T0360532

# SET THEORY
## AND ITS APPLICATIONS IN
## PHYSICS AND COMPUTING

# SET THEORY

## AND ITS APPLICATIONS IN
## PHYSICS AND COMPUTING

### Yair Shapira

Technion — Israel Institute of Technology, Israel

**World Scientific**

NEW JERSEY · LONDON · SINGAPORE · BEIJING · SHANGHAI · HONG KONG · TAIPEI · CHENNAI · TOKYO

*Published by*

World Scientific Publishing Co. Pte. Ltd.

5 Toh Tuck Link, Singapore 596224

*USA office:* 27 Warren Street, Suite 401-402, Hackensack, NJ 07601

*UK office:* 57 Shelton Street, Covent Garden, London WC2H 9HE

Library of Congress Control Number: 2022027204

**British Library Cataloguing-in-Publication Data**
A catalogue record for this book is available from the British Library.

ISBN 978-981-126-177-0 (hardcover)
ISBN 978-981-126-178-7 (ebook for institutions)
ISBN 978-981-126-179-4 (ebook for individuals)

For any available supplementary material, please visit
https://www.worldscientific.com/worldscibooks/10.1142/13009#t=suppl

Printed in Singapore

# Contents

## Part IV  The Binomial Formula and Quantum Statistical Mechanics

---

## Part V  Towards General Relativity and Quantum Mechanics

---

## Part VI  Applications in Cryptography and Error Correction

## 13  Coding–Decoding: the RSA Key Exchange . . . . . . . . . . . . . . . . . . . . 303

## Part VII  Towards Quantum Computing

## Part VIII  Appendix: Applications in C++

# Preface

Why learn set theory? This book gives the answer: because it is interesting, and also useful! Indeed, the book takes a new approach, looking at things from a fresh angle: not only theoretical, but also practical. The discussion flows in a friendly and transparent way, supplemented with a lot of examples and figures. This makes the theory much more close to home: the proofs get ever so vivid and visual, enveloped with interesting applications for students in (applied) math, physics, and engineering.

Thanks to the theory and its applications, the book could serve as a textbook in four (undergraduate) courses: three in a math department, and the fourth in a CS or physics department:

- Introduction to set theory and its applications (Chapters 1–2 and 9–12).
- Functional analysis – Han-Banach theory (Chapter 3).
- Chaos theory and stability – a geometrical point of view (Chapters 4–8).
- Cryptography with quantum computing (Chapters 12–19).

The book is self-contained, and requires no prerequisite at all. Indeed, Part I teaches set theory from scratch, including the axiom of choice, the well-ordering theorem, and Zorn's lemma. Part II uses Zorn's lemma to prove Han-Banach theorems in functional analysis. Part III uses Cantor set to introduce chaos theory from a geometrical point of view. Part IV proves the binomial formula (and other related formulas), and uses them in quantum statistical mechanics. Part V uses Zorn's lemma in general relativity and quantum mechanics, with a new insight. Part VI uses equivalence relations and modular arithmetic in cryptography and error correction. Part VII uses binary trees and hypercube to design quantum algorithms. Finally, the appendix uses the same mathematical logic and structures to design a practical computer code: dynamic tensors in C++, and a Maxwell solver in C++.

How to pronounce the title of the book? Well, this is a bit tricky: write "physics," but say "phyZics." This is a phonetic law: use the easier way to pronounce, no matter how it is written. Likewise, write "analysis" and "isomorphism," but say "analyZis" and "iZomorphiZm," and so on.

*Yair Shapira*
Haifa, Israel

# Part I

Introduction to
Set Theory

# Introduction to Set Theory

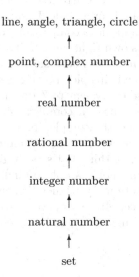

line, angle, triangle, circle

↑

point, complex number

↑

real number

↑

rational number

↑

integer number

↑

natural number

↑

set

**Fig. 0.1.** Sets are the most elementary mathematical objects. They are used to define natural and real numbers, and points in the Cartesian plane. In analytic geometry, points are then grouped in yet bigger sets: triangles, circles, etc.

To motivate the discussion, we start with a tricky concept: infinity. Fortunately, set theory helps comprehend infinity in a few forms: discrete (countable), continuous (uncountable), and even bigger: a set of functions.

Natural numbers are fundamental mathematical objects:

$$1, 2, 3, 4, \ldots.$$

They are the elementary bricks, from which more advanced numbers can be formed.

How can one tell that the natural numbers are indeed fundamental? Because they are defined in terms of axioms: Peano's axioms. Still, these axioms are based on a yet more fundamental axiom: mathematical induction.

This hints that the natural numbers are not the most elementary atoms. So, how to form them? Is there a more fundamental brick to help do this?

Fortunately, there is: the set. Indeed, in set theory, mathematical induction takes its complete form, and can help define the natural numbers one by one.

Thanks to sets, more advanced numbers can also be defined: an integer number is just a set of two elements (items): one natural number, followed by one bit, to tell us the sign. Furthermore, a rational number is just a pair of two numbers: the numerator and the denominator.

Still, the rational numbers only make a "minimal" kind of infinity: a discrete (countable) infinity. Is there a yet "bigger" infinity? Yes, there is: the real numbers already make a continuum.

Thanks to sets, we can go on and define geometrical objects as well. A pair of two real numbers make a point in the Cartesian plane. In analytic geometry, points could then be grouped in a triangle or a circle. In summary, thanks to sets, we have a complete hierarchy of more and more complex objects (Figure 0.1).

A real number is a set in its own right: an infinite sequence of rational numbers, converging to it. Still, there may be many such sequences. Which one to pick? Fortunately, the axiom of choice helps pick just one sequence. Later on, we'll prove that the axiom of choice is also equivalent to Zorn's lemma.

A pair of two real numbers make a complex number. Together, the complex numbers make a new infinite set: the complex plane. By adding the infinity point as well, we obtain another important set: the Riemann sphere.

In the world of sets, there is nothing but sets. A given set $A$ contains elements, which are sets in their own right. For example, an individual element $a \in A$ is a set in its own right, which may contain a yet inner element of the form $\alpha \in a$, which is not necessarily in the original set $A$: we may well have $\alpha \notin A$.

Don't confuse '$\in$' with '$\subset$' (inclusion). In inclusion,

$$a \in A \subset B \ \Rightarrow a \in B.$$

Geometrically, we often picture a set with elements in it. In set theory, however, this is not quite so. Strictly speaking, a set doesn't "contain:" it is just a pure abstract object that may relate to other sets in terms of '$\in$' or '$\subset$'. Still, to comprehend the theory in more concrete terms, we often think and talk about elements contained in a set, or sets that include each other.

# 1

# Sets
# and Their Cardinality

As discussed above, sets are necessary to construct composite mathematical objects. For example, a rational number is just a pair of two numbers: its numerator and denominator. Furthermore, a real number may be viewed as an infinite sequence of rational numbers that converge to it. Moreover, a complex number is just a pair of two real numbers: the real part (or the $x$-coordinate), and the imaginary part (or the $y$-coordinate). Finally, in analytic geometry, we often use infinite sets of two-dimensional points: line, angle, triangle, and circle contain those points that satisfy some algebraic condition.

The concept of set is thus in the very heart of mathematics [7, 22, 25, 36, 37, 38, 50]. This chapter introduces sets and their properties. In particular, it studies their cardinality: different kinds of infinity.

## 1.1 Russell's Paradox

### 1.1.1 The Mad-Hatter Paradox

You can learn some math even from a funny creature in *Alice in Wonderland* by Lewis Carroll. Indeed, in one episode, Alice is invited to a rather strange tea party. The host, the Mad Hatter, declares that he is going to pour tea only to those who don't pour for themselves. But then he is suddenly puzzled: should he pour for himself? If he does, then he is considered as a guest who pours for himself, so according to his own declaration he shouldn't. If, on the other hand, he doesn't, then he is considered as a guest who doesn't pour for himself, so according to his own declaration he should. In either case, this is a logical contradiction. As discussed below, this paradox is at the very heart of set theory: it is equivalent to Russell's paradox.

### 1.1.2 Sets and Containers

In the naive approach, a set is a mere container that can contain elements of any kind. Below, however, we'll see that this is not quite true: not every container is a legitimate set.

Let us introduce a few useful notations about sets. First, if the set $S$ contains the element $e$, then this is denoted by

$$e \in S$$

($e$ is in or belongs to $S$). Furthermore, it is assumed that every subset of the original set $S$ is a set in its own right. What is a subset? Well, if every element in $T$ belong to $S$ as well, then $T$ is a special kind of set: a subset of $S$. This is denoted by

$$T \subset S$$

($T$ is a subset of $S$).

So far, we haven't specified the type of the elements in the original set $S$. Actually, they could be of any type. In particular, an element $e \in S$ may well be a set in its own right. Still, this doesn't mean that $e \subset S$, because we have no information about the inner elements in $e$, so we can't check whether they belong to $S$ or not. All we know is that $e$ belongs to $S$ as an individual element. In other words, $\{e\}$ (the set whose only element is $e$) is a subset of $S$:

$$\{e\} \subset S.$$

After all, the only element in $\{e\}$ is $e$, and it indeed belongs to $S$.

### 1.1.3 Russell's Paradox

A set is not just a container of elements. In fact, not every container is a legitimate set. Consider, for example, the container $S$ that contains everything. Let us show that $S$ is not a set. Let's prove this by contradiction. Indeed, if $S$ were a set, then one could extract from it the following subset:

$$T \equiv \{A \in S \mid A \notin A\}.$$

In other words, $T$ contains only those containers that do not contain themselves as an element. Here, '$\in$' is a relation between two things in $S$, just like "to pour tea for" is a relation between two people in a tea party.

Now, is $T \in T$? In other words, does $T$ belong to itself as an element? If it does, then it is not among those $A$'s in the above definition, so it mustn't belong to $T$. If, on the other hand, it doesn't, then it is among those $A$'s in the above definition, so it must belong to $T$ as an element. In either case, this is a logical contradiction. So, we can have neither $T \in T$ nor $T \notin T$.

This is Russell's paradox [13]. It tells us that $T$ is too big to be a legitimate set, let alone the original container $S$ that is even bigger.

### 1.1.4 Fraenkel Axiom

In a container, could we still look at elements that certainly belong in it, and discuss their properties? Fortunately, we can. What we can't do is carry out operations like union (disjunction) or intersection (conjunction). Such operations could be applied to sets only, not to containers.

For example, pick some elements from the original container, and place them in a new subcontainer. If this happens to be a legitimate set, then it is indeed a subset of the original container. We can then go ahead and define all sorts of new relations between an element of the subset and any other element of the original container. This is indeed Fraenkel's axiom [17].

For example, the relation '=' tells us whether an element of the subset is the same as an element of the container. As a matter of fact, we've just used this relation implicitly: this is how we made sure that the subset is indeed included in the container. This will be useful later.

### 1.1.5 Kinds of Infinity

Russell's paradox tells us that not every kind of infinity is allowed. Some infinities may be just too big to consider. Smaller infinities, on the other hand, may still be legitimate. To construct them, we must work more slowly and patiently: better start from a small set, and advance gradually and methodically.

## 1.2 Finite Sets and Von-Neumann Numbers

### 1.2.1 The Empty Set

As discussed above, not every container is a legitimate set. For example, the container that contains everything is too big to be a set. We must therefore start from scratch: define small sets first, and then use them to construct bigger and bigger sets gradually and patiently.

What is the smallest set possible? This is the empty set: it contains no element at all. This set is denoted by $\emptyset$.

Although it may seem trivial, the empty set has a most important role in set theory. To see this, recall that 0 is the additive unit number. In other words, adding 0 changes nothing: for every number $r$,

$$r + 0 = r.$$

Now, in set theory, what operator plays the role of addition? Well, this is the union operator '$\cup$': for every two sets $A$ and $B$,

$$A \cup B \equiv \{e \mid e \in A \text{ or } e \in B\}.$$

This is a legitimate set as well. In fact, even the union of infinitely many sets is a legitimate set. This is one of the fundamental axioms in set theory.

Now, with respect to this operator, $\emptyset$ is the "unit" set. Indeed, taking the union with $\emptyset$ changes nothing: for every set $S$,

$$S \cup \emptyset = \{e \mid e \in S \text{ or } e \in \emptyset\} = \{e \mid e \in S\} = S.$$

Let's use $\emptyset$ to construct more sets.

### 1.2.2 Natural Numbers as Sets

As we have seen above, the empty set $\emptyset$ is equivalent to zero. For this reason, it can also be named '0'.

Still, the empty set is different from the nonempty set $\{\emptyset\}$. After all, this set does contain one element: $\emptyset$. For this reason, this set can be named '1'.

Furthermore, both sets serve as elements in the new set

$$\{\emptyset, \{\emptyset\}\} \, .$$

For this reason, this set can be named '2', and so on.

In fact, this is just a mathematical induction. Indeed, define

$$0 \equiv \emptyset.$$

Now, for $n = 0, 1, 2, 3, \ldots$, assume that $n$ has already been defined. (This is the induction hypothesis.) Then, define its successor:

$$\text{``}n+1\text{''} \equiv n \cup \{n\} \, .$$

Why do we use quotation marks? To indicate that there is no addition here: "$n+1$" doesn't mean $n$ plus 1, but just the successor of $n$, defined in this induction step. This way, the new natural number contains all previous ones:

$$0, 1, 2, \ldots, n \in \text{``}n+1\text{''}.$$

This can be proved by the same mathematical induction, and is often denoted by

$$0, 1, 2, \ldots, n < \text{``}n+1\text{''}.$$

This completes the definition of all natural numbers, from 0 onward.

This is the Von-Neumann definition. Later on, we'll return to it in the context of ordinals.

## 1.3 Equivalent Sets

### 1.3.1 Invertible Mapping

Two sets $S$ and $T$ are equivalent to each other if there is a one-to-one mapping $M$ from $S$ onto $T$ (Figure 1.1). Actually, we said here three things:

1. $M : S \to T$ ($M$ maps $S$ into $T$): every element $s \in S$ is mapped to some element $M(s) \in T$.
2. $M$ is one-to-one: every two distinct elements $s$ and $\hat{s}$ in $S$ are mapped to two distinct elements $M(s)$ and $M(\hat{s})$ in $T$. Numerically, $M(s)$ and $M(\hat{s})$, may have the same value: in this case, they are just duplicate copies.
3. $M$ is not only *into* but also *onto* $T$: every element $t \in T$ has an element $s \in S$ that is mapped to it:

$$M(s) = t.$$

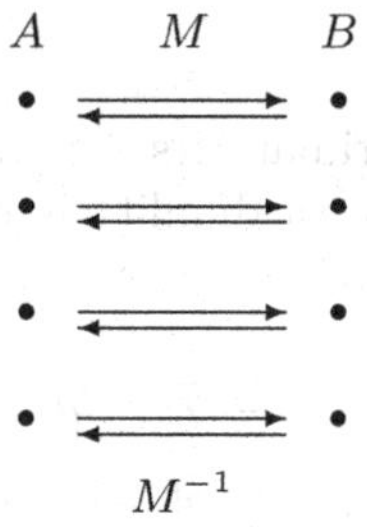

$$A \quad M \quad B$$

$$M^{-1}$$

**Fig. 1.1.** Two sets $A$ and $B$ are equivalent to each other if there is a one-to-one mapping $M$ from $A$ onto $B$. This way, each element $b \in B$ has a unique element $a \in A$ that is mapped to it. Therefore, the inverse mapping is just $M^{-1}(b) \equiv a$.

Thanks to the one-to-one property, this $s$ is unique, and can be used to define the inverse mapping

$$M^{-1} : T \to S$$

by

$$M^{-1}(t) \equiv s.$$

In the notation of Chapter 4, Section 4.6.2, this should have been written as

$$M^{-1}(\{t\}) \equiv \{s\}.$$

Here, however, we want $M^{-1}$ to be not just the origin of a set but actually a proper mapping: the inverse of $M$. This is why we drop the braces.

Consider, for example, the finite set that contains the first $n$ natural numbers:

$$T \equiv \{1, 2, 3, \ldots, n\}.$$

This is indeed a set: after all, it can be written as a union of sets:

$$T = \{1, 2, 3, \ldots, n\} = \{1\} \cup \{2\} \cup \{3\} \cup \cdots \cup \{n\} = \cup_{i=1}^{n}\{i\}.$$

How to define another set of $n$ elements? Well, for $i = 1, 2, 3, \ldots, n$, pick a new item

$$M(i) \equiv a_i.$$

Numerically, $a_i$ could have the same value as a previous item. By "new," we mean that it is distinct: not the same thing as any previous item. It might have the same value: in this case, it would be just a duplicate copy.

This guarantees that $T$ is indeed equivalent to any other finite set of $n$ elements, such as

$$S \equiv \{a_1, a_2, a_3, \ldots, a_n\}.$$

Furthermore, the inverse mapping is just

$$M^{-1}(a_i) \equiv i, \quad i = 1, 2, 3, \ldots, n.$$

## 1.3.2 Cardinality

The cardinality of a set helps estimate its size, in comparison to other sets. For finite sets, this is quite simple: the cardinality is just the total number of elements in the set. For $S$ and $T$ defined above, for example, the cardinality is just $n$. This is denoted by

$$|S| = |T| = n.$$

Later on, we'll also study the cardinality of some infinite sets. Clearly, this cardinality can no longer be a number. After all, it must be as large as any natural number. Therefore, it must be as large as infinity. How to denote it?

Well, we already have a symbol to denote infinity: $\infty$. However, this symbol is already overused. We need more specific symbols, to help distinguish between different kinds of infinities. After all, some infinities are "bigger" than others. Still, equivalent sets should have the same cardinality, denoted by the same symbol.

## 1.3.3 Ordered Sets and Sequences

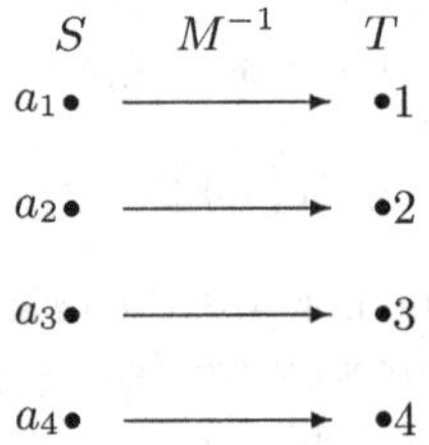

$$
\begin{array}{ccc}
S & M^{-1} & T \\
a_1 \bullet & \longrightarrow & \bullet 1 \\
a_2 \bullet & \longrightarrow & \bullet 2 \\
a_3 \bullet & \longrightarrow & \bullet 3 \\
a_4 \bullet & \longrightarrow & \bullet 4 \\
\end{array}
$$

**Fig. 1.2.** The order in $S$ is defined by the mapping $M^{-1} : S \to T$. For every two distinct elements $a, b \in S$, $a$ is before $b$ if $M^{-1}(a) < M^{-1}(b)$.

The set $T$ that contains the first $n$ natural numbers is an ordered set: for every two elements $i$ and $j$ in $T$, either $i < j$ or $i > j$ or $i = j$. Thanks to this order, we can also order the equivalent set $S$, defined in Section 1.3.1: if $a$ and $b$ are two distinct elements in $S$, then $a$ is before $b$ if $M^{-1}(a) < M^{-1}(b)$ (Figure 1.2). Thus, $S$ is ordered simply as

$$a_1, a_2, a_3, \ldots, a_n.$$

With this order, $S$ is not only a set but also a sequence of $n$ consecutive elements, denoted also by

$$S \equiv \{a_i\}_{i=1}^{n}.$$

In particular, if $n = 2$, then $S$ is a pair: a very short sequence of just two elements. In analytic geometry, pairs are most useful. For example, a pair of two real numbers make a two-dimensional point in the Cartesian plane, denoted by

$$S = (a_1, a_2).$$

Here, $a_1$ stands for the $x$-coordinate, and $a_2$ for the $y$-coordinate.

## 1.4 Infinite Set

### 1.4.1 Infinite Set

Consider the container that contains all the natural numbers:

$$\mathbb{N} \equiv \{1, 2, 3, \ldots\}.$$

(Why do we say that $\mathbb{N}$ is a container? Because, by now, we don't know that $\mathbb{N}$ is a legitimate set yet. We'll prove this soon.) More precisely, $\mathbb{N}$ is defined by mathematical induction: first, 1 is placed in $\mathbb{N}$. Then, for $n = 1, 2, 3, 4, \ldots$, if $n$ is in $\mathbb{N}$, then "$n+1$" is placed in $\mathbb{N}$ as well. Why do we use quotation marks? To indicate that there is no addition here: "$n+1$" doesn't mean $n$ plus 1, but just the successor of $n$: the next natural number that follows $n$ (Section 1.2.2).

Is $\mathbb{N}$ a set? This is not a rhetorical question. After all, in Section 1.1.3, we've already seen that a container may be too big to be a set. To make sure that $\mathbb{N}$ is a legitimate set, we have to use an axiom from set theory.

This axiom says that there exists an infinite set, denoted by $S$. Thus, every subset of $S$ is a set as well. Now, to verify that $\mathbb{N}$ is indeed a legitimate set, it is sufficient to embed it in $S$: show that $\mathbb{N}$ is equivalent to a subset of $S$.

For this purpose, define a new one-to-one mapping from $\mathbb{N}$ onto a new subset $S_0 \subset S$. First, pick an element $s_1 \in S$, and map 1 to it. Then, pick another element $s_2 \in S$, and map 2 to it, and so on.

This is just mathematical induction: assume that $1, 2, 3, \ldots, n \in \mathbb{N}$ have already been mapped to the distinct elements $s_1, s_2, s_3, \ldots, s_n \in S$, respectively. This is the induction hypothesis. Now, consider the set

$$S \setminus \{s_1, s_2, s_3, \ldots, s_n\}.$$

What is this set? Well, it contains those elements in $S$ that haven't been picked so far. Because $S$ is infinite, this set is nonempty, so we can pick some element from it, name it "$s_{n+1}$", and map "$n+1$" to it. This completes the induction step.

Here one could rightly ask: is the induction step well-defined? After all, there are many options to pick $s_{n+1}$. What option to use? Fortunately, the axiom of choice comes to our aid. We'll study it later (Chapter 2).

As a result, $\mathbb{N}$ is indeed equivalent to the infinite sequence

$$S_0 \equiv \{s_1, s_2, s_3, \ldots\} \subset S.$$

Because $S$ is a set, $\mathbb{N}$ is a set as well.

### 1.4.2 Countable Set

The cardinality of $\mathbb{N}$ is denoted by

$$|\mathbb{N}| = \aleph_0.$$

This cardinality is not a number but a mere symbol to characterize every set that is equivalent to $\mathbb{N}$. Such a set is called countable: the elements in it could be listed one by one in an infinite sequence, indexed by $1, 2, 3, \ldots$ (Figure 1.3).

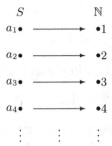

$$
\begin{array}{cc}
S & \mathbb{N} \\
a_1 \bullet \longrightarrow & \bullet 1 \\
a_2 \bullet \longrightarrow & \bullet 2 \\
a_3 \bullet \longrightarrow & \bullet 3 \\
a_4 \bullet \longrightarrow & \bullet 4 \\
\vdots \quad \vdots & \vdots
\end{array}
$$

**Fig. 1.3.** A countable set $S$ can be embedded in $\mathbb{N}$.

The cardinality $\aleph_0$ is greater than the finite cardinality $n$ defined in Section 1.3.2:

$$n < \aleph_0.$$

After all, every finite set of $n$ elements could be easily embedded in $\mathbb{N}$, or mapped by a one-to-one mapping into $\mathbb{N}$.

Furthermore, $\aleph_0$ is the minimal cardinality that is as big as any finite cardinality. Indeed, as we've seen in Section 1.4.1, every infinite set $S$ has a countable subset, so

$$|S| \geq |\mathbb{N}| = \aleph_0.$$

### 1.4.3 Is Zero a Natural Number?

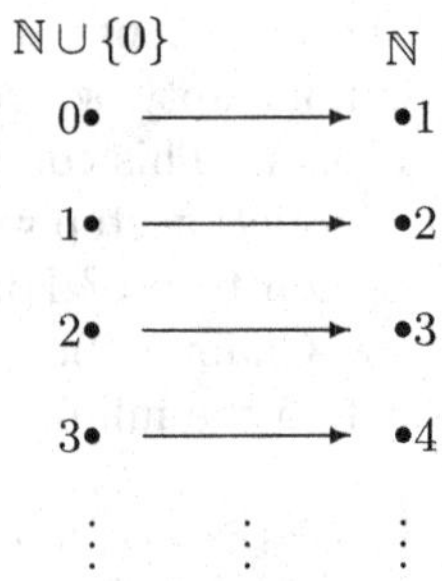

$$
\begin{array}{cc}
\mathbb{N} \cup \{0\} & \mathbb{N} \\
0 \bullet \longrightarrow & \bullet 1 \\
1 \bullet \longrightarrow & \bullet 2 \\
2 \bullet \longrightarrow & \bullet 3 \\
3 \bullet \longrightarrow & \bullet 4 \\
\vdots \quad \vdots & \vdots
\end{array}
$$

**Fig. 1.4.** The set $\mathbb{N} \cup \{0\}$ is equivalent to $\mathbb{N}$ by the one-to-one mapping $i \to i + 1$.

Usually, 0 is not considered as a natural number. This way, the natural numbers start from 1:

$$\mathbb{N} = \{1, 2, 3, \ldots\}.$$

Still, it makes little difference whether 0 is considered as a natural number or not. After all, the mathematical induction that defines the natural numbers in Section 1.2.2 could easily start from 1 rather than 0. Furthermore, the one-to-one mapping

$$i \to i + 1$$

maps $\mathbb{N} \cup \{0\}$ onto $\mathbb{N}$ (Figure 1.4). Thus, these sets are indeed equivalent to each other:

$$|\mathbb{N} \cup \{0\}| = |\mathbb{N}| = \aleph_0.$$

Thus, both $\mathbb{N} \cup \{0\}$ and $\mathbb{N}$ are good instances of a countable set. A given infinite set can now be compared to them, to check whether it is countable or not.

### 1.4.4 The Set of Integer Numbers

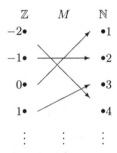

**Fig. 1.5.** The one-to-one mapping $M$ that maps $\mathbb{Z}$ onto $\mathbb{N}$: the negative numbers map to the even numbers, and the nonnegative numbers map to the odd numbers.

Let $\mathbb{Z}$ be the set of integer numbers:

$$\mathbb{Z} \equiv \{\dots, -3, -2, -1, 0, 1, 2, 3, \dots\}.$$

At first glance, $\mathbb{Z}$ seems bigger than $\mathbb{N}$. However, this is just an optical illusion. In fact, $\mathbb{Z}$ is countable as well:

$$|\mathbb{Z}| = \aleph_0.$$

To show this, let's define the mapping

$$M(i) \equiv \begin{cases} 2i + 1 & \text{if } i \geq 0 \\ 2|i| & \text{if } i < 0 \end{cases}$$

(Figure 1.5). Clearly, $M$ is a one-to-one mapping from $\mathbb{Z}$ onto $\mathbb{N}$, as required. This proves that $\mathbb{Z}$ is indeed countable:

$$|\mathbb{Z}| = |\mathbb{N}| = \aleph_0.$$

## 1.5 Infinite Grid

### 1.5.1 Product of Sets

Let $A$ and $B$ be some given sets. Their product is defined as the set of pairs with a first component from $A$ and a second component from $B$:

$$AB \equiv \{(a,b) \mid a \in A, \ b \in B\}.$$

This tells us how to multiply cardinalities:

$$|A| \cdot |B| \equiv |AB|.$$

### 1.5.2 Two-Dimensional Grid

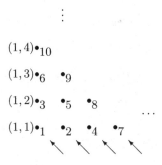

**Fig. 1.6.** The infinite two-dimensional grid $\mathbb{N}^2$ is countable: it could be listed diagonal by diagonal (as indicated by the index to the right of each grid point).

Let us use the above definition to obtain the square of $\aleph_0$:

$$\aleph_0^2 \equiv \aleph_0 \cdot \aleph_0 \equiv |\mathbb{N}| \cdot |\mathbb{N}| \equiv |\mathbb{N}^2|.$$

Here, $\mathbb{N}^2$ is the set of pairs of natural numbers:

$$\mathbb{N}^2 \equiv \{(m,n) \mid m,n \in \mathbb{N}\}.$$

This set could be viewed as an infinite two-dimensional grid (Figure 1.6).

At first glance, $\mathbb{N}^2$ seems much bigger than $\mathbb{N}$. After all, it makes not one- but two-dimensional grid. Still, this turns out to be just an optical illusion. The truth is that it has the same size (or cardinality) as $\mathbb{N}$.

Indeed, the grid points could be listed diagonal by diagonal, as in Figure 1.6. The first diagonal is quite short: it contains just one point: $(1,1)$. The second diagonal, on the other hand, is slightly longer: it contains two points: $(2,1)$ and $(1,2)$. The third diagonal, on the other hand, is yet longer: it contains three points: $(3,1)$, $(2,2)$, and $(1,3)$, and so on.

Now, place these diagonals one by one in a row. As a result, the grid points are listed in one infinite sequence, as required:

$$(1,1), \ (2,1), \ (1,2), \ (3,1), \ (2,2), \ (1,3), \ \ldots.$$

Thus, $\mathbb{N}^2$ is countable:

$$\aleph_0^2 = |\mathbb{N}^2| = \aleph_0.$$

Not convinced? In Section 1.12.2, we'll give an alternative proof. Yet another proof will also be given in the exercises below.

## 1.6 Equivalence

### 1.6.1 Mathematical Equivalence Relation

The notion of equivalence introduced above is a relation between sets. Indeed, given two sets $A$ and $B$, one could ask whether they are equivalent to each other or not:

$$|A| = |B|?$$

Still, is the name "equivalence" really appropriate here? In other words, is it indeed a mathematical equivalence relation?

To be a mathematical equivalence relation, a relation must have three desirable properties: reflexivity, symmetry, and transitivity. Fortunately, our kind of equivalence between sets indeed has these properties:

1. Reflexivity: every set $A$ is equivalent to itself by the identity mapping $I(a) \equiv a$ that maps each element to itself.
2. Symmetry: if $A$ is equivalent to $B$ by the one-to-one mapping $M$ from $A$ onto $B$, then $B$ is also equivalent to $A$ by the inverse mapping $M^{-1}$ from $B$ back onto $A$.
3. Transitivity: if $A$ is equivalent to $B$ by the one-to-one mapping $M$ from $A$ onto $B$, and $B$ is also equivalent to $C$ by the one-to-one mapping $M'$ from $B$ onto $C$, then $A$ is also equivalent to $C$ by the composite one-to-one mapping $M' \circ M$ from $A$ onto $C$:

$$(M' \circ M)(a) \equiv M'(M(a)) \in C, \quad a \in A.$$

### 1.6.2 Cantor-Bernstein's Theorem

To establish that two given sets $A$ and $B$ are equivalent to each other, we have a powerful tool: Cantor-Bernstein's theorem. It says that

$$\text{if } |A| \le |B| \text{ and } |B| \le |A|, \text{ then } |A| = |B|.$$

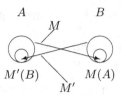

**Fig. 1.7.** The Cantor-Bernstein theorem. Assume that $A$ can be embedded in $B$ by the one-to-one mapping $M$ from $A$ onto its image $M(A) \subset B$. Assume also that $B$ can be embedded in $A$ by the one-to-one mapping $M'$ from $B$ onto its image $M'(B) \subset A$. Then, $A$ and $B$ must be equivalent to each other.

In other words,

$$|A| \leq |B| \leq |A| \Rightarrow |A| = |B|.$$

For finite sets, this is trivial. For infinite sets, on the other hand, this is not trivial at all.

Thanks to this theorem, it is now much easier to establish that $A$ and $B$ are equivalent to each other. Indeed, there is no longer any need to map $A$ onto $B$ exactly. Instead, it is sufficient to embed $A$ in $B$, and also $B$ in $A$.

To prove this theorem, we need some more notations. For any mapping $M : A \to B$, its image $M(A)$ contains its targets:

$$M(A) \equiv \{M(a) \mid a \in A\} \subset B.$$

In other words, $M(A)$ contains those elements in $B$ to which some element is mapped. In Figure 1.7, for example, $A$ is equivalent to $M(A) \subset B$ by the one-to-one mapping $M$, and $B$ is equivalent to $M'(B) \subset A$ by the one-to-one mapping $M'$.

Another useful notation stands for the set containing those elements in $A$ that are not in the image of $M'$:

$$A \setminus M'(B) \equiv \{a \in A \mid a \notin M'(B)\}.$$

In Figure 1.7, this is the big circle on the left, minus the small inner circle.

To this set, let's apply the composite mapping $M' \circ M$ $i$ times ($i \geq 0$). This gives infinitely many sets. Let's take their union:

$$\cup_{i=0}^{\infty} (M' \circ M)^i (A \setminus M'(B))$$
$$\equiv \left\{ q \mid q \in (M' \circ M)^i (A \setminus M'(B)) \text{ for some } i \geq 0 \right\}.$$

Is this a legitimate set? After all, it could take forever to check whether something is in it or not! Fortunately, one of the axioms in set theory tells us that a union of infinitely many sets is a legitimate set as well.

Consider a typical element $q$ in the above union. Where could it be? Well, $q$ could be in the first set in the union:

$$q \in (M' \circ M)^0 (A \setminus M'(B)) \equiv A \setminus M'(B)$$

(the big circle on the left in Figure 1.7, minus the small inner circle). On the other hand, $q$ could also be in the second set in the union, which is the image of the first set under $M' \circ M$:

$$q \in (M' \circ M)^1 (A \setminus M'(B)) \equiv (M' \circ M)(A \setminus M'(B)).$$

Still, there are more options: $q$ could be in the third set — the image of the image, and so on. In general, if $q$ is in the $i$th set in the above union, then one could apply $M' \circ M$ to it, and advance it to the $(i+1)$st set.

To show that $A$ and $B$ are indeed equivalent to each other, we need to form a new one-to-one mapping from $A$ onto $B$. For this purpose, let's take a geometrical approach: in Figure 1.7, let's show that the big circle on the left is equivalent to the small inner circle by the new mapping

$$K : A \to M'(B).$$

How to define $K$? As follows:

$$K(a) \equiv \begin{cases} (M' \circ M)(a) & \text{if } a \in \cup_{i=0}^{\infty} (M' \circ M)^i (A \setminus M'(B)) \\ a & \text{otherwise.} \end{cases}$$

This way, if $a$ is in the $i$th set in the infinite union, then $K$ advances it to the $(i+1)$st set. If, on the other hand, $a$ is not in the union at all, then $K$ leaves it unchanged.

Thanks to this definition, the infinite union is invariant under both $K$ and $M' \circ M$: every element $q$ in it is mapped to $K(q) = M'(M(q))$, which is in it as well. Thus, $K$ is one-to-one, as required.

Clearly, $K$ maps $A$ into $M'(B)$. Is this "onto" as well? Fortunately, it is. Indeed, every element $a \in M'(B)$ is either in the above union or not. If it is, then it can't be in the first set in it: it must lie in $(M' \circ M)^i (A \setminus M'(B))$ for some $i \geq 1$, so it is in the image of $M' \circ M$. If, on the other hand, it isn't, then it must satisfy $K(a) = a$. In either case, it is in the image of $K$, as required.

Thus, $K$ is indeed a one-to-one mapping from $A$ onto $M'(B)$. In other words, $A$ is equivalent to $M'(B)$ by the one-to-one mapping $K$. As a consequence, $A$ is also equivalent to $B$ by the composite mapping $(M')^{-1} \circ K$. This completes the proof of Cantor-Bernstein's theorem:

$$|B| \leq |A| \leq |B| \Rightarrow |A| = |B|.$$

Let's use this theorem to calculate the cardinality of $\mathbb{Q}$: the set of rational numbers.

## 1.7 Countable Set

### 1.7.1 The Set of Rational Numbers

The set of rational numbers, denoted by $\mathbb{Q}$, can now be embedded in the infinite two-dimensional grid $\mathbb{N}^2$ (Figure 1.8). Indeed, every positive rational number can

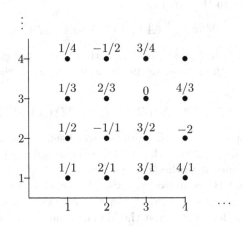

**Fig. 1.8.** How to embed $\mathbb{Q}$ in $\mathbb{N}^2$? Map each positive rational number $m/n$ to $(m, n)$. Then, map $-m/n$ to $(2m, 2n)$. Finally, map 0 to $(3, 3)$. This is indicated above each bullet in the grid.

be written as $m/n$, for some minimal natural numbers $m$ and $n$, with no common divisor. Thus, it makes sense to map $m/n$ to the point $(m, n) \in \mathbb{N}^2$. For example, $1 = 1/1$ is mapped to $(1, 1)$.

Furthermore, the negative rational number $-m/n$, is mapped to $(2m, 2n)$. For example, $-1 = -1/1$ is mapped to $(2, 2)$. Finally, zero is mapped to $(3, 3)$. Thus,

$$|\mathbb{Q}| \le |\mathbb{N}^2| = \aleph_0.$$

On the other hand, $\mathbb{N}$ is a subset of $\mathbb{Q}$, so we also have

$$|\mathbb{Q}| \ge |\mathbb{N}| = \aleph_0.$$

Thanks to Cantor-Bernstein's theorem, we therefore have

$$|\mathbb{Q}| = \aleph_0.$$

Thus, the set of the rational numbers is not really bigger than the original set of the natural numbers. Both are countable: they have the same cardinality: $\aleph_0$.

### 1.7.2 The Set of All Finite Sequences

So, every infinite subset of $\mathbb{Q}$ is countable as well. For example, consider the set $S$ containing the (arbitrarily long) finite sequences of 0's or 1's. For instance, a typical element in $S$ could be

$$01101110010000011010101 \in S.$$

Because $S$ is an infinite set, we have from Sections 1.4.1–1.4.2 that

$$|S| \geq |\mathbb{N}| = \aleph_0.$$

On the other hand, $S$ could be easily embedded in $\mathbb{Q}$: given a finite sequence in $S$, just place it behind the decimal point. Then, replace each '0' by '2'. This makes a distinct (finite) decimal fraction. For example,

$$0110111001000001101010 1 \rightarrow 0.2112111221222221121212 1 \in \mathbb{Q}.$$

Thus,

$$|S| \leq |\mathbb{Q}| = \aleph_0.$$

These two inequalities can now combine in Cantor-Bernstein's theorem to conclude that

$$|S| = \aleph_0.$$

Not convinced? In the exercises, below, we'll give yet another proof.

Why is $S$ so "small?" Because it contains finite sequences only. The set of infinite sequences, on the other hand, is genuinely bigger: its cardinality is greater than $\aleph_0$, so it is no longer countable. After all, an infinite sequence differs from a finite sequence, just like an infinite universe differs from our finite universe (Chapter 6). This will be discussed below.

## 1.8 Subsets

### 1.8.1 Characteristic Function

A function can be viewed as a mapping from one set to another: a machine that takes an input (or argument) from one set to produce an output: a specific element in the other set. How to define a function $f$? Well, let $x$ be some input element. Then, specify $f(x)$: the output element, produced by $f$ (or the target element, to which $x$ is mapped by $f$).

The most elementary function is the characteristic (or indicator) function: each element is mapped either to 0 or to 1. In other words, a characteristic function $f$ maps a given set $S$ into the small set that contains two elements only — 0 and 1:

$$f : S \rightarrow \{0,1\}.$$

This way, each element $s \in S$ produces an output $f(s)$: either 0 or 1, depending on the specific definition of the function.

### 1.8.2 Characteristic Function as a Subset

Consider a characteristic function as above:

$$f : S \rightarrow \{0,1\}.$$

Now, look at those elements that are mapped to 1. Together, they make a subset $T \subset S$. In Chapter 4, Section 4.6.2, we already gave a name to such a subset: the origin of $\{1\}$ under $f$:

$$T \equiv f^{-1}(\{1\}) \equiv \{s \in S \mid f(s) = 1\}.$$

This way, $T$ mirrors $f$: it contains all the information in $f$.

At the same time, $f$ also defines the complementary subset — those elements that are mapped to 0:

$$S \setminus T \equiv\equiv \{s \in S \mid s \notin T\} = \{s \in S \mid f(s) = 0\} = f^{-1}(\{0\}).$$

Thus, $f$ actually splits the original set $S$ into two disjoint subsets: $T$ and $S \setminus T$:

$$S = T \cup (S \setminus T).$$

This also works the other way around: given a subset $T \subset S$, it defines a unique characteristic function $f$ as above. This way, $f$ mirrors $T$: it contains all the information in $T$.

### 1.8.3 The Power Set

Subsets like $T$ could also serve as elements in a new set — the power set of $S$, denoted by $P(S)$:

$$P(S) \equiv \{T \mid T \subset S\}.$$

This new set mirrors the set of characteristic functions on $S$. For this reason, both have the same cardinality:

$$|\{f \mid f : S \to \{0,1\}\}| = |P(S)|.$$

Below, we discuss some properties of the set of characteristic functions. Thanks to the above mirroring, the power set shares the same properties as well.

## 1.9 Functions

### 1.9.1 Set of Functions

The set of characteristic functions on $S$ could be modeled as a "list" of $|S|$ duplicate copies of $\{0, 1\}$:

$$\{0, 1\}, \ \{0, 1\}, \ \{0, 1\}, \ \ldots \quad (|S| \text{ times}).$$

In this list, each copy of $\{0, 1\}$ is associated with a particular element in $S$.

A characteristic function $f$ assigns either 0 or 1 to each individual element $s \in S$. In terms of the above model, $f$ picks either 0 or 1 from that copy of $\{0, 1\}$ that is associated with $s$. The picked value serves then as the output $f(s)$. The list of all these outputs makes a unique sequence of 0's or 1's, of length $|S|$.

In Section 1.5.1, we've already defined the product of two sets. In these terms, the above model could be viewed as a long "product:"

$$\{0,1\} \cdot \{0,1\} \cdot \{0,1\} \cdots \quad (|S| \text{ times}).$$

In other words, this is just the $|S|$th "power" of $\{0,1\}$. For this reason, the set of characteristic functions is often denoted by

$$\{0,1\}^S \equiv \{f \mid f : S \to \{0,1\}\}.$$

After all, this is just a natural extension of the notation in Sections 1.5.1–1.5.2.

### 1.9.2 Cardinality of Set of Functions

Could the original set $S$ be embedded in $\{0,1\}^S$? This is easy enough: each individual element $s \in S$ could be mapped to its own characteristic function, which is nonzero at $s$ only:

$$M(s)(t) \equiv \begin{cases} 1 & \text{if } t = s \\ 0 & \text{if } t \in S \text{ and } t \neq s. \end{cases}$$

This proves that

$$|\{0,1\}^S| \geq |S|.$$

Is this inequality strict? In other words, is it true that

$$|\{0,1\}^S| > |S|?$$

In other words, is $S$ *not* equivalent to $\{0,1\}^S$?

This is a negative conjecture. Therefore, it should better be proved by contradiction. Assume momentarily that there were a one-to-one mapping $M$ from $S$ onto $\{0,1\}^S$. For each element $s \in S$, $M(s)$ would then be a characteristic function from $S$ to $\{0,1\}$. In particular, when it takes the input $s$, this function produces the output $M(s)(s)$, which is either 0 or 1.

Now, let's design a new characteristic function $f : S \to \{0,1\}$ that disagrees with this output:

$$f(s) \equiv 1 - M(s)(s) \neq M(s)(s), \quad s \in S.$$

In other words, $f$ disagrees with $M(s)$ on $s$:

$$f(s) \neq M(s)(s).$$

Now, to be the same, two functions must agree on all possible inputs. Since we've found one disagreement, these two functions can no longer be the same:

$$f \neq M(s).$$

As a consequence, $f$ can never be in the image of $M$. This contradicts our momentary assumption that $M$ was onto $\{0,1\}^S$. Thus, our momentary assumption must have been false. So,

$$|\{0,1\}^S| > |S|,$$

as asserted.

We can now define the power of two cardinalities as follows: for any two sets $A$ and $B$,

$$|A|^{|B|} \equiv |A^B|.$$

With this new definition,

$$2^{|S|} \equiv |\{0,1\}|^{|S|} \equiv |\{0,1\}^S| > |S|.$$

Let's use this to design an uncountable set.

## 1.10 Uncountable Sets

### 1.10.1 The Set of Infinite Sequences

Let's look at a more concrete example. In the above, let $S$ be the set of natural numbers:

$$S \equiv \mathbb{N}.$$

In this special case, our model is

$$\{0,1\}^{\mathbb{N}} = \{0,1\}, \ \{0,1\}, \ \{0,1\}, \ \ldots \quad (\aleph_0 \text{ times}).$$

This set contains the infinite sequences of 0's and 1's, such as

$$a_1, a_2, a_3, \ldots,$$

where $a_i$ is either 0 or 1 $(i \geq 1)$.

Furthermore, this sequence can also be viewed as a characteristic function of the form

$$f : \mathbb{N} \to \{0,1\},$$

with $f(i) = a_i$.

Thus, the above set of infinite sequences could also be viewed as a set of functions: the set of characteristic functions on $\mathbb{N}$. Moreover, it could also be viewed as a set of subsets: the set of subsets of $\mathbb{N}$.

We're now ready to see that this set is too big to be countable. On the contrary: it is uncountable.

### 1.10.2 Uncountable Set

Indeed, thanks to the theorem in Section 1.9.2,

$$\left| \{0,1\}^{\mathbb{N}} \right| = 2^{\aleph_0} > \aleph_0.$$

Thus, we've just designed an uncountable set, with a new cardinality, denoted by $\aleph$:

$$\aleph \equiv 2^{\aleph_0} > \aleph_0.$$

This new cardinality is most useful: it is also the cardinality of the real axis $\mathbb{R}$. To see this, let's look at the unit interval.

## 1.11 Real Numbers

### 1.11.1 The Unit Interval

To calculate the cardinality of the entire real axis $\mathbb{R}$, we must first study a small part of it: the unit interval. Recall that it comes in three forms. In Chapter 4, Section 4.1.1, we've already met the closed (and open) unit interval. In Chapter 5, Section 5.2.2, on the other hand, we also met the half-open unit interval.

So far, we've used braces to denote the small set

$$\{0,1\},$$

which contains two elements only: 0 and 1. The real numbers in between, on the other hand, are not included.

To denote the closed unit interval, on the other hand, we use a different kind of delimiters: square bracket. This way, we include not only 0 and 1 but also all real numbers in between:

$$[0,1] \equiv \{x \in \mathbb{R} \mid 0 \le x \le 1\}.$$

Here, a square bracket includes an endpoint. A round parenthesis, on the other hand, excludes an endpoint. For this reason, the open unit interval that excludes the endpoints is denoted by

$$(0,1) \equiv \{x \in \mathbb{R} \mid 0 < x < 1\}.$$

Don't get confused! In the Cartesian plane, $(0,1)$ stands for just one point: the $x$-coordinate is 0, and the $y$-coordinate is 1. In the context of the real axis, on the other hand, there is no $y$-coordinate at all. For this reason, in the real axis, $(0,1)$ can be used for another purpose: to denote the open unit interval, as above.

So, the open unit interval is obtained from the closed unit interval by dropping the endpoints:

$$(0,1) = [0,1] \setminus \{0,1\}.$$

Conversely, the closed unit interval could be obtained from the open unit interval by adding the endpoints:

$$[0,1] = (0,1) \cup \{0,1\}.$$

### 1.11.2 Cardinality of The Unit Interval

Now, let's embed $\{0,1\}^{\mathbb{N}}$ in the closed unit interval. For this purpose, pick some infinite sequence of 0's and 1's. Let's place it behind the decimal point. This produces an infinite decimal fraction, in base 10.

This is indeed a one-to-one mapping into $[0,1]$. Thus, we have established that

$$|[0,1]| \ge |\{0,1\}^{\mathbb{N}}| = 2^{\aleph_0} = \aleph.$$

On the other hand, let's embed the closed unit interval in $\{0,1\}^{\mathbb{N}}$. For this purpose, pick a real number in $[0,1]$. Look at its binary form: in base 2 rather than base 10. In this form, it has an infinite sequence of 0's and 1's behind the point.

If the real number is also rational, then it may have two binary forms: one ending with infinitely many 0's, and the other ending with infinitely many 1's. In such a case, pick the latter.

This is indeed a one-to-one mapping. So, we've also established that

$$|[0,1]| \le |\{0,1\}^{\mathbb{N}}| = 2^{\aleph_0} = \aleph.$$

From Cantor-Bernstein's theorem, we now have

$$|[0,1]| = |\{0,1\}^{\mathbb{N}}| = 2^{\aleph_0} = \aleph,$$

as asserted.

### 1.11.3 The Real Axis and Its Cardinality

We're now ready to calculate the cardinality of the entire real axis $\mathbb{R}$. On one hand, $[0,1] \subset \mathbb{R}$, so

$$|\mathbb{R}| \geq |[0,1]| = \aleph.$$

On the other hand, the function

$$\tan\left(\pi\left(x - \frac{1}{2}\right)\right)$$

is a one-to-one mapping from the open unit interval $(0,1)$ onto $\mathbb{R}$, so

$$|\mathbb{R}| = |(0,1)| \leq |[0,1]| = \aleph.$$

(Here, instead of the tangent function, we could also use the hyperbolic or cotangent function.) From Cantor-Bernstein's theorem, we now have

$$|\mathbb{R}| = |[0,1]| = \aleph,$$

as asserted.

## 1.12 Higher Dimension

### 1.12.1 The Unit Square and Its Cardinality

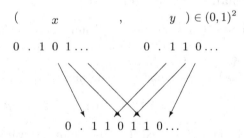

**Fig. 1.9.** The point $(x,y)$ in the open unit square is mapped to a new decimal fraction, with odd-numbered digits from $x$, and even-numbered digits from $y$.

At first glance, the Cartesian plane $\mathbb{R}^2$ seems much bigger than the real axis $\mathbb{R}$. It turns out, however, that this is just an optical illusion: both have the same cardinality $\aleph$.

To see this, recall the open unit interval:

$$(0,1) \equiv \{x \in \mathbb{R} \mid 0 < x < 1\}$$

(Section 1.11.1). Let's square it up! This gives the open unit square:

$$(0,1)^2 \equiv (0,1) \times (0,1) = \{(x,y) \in \mathbb{R}^2 \mid 0 < x, y < 1\}.$$

Let's see that these two are equivalent to each other. Let's start from the easy bit — embed the former in the latter, say in the middle of it:

$$|(0,1)| = \left|\left\{\left(x, \frac{1}{2}\right) \mid 0 < x < 1\right\}\right| \leq |(0,1)^2|.$$

Now, let's do the more difficult bit: embed the latter in the former. For this purpose, consider some point $(x,y) \in (0,1)^2$. Look at its first coordinate: $x$. View it as a binary fraction (in base 2), with infinitely many 0's and 1's behind the point.

If $x$ is rational, then it may have two binary forms: one ending with infinitely many 0's, and the other ending with infinitely many 1's. In this case, pick the former.

The same can be done for $y$ as well. So, we have two infinite sequences of 0's and 1's. Now, merge them into one sequence, with odd-numbered digits picked one by one from $x$, and even-numbered digits from $y$ (Figure 1.9). Place this new sequence behind the decimal point, to make a new decimal fraction (in base 10 rather than 2).

This is indeed a one-to-one mapping: after all, both $x$ and $y$ could be reconstructed from the new decimal fraction. So,

$$|(0,1)^2| \leq |(0,1)| = \aleph.$$

From Cantor-Bernstein's theorem, we therefore have

$$|(0,1)^2| = |(0,1)| = \aleph,$$

as asserted.

## 1.12.2 Infinite Grid — a New Proof

In the above, $x$ and $y$ are viewed as binary fractions: each has an infinite sequence of binary digits. Now, instead of an infinite sequence, let's look at an (arbitrarily long) finite sequence of binary digits. In base 2, it makes a natural number in $\mathbb{N}$. Two such sequences make a pair of the form $(j, i) \in \mathbb{N}^2$.

Now, let's merge these sequences with each other, to make a new natural number in base 2. This would embed $\mathbb{N}^2$ in $\mathbb{N}$, giving a new proof to the result in Section 1.5.2.

But be careful! $i$ and $j$ must have nearly the same length (number of binary digits). More precisely, as a binary number in base 2, $j$ should be as long as $i$. If it isn't, then leading zeroes must be added to it on the left, to make it as long as $i$.

$i$, on the other hand, may be slightly shorter than $j$: by one digit at most. If $i$ is shorter than that, then leading zeroes must be added to it on the left, to make it nearly as long as $j$.

In their final form (with leading zeroes if necessary), both $i$ and $j$ are ready to merge into a new binary number. This is done as follows. In the new number, index the binary digits by $1, 2, 3, \ldots$. More specifically, index the rightmost digit by 1, the next digit on its left by 2, and so on, until the leftmost digit. In this binary representation, the odd-indexed digits come from the binary representation of $j$, and the even-indexed digits come from the binary representation of $i$. This defines the new number uniquely, as required.

This way, the original pair $(j, i) \in \mathbb{N}^2$ is now mapped to a new number, which contains at least two binary digits. For example, if $i = j = 1$, then the new number is 3 (the binary number 11).

Is this mapping one-to-one? Fortunately, it is. Indeed, from the new number, both $i$ and $j$ could be reconstructed. For this purpose, one must first count the binary digits in the new number. If this count is even, then $i$ is as long as $j$, so leading zeroes may drop from $j$, to get it to its original form. If, on the other hand, the above count is odd, then $i$ is shorter than $j$ by one digit, so leading zeroes may drop from $i$, to get it to its original form. In either case, both $i$ and $j$ are uncovered uniquely in their original form, as required.

This proves that

$$|\mathbb{N}^2| \leq |\mathbb{N}|,$$

in agreement with Figure 1.6 above.

### 1.12.3 The Cartesian Plane

The entire Cartesian plane $\mathbb{R}^2$ is of the same cardinality: $\aleph$. Indeed, the open unit square could map onto it by

$$(x, y) \rightarrow \left( \tan \left( \pi \left( x - \frac{1}{2} \right) \right), \tan \left( \pi \left( y - \frac{1}{2} \right) \right) \right).$$

Thus,

$$|\mathbb{R}^2| = |(0, 1)^2| = \aleph,$$

as asserted.

Geometrically, the complex plane $\mathbb{C}$ is just the same as the Cartesian plane. Therefore,

$$|\mathbb{C}| = |\mathbb{R}^2| = \aleph$$

as well.

### 1.12.4 Multidimensional Space

Let's extend this to a yet higher dimension. For this purpose, consider the multidimensional space $\mathbb{R}^n$, where $n$ is a fixed natural number. What is this space? Well, it contains $n$-dimensional vectors, with $n$ real components (coordinates):

$$\mathbb{R}^n \equiv \{(v_1, v_2, v_3, \ldots, v_n) \mid v_1, v_2, v_3, \ldots, v_n \in \mathbb{R}\}.$$

As in Section 1.9.1 above, this space can be modeled as a list of $n$ duplicate copies of $\mathbb{R}$:

$$\mathbb{R}, \mathbb{R}, \mathbb{R}, \ldots, \mathbb{R} \quad (n \text{ times}).$$

Is this space big? Not at all: its cardinality is $\aleph$ as well! After all, the proof in Figure 1.9 could be easily extended to it as well.

This works well for a finite $n$. But what about the infinite-dimensional space? This space is denoted by

$$\mathbb{R}^\mathbb{N} \equiv \{f \mid f : \mathbb{N} \to \mathbb{R}\},$$

and modeled as an infinite list of duplicate copies of $\mathbb{R}$:

$$\mathbb{R}, \ \mathbb{R}, \ \mathbb{R}, \ \ldots \quad (\aleph_0 \text{ times}).$$

In this case, our proof no longer works. Thus, we better develop a more general theory, suitable to both finite- and infinite-dimensional spaces at the same time.

## 1.13 The Power Rule

### 1.13.1 Geometrical Point of View

$$A^B : \text{------} \ A \quad A \quad A \ \cdots \ (|B| \text{ times})$$

$$A^B : \text{------} \ A \quad A \quad A \ \cdots \ (|B| \text{ times})$$

$$A^B : \text{------} \ A \quad A \quad A \ \cdots \ (|B| \text{ times})$$

$$\vdots \qquad\qquad \vdots$$

$$(|C| \text{ times})$$

**Fig. 1.10.** In each row, there are $|B|$ duplicate copies of $A$. Thus, each row models $A^B$. There are $|C|$ such rows. Together, they model $(A^B)^C$. The same model works for $A^{BC}$ as well.

Let $A$, $B$, and $C$ be some given sets. In Section 1.9.1, we've already defined the power $A^B$: each individual element in it is a function from $B$ into (not necessarily onto) $A$. Thus, the power of the power is

$$\left(A^B\right)^C \equiv \{f \mid f : C \to A^B\}.$$

Recall also that the product $BC$ is defined in Section 1.5.1. Thus, we also have the power

$$A^{BC} \equiv \{f \mid f : BC \to A\}.$$

Let's show that the former is equivalent to the latter. For this purpose, let's design a one-to-one mapping $M$ from the former onto the latter. This will indeed prove that

$$\left|\left(A^B\right)^C\right| = \left|A^{BC}\right|.$$

For finite sets, this is well-known. Here, however, we extend this to infinite sets as well.

To define $M$, consider a particular function $f : C \to A^B$. For each individual element $c \in C$, $f(c)$ is by itself a function from $B$ to $A$:

$$f(c)(b) \in A, \quad b \in B.$$

In this function, to obtain the final output in $A$, we must pick two concrete inputs: $c \in C$ and $b \in B$. Roughly speaking, to determine the unique output, $f$ uses $|BC|$ degrees of freedom (Figure 1.10). Fortunately, the same is true in $A^{BC}$ as well. Thus, $f$ could be mirrored by the new function $M(f)$, defined in $BC$:

$$M(f)((b,c)) \equiv f(c)(b), \quad (b,c) \in BC.$$

This way, $M$ is one-to-one. Indeed, let $f, g \in (A^B)^C$ be two distinct functions from $C$ to $A^B$, which disagree with each other on at least one element $c \in C$. Thus, as a function from $B$ to $A$, $f(c)$ must disagree with $g(c)$ on at least one element $b \in B$:

$$f(c)(b) \neq g(c)(b).$$

Thus, we've just found a pair $(b,c) \in BC$ on which $M(f)$ disagrees with $M(g)$:

$$M(f)(b,c) = f(c)(b) \neq g(c)(b) = M(g)(b,c).$$

This is enough to conclude that they are not the same function:

$$M(f) \neq M(g).$$

Thus, $M$ is indeed one-to-one, as asserted.

Furthermore, $M$ maps $(A^B)^C$ not only into but also *onto* $A^{BC}$. Indeed, given a function in $A^{BC}$, it is easy to reconstruct that function in $(A^B)^C$ that is mapped to it. This completes the proof of the power rule:

$$\left|\left(A^B\right)^C\right| = |A^{BC}|.$$

Let's go ahead and use it.

### 1.13.2 Finite and Infinite Dimension

Thanks to the power rule, we can now study both finite- and infinite-dimensional spaces at the same time. It turns out that both are rather small. Indeed, for the former,

$$|\mathbb{R}^n| = \left|\left(\{0,1\}^{\mathbb{N}}\right)^n\right| = \left|\{0,1\}^{n\mathbb{N}}\right| = \left|\{0,1\}^{\mathbb{N}}\right| = \aleph.$$

Similarly, for the latter,

$$|\mathbb{R}^{\mathbb{N}}| = \left|\left(\{0,1\}^{\mathbb{N}}\right)^{\mathbb{N}}\right| = \left|\{0,1\}^{\left(\mathbb{N}^2\right)}\right| = \left|\{0,1\}^{\mathbb{N}}\right| = \aleph.$$

(See Section 1.5.2 and the exercises at the very end of this chapter.) In summary,

$$\aleph^{\aleph_0} = \aleph^n = \aleph.$$

So, these spaces are as small as the real axis. Is there a genuinely bigger set? Fortunately, there is: the exponent set, containing those functions defined on the real axis.

### 1.13.3 Yet Greater Cardinalities

So far, we didn't see any cardinality greater than $\aleph$. In fact, even the infinite-dimensional space $\mathbb{R}^{\mathbb{N}}$ is not really bigger than the real axis. Fortunately, we can still design a yet bigger set. Indeed, thanks to the theorem in Section 1.9.2, we can move on to the exponent set, containing the characteristic functions defined on the real axis:

$$2^{\aleph} = \left|\{0,1\}^{\mathbb{R}}\right| > |\mathbb{R}| = \aleph.$$

Is there a yet bigger set, with a yet greater cardinality? Well, let's try the set of real functions:

$$\mathbb{R}^{\mathbb{R}} = \{f \mid f : \mathbb{R} \to \mathbb{R}\}.$$

This seems bigger, but is not. Indeed, from the power rule and the exercises at the very end of this chapter, we have that

$$\aleph^{\aleph} = \left|\mathbb{R}^{\mathbb{R}}\right| = \left|\left(\{0,1\}^{\mathbb{N}}\right)^{\mathbb{R}}\right| = \left|\{0,1\}^{\mathbb{N}\mathbb{R}}\right| \le \left|\{0,1\}^{\left(\mathbb{R}^2\right)}\right| = \left|\{0,1\}^{\mathbb{R}}\right| = 2^{\aleph}.$$

Thus, to have a greater cardinality, we must turn to the exponent:

$$2^{\left(2^{\aleph}\right)} > 2^{\aleph}.$$

Is there a yet greater cardinality? Well, let's try $(2^{\aleph})^{(2^{\aleph})}$:

$$(2^{\aleph})^{(2^{\aleph})} = 2^{\left(\aleph \cdot 2^{\aleph}\right)} \le 2^{\left(2^{\aleph} \cdot 2^{\aleph}\right)} = 2^{\left((2^{\aleph})^2\right)} = 2^{\left(2^{2\aleph}\right)} = 2^{\left(2^{\aleph}\right)}.$$

Thus, to have a greater cardinality, we must again turn to the exponent:

$$2^{\left(2^{\left(2^{\aleph}\right)}\right)} > 2^{\left(2^{\aleph}\right)},$$

and so on.

## 1.14 Null Set

### 1.14.1 Null Set

What is a null set? It is a subset of $\mathbb{R}$ that is really small in size: it has zero size (or measure). Indeed, for every (arbitrarily small) $\varepsilon > 0$, it could be covered by open intervals of total length as small as $\varepsilon$. As we'll see below, being small in this sense is not necessarily the same as being small in terms of cardinality.

### 1.14.2 Countable Set

Is $\mathbb{N}$ a null set? Well, consider a fixed $\varepsilon > 0$. Now, cover $\mathbb{N}$ as in Figure 1.11: the first number, 1, is covered by the open interval

$$\left(1 - \frac{\varepsilon}{4}, 1 + \frac{\varepsilon}{4}\right).$$

The second number, 2, is covered by a slightly shorter interval:

Fig. 1.11. $\mathbb{N}$ is a null set. Indeed, for arbitrarily small $\varepsilon > 0$, the natural numbers could be covered by open intervals of total length as small as $\varepsilon$.

$$\left(2 - \frac{\varepsilon}{8}, 2 + \frac{\varepsilon}{8}\right).$$

The third number, 3, is covered by a yet shorter interval:

$$\left(3 - \frac{\varepsilon}{16}, 3 + \frac{\varepsilon}{16}\right),$$

and so on. What is the total length of all such intervals? It is as small as

$$\frac{\varepsilon}{2} + \frac{\varepsilon}{4} + \frac{\varepsilon}{8} + \cdots = \frac{\varepsilon}{2} \sum_{i=0}^{\infty} \left(\frac{1}{2}\right)^i = \frac{\varepsilon}{2} \cdot \frac{1}{1 - \frac{1}{2}} = \varepsilon.$$

Thus, $\mathbb{N}$ is indeed a null set: it has been covered by open intervals of total length as small as $\varepsilon$, as required.

In terms of cardinality, $\mathbb{N}$ is much bigger than the empty set. Still, in terms of measure, it is as small. Indeed, in the above proof, $\varepsilon$ could be as small as you like.

The same is true for every countable subset of $\mathbb{R}$. For example, $\mathbb{Q}$ is a null set as well. In the exercises below, we'll meet yet another null set: Cantor set. Still, Cantor set is not so small in terms of cardinality. On the contrary: its cardinality is $\aleph$.

## 1.15 Exercises: De-Morgan Laws and Mathematical Logic

### 1.15.1 Logical "Or"

1. Set theory mirrors mathematical logic. In what sense? Hint: see below.
2. Indeed, the logical "or" operator mirrors the union operator. In what sense?

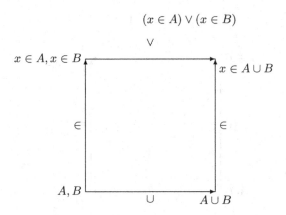

**Fig. 1.12.** Commutative diagram: in set theory, go rightwards (the union operator), and then upwards ('∈'). In mathematical logic, on the other hand, go the other way around: first upwards ('∈'), and then rightwards (the logical "or" operator), to obtain the same thing.

3. Show that, for every two sets $A$ and $B$, the following statements are the same:

$$(x \in A \ \text{or} \ x \in B) \ \Leftrightarrow \ x \in A \cup B.$$

4. Illustrate this geometrically. Hint: Figure 1.12 illustrates the commutative diagram.

### 1.15.2 Logical "And"

1. Likewise, the logical "and" operator mirrors the intersection operator. In what sense?
2. Show that, for every two sets $A$ and $B$, the following statements are the same:

$$(x \in A \ \text{and} \ x \in B) \ \Leftrightarrow \ x \in A \cap B.$$

3. Illustrate this geometrically. Hint: Figure 1.13 illustrates the commutative diagram.

### 1.15.3 Mathematical Equivalence Relation

1. What is a set? Hint: something that satisfies the axioms of set theory.
2. What is an element in a set?
3. Interpret a set and its elements as pure abstract objects, with the relation '∈' between them: $e \in A$ means that $e$ is an element of $A$.
4. Show that, in general, '∈' is neither reflexive nor symmetric nor transitive. Hint: see Section 1.6.1, and design counter examples.

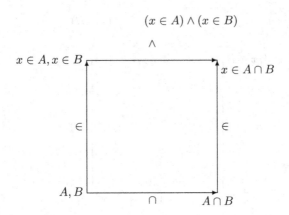

**Fig. 1.13.** Commutative diagram: in set theory, go rightwards (the intersection operator), and then upwards ('$\in$'). In mathematical logic, on the other hand, go the other way around: first upwards ('$\in$'), and then rightwards (the logical "and" operator), to obtain the same thing.

5. Conclude that '$\in$' is not a mathematical equivalence relation.
6. Interpret sets as pure abstract objects, with the relation '$\subset$' between them: $A \subset B$ means that every element $e \in A$ satisfies $e \in B$ as well:

$$A \subset B \Leftrightarrow (e \in A \Rightarrow e \in B).$$

7. Show that '$\subset$' is reflexive: for every set $A$, $A \subset A$.
8. Show that '$\subset$' is also transitive: for every three sets $A$, $B$, and $C$, if $A \subset B$ and $B \subset C$, then $A \subset C$ as well.
9. Show that, in general, '$\subset$' is not symmetric: $A \subset B$ doesn't necessarily imply that $B \subset A$. Hint: design a counter example, in which $A \subset B$ but $A \neq B$.
10. Conclude that '$\subset$' is *not* a mathematical equivalence relation.
11. Show that $A = B$ if and only if $A \subset B$ and $B \subset A$ at the same time.
12. Show that '$=$' is reflexive: for every set $A$, $A = A$.
13. Show that '$=$' is also symmetric: for every two sets $A$ and $B$, $A = B$ implies that $B = A$ as well.
14. Show that '$=$' is also transitive: for every three sets $A$, $B$, and $C$, if $A = B$ and $B = C$, then $A = C$ as well.
15. Conclude that, unlike '$\subset$', '$=$' is a mathematical equivalence relation.

### 1.15.4 Union and Intersection

1. For every two sets $A$ and $B$, define their union (or disjunction) by

$$A \cup B \equiv \{a \mid a \in A \text{ or } a \in B\}.$$

2. Is this a legitimate set as well? Hint: it sure is. In fact, even the union of infinitely many sets is a legitimate set as well. This is indeed one of the fundamental axioms in set theory.
3. Furthermore, define also their intersection (or conjunction) by

$$A \cap B \equiv \{a \mid a \in A \text{ and } a \in B\}.$$

4. Is this a legitimate set? Hint: it is a subset of $A$.
5. Show that the intersection of any number of sets is a legitimate set as well. Hint: use mathematical induction.

### 1.15.5 Associative Laws

1. Let $A$, $B$, and $C$ be some sets. Show that

$$A \cup (B \cup C) = (A \cup B) \cup C.$$

2. This is a sort of associative law: parentheses are not really needed, and could be dropped.
3. What does this mean in terms of mathematical logic? Hint: see Figure 1.12.
4. Furthermore, show also that

$$A \cap (B \cap C) = (A \cap B) \cap C.$$

5. This is a sort of associative law: parentheses are not really needed, and could be dropped.
6. What does this mean in terms of mathematical logic? Hint: see Figure 1.13.

### 1.15.6 Distributive Laws

1. Next, show also that

$$A \cap (B \cup C) = (A \cap B) \cup (A \cap C).$$

Hint: consider an element in $A$ that also lies in $B$ or in $C$.
2. This is a sort of distributive law: it tells us how to open parentheses.
3. What does this mean in terms of mathematical logic?
4. Moreover, show also that

$$A \cup (B \cap C) = (A \cup B) \cap (A \cup C).$$

Hint: look at an element $x$. Consider two possibilities: either $x \in A$, or $x \in B \cap C$.
5. This is a sort of distributive law: it tells us how to open parentheses.
6. What does this mean in terms of mathematical logic?

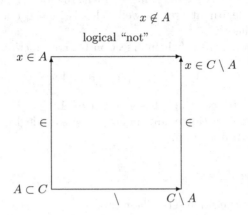

**Fig. 1.14.** In set theory, go rightwards (the '\' operator), and then upwards ('∈'). In mathematical logic, on the other hand, go the other way around: first upwards ('∈'), and then rightwards (the logical "not" operator).

### 1.15.7 Logical "Not"

1. Next, assume that $C$ is a bigger set that includes $A$:

$$A \subset C.$$

2. Let $x$ be an element of $C$, but not of $A$:

$$x \in C, \quad \text{but} \quad x \notin A.$$

3. Use the '\' operator to write this more concisely. Hint:

$$x \in C \setminus A.$$

4. Illustrate this geometrically. Hint: see Figure 1.14.

### 1.15.8 De-Morgan's Laws

1. Next, assume that both $A$ and $B$ are included in $C$:

$$A \subset C, \ B \subset C.$$

2. Under these circumstances, show that

$$C \setminus (A \cup B) = (C \setminus A) \cap (C \setminus B)$$
$$C \setminus (A \cap B) = (C \setminus A) \cup (C \setminus B).$$

Hint: to prove the former law, consider an element in $C$ that is neither in $A$ nor in $B$. To prove the latter, on the other hand, consider an element in $C$ that is either not in $A$ or not in $B$.

3. How to open parentheses that follow the '\' operator? Hint: in the above laws, upon opening parentheses, the inner symbol gets reversed, upside down: '∪' converts to '∩', and '∩' converts back to '∪'.
4. These are De-Morgan's laws.
5. What do they mean in terms of mathematical logic? Hint: mirror the '\' operator by the logical "not" operator (Figure 1.14).

### 1.15.9 Bell's Inequality

1. Let $A$, $B$, and $C$ be some sets. Show that

$$A \setminus C \subset (A \setminus B) \cup (B \setminus C).$$

Hint: pick some element $a \in A \setminus C$. This means that $a \in A$, but $a \notin C$. Now, is $a \in B$? If it is, then $a \in B \setminus C$, as required. If, on the other hand, it isn't, then $a \in A \setminus B$, as required.
2. This is Bell's inequality.

### 1.15.10  Von-Neumann Numbers

1. In Section 1.2.2, why do we write "$n + 1$" in quotation marks? Hint: this is no addition. This is just the next natural number: the successor of $n$.
2. Use the original definition to show that "$n + 1$" $\neq n$.
3. Furthermore, show that

$$1, 2, 3, \ldots, n \subset \text{"}n + 1\text{."}$$

Hint: use mathematical induction.
4. Moreover, show that

$$1, 2, 3, \ldots, n \in \text{"}n + 1\text{."}$$

Hint: use mathematical induction.
5. Let $m, n \in \mathbb{N}$ be two natural numbers. Use the above to define a new relation: $m \leq n$ if $m \subset n$.
6. Alternatively, define this relation in a slightly different way: $m < n$ if $m \in n$.
7. Show that $m < n$ if (and only if) $m \leq n$ or $m = n$. Hint: use the original definitions, and mathematical induction on $n - m = 1, 2, 3 \ldots$.

### 1.15.11  Union of Countable Sets

1. Show that the set of even numbers is equivalent to $\mathbb{N}$.
2. Conclude that this set is countable (has cardinality $\aleph_0$).
3. Show that the set of odd numbers is equivalent to $\mathbb{N}$ as well.
4. Conclude that this set is countable as well.
5. Let $A$ and $B$ be two sets. What is their product $AB$? Hint: it contains pairs of the form $(a, b)$, for every $a \in A$ and $b \in B$ (Section 1.5.1).
6. For example, assume that $A$ contains one element only: $A \equiv \{a\}$. In this case, what is $AB$? Hint: it contains pairs of the form $(a, b)$, for every $b \in B$.

7. For example, assume that $A \equiv \{n\}$ (where $n$ is some fixed natural number) and $B = \mathbb{N}$. In this case, what is $AB$? Hint: in this case, $AB = \{n\}\mathbb{N}$ contains pairs of the form $(n, i)$, for every $i \in \mathbb{N}$.

8. Show that this set is countable:
$$|\{n\}\mathbb{N}| = |\mathbb{N}|.$$

9. Now, take the union over all $n \in \mathbb{N}$. What do you get? Hint:
$$\cup_{\{\{n\}\mathbb{N} \mid n \in \mathbb{N}\}} \equiv \cup_{n=1}^{\infty}\{n\}\mathbb{N} = \mathbb{N} \cdot \mathbb{N} = \mathbb{N}^2.$$

10. Consider the $n$-dimensional infinite grid. What is its cardinality? Hint: for $n = 2$, we already have two proofs that
$$\left|\mathbb{N}^2\right| = |\mathbb{N}| = \aleph_0$$

(Sections 1.5.2 and 1.12.2). Extend these proofs to a bigger $n$ as well:
$$|\mathbb{N}^n| = |\mathbb{N}| = \aleph_0.$$

Better yet, prove this by mathematical induction.

11. In Section 1.7.2, we studied the set of finite binary sequences, made of 0's and 1's. But what if they were made not only of 0's and 1's but also of any natural number? Show that the cardinality would still be $\aleph_0$. Hint: thanks to the previous exercises,
$$\left|\cup_{n \in \mathbb{N}}\mathbb{N}^n\right| = \left|\cup_{n \in \mathbb{N}}\{n\}\mathbb{N}\right| = \left|\mathbb{N}^2\right| = \aleph_0.$$

## 1.15.12  Prime Factors — Unique Representation

1. Prove the above yet more directly. For this purpose, define a prime number: a natural number (greater than 1) that can be divided (evenly, with no remainder) only by 1 and itself.

2. Prove that there are infinitely many prime numbers. Hint: by contradiction: if there were only a few prime numbers, then take the product of all of them, and add 1, to design a new (yet bigger) prime number.

3. Let $m$ be a natural number, greater than 1. Does $m$ have at least one prime factor that divides it (evenly, with no remainder)? Hint: if $m$ is prime, then $m$ itself is its own prime "factor." Otherwise, $m$ could be written as the product of two smaller numbers. By mathematical induction, they must have a prime factor.

4. Does $m$ have a unique representation as the product of its own prime factors? Hint: if $m$ is prime, then $m$ itself is its own (unique) prime "factorization." Otherwise, then $m$ must have a prime factor $p$. Apply the induction hypothesis to $m/p$.

5. Show more directly that this representation is unique. Hint: if there were two, then each factor of the form $p^l$ (for some power $l$) in one of them must also divide the other (evenly, with no remainder).

6. Map each finite sequence of natural numbers to a new natural number: product of powers of prime numbers:
$$(k, l, m, n, \ldots) \rightarrow 2^k 3^l 5^m 7^n \cdots.$$

7. Is this mapping one-to-one? Hint: we've already shown twice that the prime factorization is unique.
8. Alternatively, map each finite sequence of natural numbers to a new decimal number:

$$(k, l, m, n, \ldots) \rightarrow k7l7m7n7\ldots.$$

On the right, $k$, $l$, $m$, and $n$ are in their binary form, and the fictitious digit '7' separates them from each other, to yield a unique decimal number in $\mathbb{N}$.
9. Is this mapping one-to-one?
10. Conclude that the set of finite sequences of natural numbers is indeed countable:

$$\left| \cup_{\{\mathbb{N}^n \mid n \in \mathbb{N}\}} \right| \equiv \left| \cup_{n=1}^{\infty} \mathbb{N}^n \right| \leq |\mathbb{N}| = \aleph_0.$$

11. Conclude that, in general, the union of $\aleph_0$ (disjoint) countable sets is countable as well.

## 1.15.13 Uncountable Null Set

1. The above exercises focus on countable sets and their union. Now, let's consider uncountable sets as well.
2. Give an example of an uncountable set.
3. Prove that it is indeed uncountable.
4. Could an uncountable set be equivalent to $\mathbb{N}$?
5. Could an uncountable set be equivalent to a countable set?
6. Let $a$ and $b$ be two real numbers satisfying $a < b$. What is the cardinality of the closed interval

$$[a, b] \equiv \{x \in \mathbb{R} \mid a \leq x \leq b\}?$$

7. Show that the answer is

$$|[a, b]| = |[0, 1]|.$$

Hint: use the linear mapping

$$x \rightarrow a + (b - a)x.$$

8. Similarly, show that the same is true for the open interval as well:

$$|(a, b)| = |(0, 1)|.$$

9. Conclude that all closed, open, and half-open intervals (with or without their endpoints) have the same cardinality:

$$|[a, b]| = |(a, b)| = |(a, b]| = |[a, b)| = \aleph.$$

Hint: map $[0, 1]$ into a small inner interval, and use Cantor-Bernstein's theorem (Section 1.6.2).
10. Similarly, show that all squares (either closed or open or half-open) are equivalent to each other.
11. Use the mapping $x \rightarrow \tan(x)$ to show that

$$\left| \left( -\frac{\pi}{2}, \frac{\pi}{2} \right) \right| = |\mathbb{R}|.$$

12. Show that the half-open interval $[0, 2\pi)$ is equivalent to a circle. Hint: map $x \to \theta$.

13. Similarly, show that the open interval $(0, 2\pi)$ is equivalent to a circle minus one point.

14. Use the above to show once again that

$$|(0, 2\pi)| = |\mathbb{R}|.$$

15. Use the mapping $x \to \cotan(x)$ to show once again that

$$|(0, \pi)| = |\mathbb{R}|.$$

16. Modify the hyperbolic mapping to define the new mapping

$$x \to \begin{cases} \frac{1}{1-x} - 1 & \text{if} \quad 0 \le x < 1 \\ \\ -\frac{1}{1+x} + 1 & \text{if} \ -1 < x < 0. \end{cases}$$

17. Show that this maps the open interval $(-1, 1)$ onto the entire real axis.

18. Show that this mapping is monotonically increasing.

19. Conclude that it preserves the standard order of real numbers.

20. Conclude that it is one-to-one.

21. Conclude that it is invertible.

22. Write the inverse mapping explicitly.

23. Use the above to show once again that

$$|(-1, 1)| = |\mathbb{R}|.$$

24. What's special about this mapping? Hint: it is algebraic rather than trigonometric. Therefore, it maps each rational number to a rational number.

25. Conclude that, in the open interval $(-1, 1)$, there are as many rational numbers as in $\mathbb{Q}$:

$$|\mathbb{Q} \cap (-1, 1)| = |\mathbb{Q}|.$$

26. Prove this also indirectly, using Cantor-Bernstein's theorem.

27. Show that a square is equivalent to a sphere. Hint: fold the square in such a way that its boundary shrinks to a single point.

28. Conclude that every square (either closed or open or half-open) is equivalent to the entire Cartesian plane.

29. Show that the unit interval is uncountable. Hint: see Section 1.11.2.

30. Show that the real axis has the same cardinality as well:

$$|\mathbb{R}| = \aleph.$$

31. Show that the unit square is equivalent to the unit interval. Hint: see Section 1.12.1.

32. Show that the Cartesian plane has the same cardinality as well:

$$|\mathbb{R}^2| = \aleph.$$

33. Prove this in yet another way. Hint: In Section 1.13.2, set $n = 2$.

34. Extend this to every finite-dimensional space as well:

$$|\mathbb{R}^n| = \aleph.$$

Hint: use mathematical induction.

35. Show that this is also true for an infinite-dimensional space:

$$|\mathbb{R}^{\mathbb{N}}| = \aleph.$$

Hint: see Section 1.13.2.

36. What is a function?

37. Interpret a function as a mapping.

38. Interpret a function as a machine that takes an input to produce an output.

39. Consider the set of characteristic functions, defined on the unit interval. What is its cardinality? Is it greater than $\aleph$? Hint: from Section 1.9.2,

$$\left|\{0,1\}^{[0,1]}\right| > |[0,1]|.$$

40. What's the connection between a characteristic function and a subset?

41. What is the power set? Hint: the set of subsets.

42. What is its cardinality? Is it greater than that of the original set? Hint: yes, it is (Section 1.9.2).

43. Recall that $\mathbb{Q}$ is the set of rational numbers. Is it countable?

44. Is it a null set? Hint: see Sections 1.7.1 and 1.14.1.

45. Must a null set be countable?

46. Is there a null set that is uncountable?

47. Is Cantor set a null set? Hint: in Figure 5.1, drop not the entire subinterval, but a slightly shorter one: shorter by $\varepsilon/4$ from either side. Likewise, in Figure 5.2, drop not the entire subsubinterval, but a slightly shorter one: shorter by $\varepsilon/16$ from either side, and so on. In the end, the remaining intervals have a total length as small as $\varepsilon$, yet they cover Cantor's set completely.

48. Is Cantor's set countable? Hint: no — see exercises at the end of Chapter 5.

49. What is its cardinality? Hint: $\aleph$ — see exercises at the end of Chapter 5.

50. Conclude that a null set is not necessarily countable.

51. Now, look at another version of Cantor set, constructed in Figures 5.3–5.4. Is it a null set?

52. Is it countable?

53. What is its cardinality?

54. Conclude once again that a null set is not necessarily countable.

55. Show that the mapping $M$ defined in Section 1.13.1 indeed maps $(A^B)^C$ onto $A^{BC}$.

56. Let $A$, $B$, and $C$ be three given sets. Assume that $B$ and $C$ are equivalent to each other by the one-to-one mapping $M$ from $B$ onto $C$. Construct a one-to-one mapping from $A^B$ onto $A^C$. Hint: each function of the form $f : B \to A$ is mapped to the composite function $f \circ M^{-1} : C \to A$.

57. Conclude that

$$|B| = |C| \Rightarrow |A^B| = |A^C|.$$

58. Show also that
$$|B| \leq |C| \Rightarrow |A^B| \leq |A^C|.$$

Hint: in this case, $M$ maps $B$ onto the subset $M(B) \subset C$, so each function of the form $f : B \to A$ could be mapped to a function that agrees with $f \circ M^{-1}$ on $M(B)$, and is constant in $C \setminus M(B)$.

59. Conclude that the set of integer functions
$$\mathbb{Z}^{\mathbb{Z}} \equiv \{f \mid f : \mathbb{Z} \to \mathbb{Z}\}$$

is equivalent to the real axis. Hint:

$$\aleph = 2^{\aleph_0} \leq \aleph_0^{\aleph_0} = |\mathbb{Z}^{\mathbb{Z}}| \leq \aleph^{\aleph_0} = \left(2^{\aleph_0}\right)^{\aleph_0} = 2^{\aleph_0 \cdot \aleph_0} = 2^{\aleph_0} = \aleph.$$

60. Let $A$ be some infinite set. Is
$$|A^2| = |A|?$$

Hint: if $|A| = \aleph_0$ or $|A| = \aleph$, then we already know that this is indeed true. If, on the other hand, $A$ has a different cardinality, then we don't know this as yet. Only at the end of the next chapter will we discover the answer.

# 2

## Ordinals
## and Zorn's Lemma

We are now ready to introduce an order between elements in the set. For natural numbers, $m < n$ could be the standard order: $m$ is smaller than $n$. Still, this is not the only option: one could also define a nonstandard order, completely different from this order. In fact, this could be done in every set.

Once a set is well-ordered, it makes an ordinal: an example that tells us how the order looks like. One could then talk about sets of ordinals.

We've already seen that all possible sets could never be placed in one set: there are too many. This is also true for ordinals: all possible ordinals could never be placed in one set: there are too many. At best, they could be contained in a container that is *not* a set: on a container, one can never carry out operations like union or intersection.

This is called the ordinal paradox. Fortunately, in their container, the ordinals are still well-ordered: we say that one ordinal is "smaller" than another if it could serve as an initial segment in it.

Thanks to the ordinal paradox, we can introduce three (equivalent) fundamental principles [3]: the axiom of choice, the well-ordering theorem, and Zorn's lemma. In fact, each of these three could serve as an axiom, from which the other two follow. Finally, we also use Zorn's lemma to add and multiply cardinalities with each other. Later on, we'll use Zorn's lemma in functional analysis and physics as well.

## 2.1 Partial Order

### 2.1.1 Partial vs. Complete Order

So far, we never imposed any order on the set. Indeed, we never assumed that one element is smaller or bigger than another. In this chapter, on the other hand, this is going to change: we're going to define all sorts of orders (standard or not), and see how they look like. Let's start with a familiar example.

Every two natural numbers $m, n \in \mathbb{N}$ are ordered: either $m \leq n$ or $n \leq m$. This is indeed a complete order.

A more general set $A$, on the other hand, may be not as perfectly ordered. It may have just a partial order: two elements $a, b \in A$ *may* be comparable or not. If they are, say $a \leq b$, then we say that $a$ is before $b$, or "smaller" than $b$, in some (nonstandard) sense, to be specified later.

### 2.1.2 Properties of Partial Order

What could this sense be? Well, we don't know as yet. We only know that it must have three properties. The first and the second are already familiar to us: they are the same as in mathematical equivalence relation (Chapter 1, Section 1.6.1). The third, on the other hand, is new. In all three, '$\leq$' means "smaller than or equal to," not necessarily in the usual sense, but possibly in a new sense, to be specified later:

1. Reflexive:

$$a \leq a, \quad a \in A.$$

2. Transitive:

$$a \leq b, \ b \leq c \ \Rightarrow \ a \leq c, \quad a, b, c \in A.$$

3. "Anti-symmetric:"

$$a \leq b, \ b \leq a \ \Rightarrow \ a = b, \quad a, b \in A.$$

Why the quotation marks? Because, later on, we'll also introduce a stricter kind of antisymmetry, with no quotation marks.

Could our set contain two different copies of the same thing? In other words, could two elements $a$ and $b$ have the same "value," and obey the same order rules? Well, in such a case, they must be comparable, and satisfy both $a \leq b$ and $b \leq a$ at the same time. Thanks to "antisymmetry," we then also know that $a = b$: $a$ and $b$ must be one and the same element. This makes life easier: our original set may contain no duplicate copies.

### 2.1.3 Example: The Power Set

Consider now a general set $S$, with no order at all. The subsets of $S$ could now be placed in a new set — the power set of $S$:

$$P(S) \equiv \{T \mid T \subset S\}.$$

In Chapter 1, Sections 1.8.2–1.8.3, we've already seen the power set, and established that it is equivalent to the set of characteristic functions on $S$:

$$|P(S)| = 2^{|S|}.$$

How to order $P(S)$? Well, what is an individual element in $P(S)$? It is actually a set in its own right: a subset of $S$. Such subsets could be ordered in terms of inclusion:

$$T \leq U \ \text{ if } \ T \subset U, \quad T, U \subset S.$$

Is this a legitimate order? Well, it is indeed reflexive, transitive, and "anti-symmetric," as required. (Check!)

Still, this order is only partial, not complete. Indeed, one could easily design two subsets that are not included in one another.

Still, there are two special subsets that are comparable to every other subset. These are the empty set $\emptyset$, and the entire set $S$. Indeed, every subset $T \subset S$ is in between $\emptyset$ and $S$ in terms of inclusion:

$$\emptyset \subset T \subset S.$$

Thus, in $P(S)$, $\emptyset$ is the smallest element, and $S$ is the greatest element.

Still, this is just an example. What about the general case? Is there always a greatest element? Not necessarily.

### 2.1.4 Maximal Element and Greatest Element

Consider again the general (partially ordered) set $A$. What is the greatest element in $A$? If exists, then it is greater than every other element in $A$.

Is it unique? Well, it must be: if there were two greatest elements $a, b \in A$, then we'd have both $a \le b$ and $b \le a$ at the same time, leading to $a = b$.

A maximal element, on the other hand, is different. If exists, then it has no greater element in $A$. It is not necessarily the greatest: there could be many elements not comparable to it at all.

It is not necessarily unique: there could be a few distinct maximal elements in $A$, not comparable to each other at all. The greatest element, if exists, must be maximal.

A minimal element and the smallest element (if exist) are defined in a similar way. The smallest element is also called the first element.

## 2.2 Chains and Their Examples

### 2.2.1 Chain and Its Upper Bound

As discussed above, some elements in $A$ may be not comparable at all. Still, let's pick some elements that are. This would make a new chain. For example,

$$a_1 \le a_2 \le a_3 \le \cdots.$$

In general, a chain is a completely-ordered subset: a subset of $A$, in which all elements are comparable to one another.

A chain may have an upper bound: an element as great as all elements in the chain. The upper bound may belong to the chain, or not. The greatest element in $A$, if exists, is a good upper bound. Still, there may be a yet smaller upper bound.

### 2.2.2 Example: Subsets Ordered by Inclusion

In Section 2.1.3, for example, a chain in $P(S)$ takes the form of a list of bigger and bigger subsets of $S$:

$$c_1 \subset c_2 \subset c_3 \subset \cdots \subset S.$$

Here, each $c_i$ is a subset of $S$, or an element of $P(S)$. Together, the entire chain is a subset of $P(S)$:

$$C \equiv \{c_1, c_2, c_3, \ldots\} \subset P(S).$$

The entire chain $C$ is not necessarily countable: the index is not necessarily an integer number, but could be a fraction as well. Still, for simplicity, we often stick to the standard index $i = 1, 2, 3, \ldots$.

Fortunately, we already have an upper bound: $S$ itself. After all, $S$ is the greatest element in $P(S)$. Still, is there a yet smaller upper bound? Yes, there is: the union of all the $c_i$'s:

$$\cup_C \equiv \{q \in c \mid c \in C\}.$$

This union contains all the inner elements in the $c_i$'s.

Is this union in $C$? Well, it could be. For example, if $C$ is finite, then this union is just the last member in $C$.

Still, as discussed above, $C$ could be infinite, and even uncountable. In such a case, the above union could be left outside $C$, and belong to $P(S) \setminus C$.

### 2.2.3 Example: Extended Functions

So far, we've ordered subsets in terms of inclusion. Let's use this to order functions as well.

Consider one-to-one functions $f$ and $g$, defined on some subset of $S$. When do we say that $f \leq g$? Well, for this, $g$ should extend $f$.

For example, $f$ could be defined on $c_1$, and $g$ on $c_2$ (where $c_1 \subset c_2 \subset S$). Still, this is not enough: to serve as a legitimate extension, $g$ must also agree with $f$ on $c_1$:

$$f \leq g \text{ if } g \mid_{c_1} \equiv f.$$

Is this a legitimate order? Well, it is reflexive, transitive, and "anti-symmetric," as required. (Check!) Still, it is not complete, but only partial. To see this, just pick two functions that disagree: they can never extend one another any more.

Is there a maximal function? Well, every one-to-one function defined on $S$ is maximal: it can never be extended any more. Is this function the greatest? No! After all, many other functions disagree with it, and can never be extended by it. Thus, there is no greatest function at all.

Consider now a chain $C$ as in Section 2.2.2: a list of bigger and bigger subsets of $S$. Let's use it to define new functions: for each index $i$, let $f_i$ be a one-to-one function defined on $c_i \subset S$. Assume that each function extends the previous ones:

$$f_1 \leq f_2 \leq f_3 \leq \cdots.$$

This makes a new chain of functions.

Does it have an upper bound? Well, let's design one. For this purpose, define a new function $f$ on the union $\cup_C$, defined in Section 2.2.2.

Recall that $\cup_C$ contains the inner elements in all the $c_i$'s. Therefore, each element $q \in \cup_C$ must belong to some $c_i \subset S$. So, one should better define $f$ to agree with this:

$$f(q) \equiv f_i(q).$$

Since the functions extend each other, there is no conflict in this definition.

Still, we are not done yet. After all, our functions are special: we allow one-to-one functions only. Like the $f_i$'s, $f$ must be one-to-one as well. Is it? To check on this, pick two distinct elements $q \neq r \in \cup_C$. They must belong to some subsets $c_i$ and $c_j$, respectively. Without loss of generality, assume that $c_j \subset c_i \subset S$. Since $f_i$ is one-to-one, we have

$$f(q) = f_i(q) \neq f_i(r) = f(r),$$

as required. So, $f$ is indeed a good upper bound. This will be useful later.

### 2.2.4 Functions Onto Their Squared Domain

So far, we didn't specify the image of the function. Let us now require that the image is the square of the domain where the function is defined. In the above chain, for example, $f_i$ is a one-to-one function from $c_i$ *onto* $c_i^2$. Is the above upper bound still valid? In other words, is $f$ a one-to-one function from $\cup_C$ *onto* $(\cup_C)^2$?

Yes, it is! Indeed, we already know that $f$ is one-to-one. To prove the "onto" bit as well, consider a pair of the form $(x, y) \in (\cup_C)^2$. In other words, $x \in \cup_C$, so $x \in c_m$ for some $m$, and $y \in \cup_C$ as well, so $y \in c_n$ for some $n$. Without loss of generality, assume that $c_n \subset c_m$. Because $f_m$ is onto $c_m^2$, $(x, y)$ is in the image of $f_m$, so it must be in the image of $f$ as well, as asserted. This proves that $f$ is indeed a legitimate upper bound, as required.

### 2.2.5 Choice Functions

So far, we've considered one-to-one functions only. Next, let's drop this requirement altogether. Instead, let's assume that the functions are choice functions that pick an inner element.

For this purpose, assume now that the elements in $S$ are (nonempty) sets in their own right. This is also true for each subset $c_i \subset S$: its elements are now nonempty sets. So, it is possible to pick an inner element from each. This is indeed what $f_i$ does: it maps each element in $c_i$ to a yet inner element:

$$f_i : c_i \to \cup_{c_i},$$

and

$$f_i(s) \in s, \quad s \in c_i.$$

This is why $f_i$ is called a choice function: from each element in $c_i$, it picks an inner element.

Clearly, even in the new circumstances, the $f_i$'s still make a legitimate chain of functions. After all, they still agree with each other, and extend each other. Now, however, we have a different requirement: we no longer care whether the $f_i$'s are one-to-one or onto. Instead, we only require them to be choice functions, as discussed above.

Now, define $f$ as before. Is $f$ a legitimate upper bound? In other words, is $f$ a choice function as well? It sure is! Indeed, every element $q \in \cup_C$ must belong to $c_i$ for some $i$, so

$$f(q) = f_i(q) \in q,$$

as required. This will be useful later.

## 2.3 Strict Partial Order and Complete Order

### 2.3.1 Strict Partial Order

In Section 2.1.1, we've introduced a "weak" partial order: '$\leq$'. From it, we can now derive a strict partial order:

$$a < b \quad \text{if:} \quad a \leq b \text{ and } a \neq b.$$

This doesn't mean that $a$ is smaller than $b$ in the usual sense, but only in the new (possibly nonstandard) sense.

The original weak order enjoys three basic properties (Section 2.1.2). Let's use them to prove three new properties for the new strict order:

1. Transitive:
$$a < b, \; b < c \Rightarrow a < c, \quad a, b, c \in A.$$

   Indeed, if $a < b$ and $b < c$, then in particular $a \leq b$ and $b \leq c$, so $a \leq c$. Now, could $a = c$? Of course not! Indeed, by contradiction: if $a = c$, then we'd have $a \leq b \leq c = a$, so $a = b$, in violation of our original assumption that $a < b$. Thus, we must have $a \neq c$. In summary, $a < c$, as asserted.

2. Anti-reflexive:
$$a \not< a, \quad a \in A.$$

   After all, $a < a$ would imply $a \neq a$, which is of course impossible.

3. Anti-symmetric:
$$a < b \Rightarrow b \not< a, \quad a, b \in A.$$

   After all, $a < b < a$ would imply $a < a$, which is impossible.

So far, we've started from a given weak order, and derived a new strict order from it. Next, let's work the other way around: assume that a strict order '<' is already given, with the above properties: transitive, anti-reflexive, and anti-symmetric (without quotation marks). From it, let's derive a new weak order:

$$a \leq b \quad \text{if:} \quad \text{either } a < b \text{ or } a = b.$$

Here, we assume that there are no duplicate copies. This way, $a = b$ means not only same value but also one and the same element.

Does the new order '$\leq$' enjoy the desirable properties in Section 2.1.2? Well, thanks to the original properties of '<', it indeed does:

1. It is transitive. Indeed, if $a \leq b \leq c$, then either $a < b < c$ or $a < b = c$ or $a = b < c$ or $a = b = c$. In either case, $a \leq c$, as required.

2. It is reflexive: since $a = a$, $a \leq a$ as well.

3. It is "anti-symmetric" (with quotation marks, as in Section 2.1.2). After all, $a \leq b \leq a$ means that either $a < b < a$ (impossible) or $a = b < a$ (impossible) or $a < b = a$ (impossible) or $a = b = a$ (must).

In summary, one could work with either kind of order: either weak, or strict.

## 2.3.2 Complete Order

We started this chapter with a (weak) partial order (Sections 2.1.1–2.1.2). This way, if you picked two elements, then they could be comparable, or not.

Now, let's assume that the order is also complete: every two elements $a$ and $b$ are comparable to each other: either $a \leq b$ or $b \leq a$. For this reason, in a complete order, a maximal element, if exists, is also the greatest.

In the example in Section 2.1.3, it is easy to make the order complete: just restrict the discussion to the chain $C$ (Section 2.2.2). If the upper bound belongs to $C$ as well, then it is maximal: the greatest element in $C$.

Likewise, once restricted to the chain in Section 2.2.3, the partial order defined there becomes complete. Furthermore, if the upper bound is in the chain, then it is not only maximal but also the greatest function in the entire chain.

In what follows, unless specified otherwise, we'll assume that the order is complete.

## 2.4 Well Order

### 2.4.1 Container and Its Subset

Later on, we'll discuss an order not only in a set but also in a more general container. For this purpose, let's look at a container, and the possible relations in it.

Let $A$ be a container (not necessarily a set). This way, $A$ can't take part in operations like union or intersection. Still, we can talk about individual elements that certainly belong in $A$.

For example, pick some elements from $A$, and place them in a new subcontainer. If this happens to be a legitimate set, then it is indeed a subset of $A$. We can then go ahead and define all sorts of new relations between an element of the subset and any other element of $A$. This is indeed Fraenkel's axiom [17].

For example, the relation '=' tells us whether an element of the subset is the same as an element of $A$. As a matter of fact, we've just used this relation implicitly: this is how we made sure that the subset is included in $A$ in the first place. This will be useful later.

### 2.4.2 Well Order

Assume that $A$ is completely-ordered. Is $A$ also well-ordered? To be well-ordered, $A$ must have the following property:

> Every nonempty subset $S \subset A$ must have a first element, smaller than all other elements in $S$ (in terms of the above order).

(Clearly, in each such $S$, the first element is unique.) In this case, we say that this is a well order. (Here, the word "well" serves as an adjective, not an adverb.)

For example, $\mathbb{N}$ is well-ordered. $\mathbb{R}$, on the other hand, is not. Indeed, every open interval has no first element at all. Likewise, $\mathbb{Q}$ is not. For example, the harmonic sequence

$$\frac{1}{2}, \frac{1}{3}, \frac{1}{4}, \frac{1}{5}, \dots$$

has no first element at all. Still, this is true in terms of the standard order. Later on, we'll see that both $\mathbb{R}$ and $\mathbb{Q}$ could be ordered in a (nonstandard) well order. Before doing this, let's see some interesting properties of a well order.

### 2.4.3 Next Element

Assume that $A$ is well-ordered. Consider some element $x \in A$ that is not maximal. What is the next element right after $x$?

Assume first that $A$ is a legitimate set. Consider the subset of those elements that are strictly greater than $x$:

$$\{y \in A \mid y > x\} \subset A.$$

Because $x$ is not maximal, this is a nonempty subset of $A$. As such, it must have a unique first element $\hat{x}$. This is the next element, or the element that follows $x$: there is no intermediate element between $x$ and $\hat{x}$.

What if $A$ is not a set, but a mere container? In this case, the above is not necessarily a set either! Fortunately, it could still be fixed, provided that the elements in $A$ are sets in their own right, ordered by inclusion, as in Sections 2.1.3 and 2.2.2.

In this case, since $x$ is not maximal, we can pick some element $z > x$. Since '$\leq$' means inclusion, $z > x$ could also be written as $x \subset z$, $x \neq z$. Now, look at the new set

$$\{y \in A \mid x < y \leq z\} \subset P(z).$$

This is a legitimate set, as required. Indeed, what are the individual elements in it? These are subsets of $z$, or elements in $P(z)$. Together, they indeed make a legitimate set: a subset of $P(z)$.

Here one may ask: who says that $P(z)$ is a legitimate set? After all, we only know that $z$ is a legitimate set, not $P(z)$! Fortunately, there is no problem: one of the fundamental axioms of set theory tells us that $P(z)$ is a legitimate set as well.

Now, what can we say about the above subset of $P(z)$? It is nonempty: it contains at least one element — $z$ itself. Thus, the first element in it can indeed serve as $\hat{x}$, the element that follows $x$ in $A$. In summary, although $A$ is not necessarily a legitimate set, $x$ does have its next element in it.

## 2.5 Isomorphism and Ordinals

### 2.5.1 Isomorphism

In Chapter 1, we've seen that two sets could be equivalent to each other. Here, we extend this into a yet more powerful relation: isomorphism. This means that the sets also mirror each other in terms of order.

Let $A$ and $B$ be two sets. Assume that both are ordered completely, but not necessarily well. Assume that there is a one-to-one function $J$ from $A$ onto $B$ that preserves order:

$$x, y \in A, \; x < y \; \Rightarrow \; J(x) < J(y) \text{ in } B.$$

Then we say that $J$ is an isomorphism, and that $A$ and $B$ are isomorphic to each other:

$$A \cong B.$$

Clearly, '$\cong$' is a mathematical equivalence relation: reflexive, symmetric, and transitive. (Check!) Thus, the isomorphism preserves the original structure of the set:

its order, and the properties derived from it. In particular, it preserves the minimal and maximal elements (if exist), and the next element.

Thus, to examine a particular (complete) order, it is sufficient to look at just one example: a particular set that supports this order. This is quite economic: this set could then represent all those sets that are isomorphic to it. Once such set is picked, all other sets that are isomorphic to it don't have to be considered any more. After all, they don't give us any new information. This saves a lot of time and effort in developing the theory.

### 2.5.2 Ordinals

In particular, if this set is well-ordered, then so are all those sets that are isomorphic to it. This set is then called an ordinal.

In summary, an ordinal is a well-ordered set, picked arbitrarily from those sets that are isomorphic to each other. Once picked, the ordinal is very useful: it serves as an example to represent all those sets that are isomorphic to it. After all, they all share the same structure, mirrored by the ordinal.

The ordinal is not defined uniquely. After all, it could be replaced by any other set that is isomorphic to it. In other words, the ordinal is only defined up to isomorphism.

### 2.5.3 Initial Segment

Let $A$ be a well-ordered set. (Actually, $A$ could even be a mere container, provided that the elements in it are sets in their own right, ordered by inclusion.) Let $x \in A$ be some element in $A$. The initial segment $A_x$ contains those elements that are strictly smaller than $x$:

$$A_x \equiv \{y \in A \mid y < x\}.$$

It turns out that $A$ could never be isomorphic to its own initial segment:

$$A \ncong A_x.$$

Indeed, by contradiction: if there were an isomorphism $J : A \to A_x$, then we could design an infinite (monotonically decreasing) chain of the form

$$\{x > J(x) > J(J(x)) > J^3(x) > \cdots\} \subset A,$$

with no first element at all, in violation of the well-order property.

Consider now a yet greater element $y > x$ (if exists). In the above argument, we could replace the original set $A$ by $A_y$. Thus, two different initial segments could never be isomorphic to each other:

$$x \neq y \Rightarrow A_x \ncong A_y.$$

For this reason, the initial segments could also be viewed as ordinals: because they are never isomorphic to each other, they stand for different ordinals.

Thus, $A$ is isomorphic to the set of its own initial segments, ordered by inclusion:

$$A \cong \{A_x \mid x \in A\},$$

by the isomorphism

$$x \to A_x, \quad x \in A.$$

Is this an isomorphism? Yes, it is: one-to-one, onto, and order preserving. (Check!)

This is most useful: without loss of generality, we can now assume that $A$ is a set of ordinals, ordered by inclusion. After all, it is isomorphic to the set of its own initial segments, which are indeed ordered by inclusion.

## 2.6 Mathematical Induction: Von-Neumann Numbers

### 2.6.1 Geometrical Model

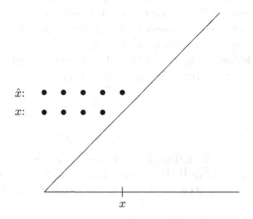

**Fig. 2.1.** Constructing $A$ from scratch: $x$ is a set in its own right. The next element $\hat{x}$ is a slightly bigger set: it contains just one more element. What is this new element? $x$ itself!

The above conclusion is quite useful. Thanks to it, the discussion gets much simpler: we can now assume that $A$ contains inner ordinals, ordered by inclusion.

Let's use this new point of view to have some geometrical intuition. Since $A$ is well-ordered, we already know that each (nonmaximal) element $x \in A$ has a next element $\hat{x} \in A$. How does $\hat{x}$ look like? To see this, let's try to define it explicitly. First, let's try a geometrical approach:

$$\hat{x} \equiv A_x \cup \{x\}.$$

In this approach, what is $\hat{x}$? Well, it contains all previous elements in $A$: not only $x$ but also those smaller than $x$ (Figure 2.1).

What is the geometrical meaning of this? Well, $x$ marks the right endpoint of the "interval" $A_x$. $\hat{x}$, on the other hand, marks a slightly longer "interval" that contains not only $A_x$ but also its right endpoint: $x$.

Could this be improved? After all, we've already established that $x$ and $A_x$ stand for one and the same thing. So, in the above definition, why not replace $A_x$ by $x$:

$$\hat{x} \equiv x \cup \{x\}.$$

This way, $x \subset \hat{x}$, so they are ordered by inclusion, as required. In fact, in terms of inclusion, $\hat{x}$ indeed follows $x$: as a set, it is slightly bigger — it contains all the elements of $x$, plus one new element. What could this element be? It must be a new element that is not in $x$. The best candidate for this is $x$ itself. After all, $x$ is not an element in $x$, so it is indeed a good candidate to serve as the new element in $\hat{x}$. This way, we have yet another attractive property: not only $x \subset \hat{x}$, but also $x \in \hat{x}$.

But what if, God forbid, $x \in x$? Fortunately, this could never happen. After all, as an ordinal in its own right, $x$ could never be isomorphic to its own initial segment (Section 2.5.3). We'll come back to this later.

Thanks to the above, $A$ could actually be constructed from scratch, step by step. This mirrors a familiar process: mathematical induction.

### 2.6.2 Natural Numbers as Ordinals

As a matter of fact, we've already used this kind of mathematical induction to define the natural numbers in the first place. This is indeed the Von-Neumann definition. In Chapter 1, Section 1.2.2, it is done in terms of sets only. Here, on the other hand, it is done in terms of ordinals, defined up to isomorphism only:

$$0 \equiv \emptyset$$
$$1 \equiv 0 \cup \{0\} = \emptyset \cup \{\emptyset\} = \{\emptyset\}$$
$$2 \equiv 1 \cup \{1\} = \{\emptyset, \{\emptyset\}\}$$
$$3 \equiv 2 \cup \{2\} = \{\emptyset, \{\emptyset\}, \{\emptyset, \{\emptyset\}\}\}$$
$$\cdots \quad \cdots$$
$$n + 1 \equiv n \cup \{n\}$$
$$\cdots \quad \cdots$$

and so on. This way, $n$ is the same as the initial segment that contains exactly $n$ numbers:

$$n \equiv \{0, 1, 2, \ldots, "n-1"\}.$$

Why the quotation marks? Because this is no subtraction: "$n-1$" is just the number prior to $n$. Thus, the natural numbers can now be interpreted as ordinals, ordered by inclusion:

$$0 \subset 1 \subset 2 \subset 3 \subset \cdots.$$

Let's go ahead and define arithmetic operations between these numbers. This is better done in a more general context: adding well-ordered sets.

## 2.7 Adding Ordinals

### 2.7.1 Adding Well-Ordered Sets

Let $A$ and $B$ be two well-ordered sets. How to add them to each other?

**Fig. 2.2.** In $A + B$, the elements of $A$ are listed one by one in the bottom row, before the elements of $B$, listed one by one in the top row.

Geometrically, the sum $A + B$ makes two horizontal lines (Figure 2.2). The elements of $A$ are listed one by one in the bottom line, before the elements of $B$, listed one by one in the top line.

How to do this more precisely? Define

$$A + B \equiv (A \times \{0\}) \cup (B \times \{1\}).$$

This introduces a new bit: either 0 or 1. To each element $a \in A$, attach the bit 0. This makes the new pair $(a, 0)$, indicating that these pairs have high priority. To each element $b \in B$, on the other hand, attach the bit 1. This makes the new pair $(b, 1)$, indicating that these pairs have low priority. Later on, we'll make sure that the new set $A + B$ is indeed well-ordered.

Geometrically, the new bit helps order the pairs in two horizontal rows: a pair with bit 0 is always before a pair with bit 1. Pairs with the same bit, on the other hand, are in the same row, where they are ordered as before: either as in $A$, or as in $B$.

This is a kind of lexicographic order: first, look at the second coordinate: the bit. Only if it is the same, look at the first coordinate, and order.

In a standard lexicographic order, as in an English dictionary, one first looks at the first letter, on the left. Here, on the other hand, things are the other way around. This is an "anti-lexicographic" order, as in a Hebrew dictionary: first, look at the second coordinate, on the right. In this sense, the bit on the right is "stronger:" it is looked at first. Only if both pairs have the same bit do we look at the first coordinate on the left, to decide which pair is prior.

### 2.7.2 Adding Ordinals

Things get even simpler when $A$ and $B$ are interpreted as ordinals, defined up to isomorphism only. Von-Neumann's natural numbers in Section 2.6.2, for example, sum as in standard arithmetic:

$$3 + 4 \cong 7.$$

(Prove!)

Note that our natural numbers are now ordinals, ordered by inclusion. Is there a yet bigger ordinal that includes them all? Fortunately, there is: their infinite union, denoted by $\omega$:

$$\omega \equiv \cup_{\mathbb{N}} = \cup_{n=1}^{\infty} n = \{m \in n \mid n \in \mathbb{N}\} = \{m \mid m \in \mathbb{N} \cup \{0\}\} = \mathbb{N} \cup \{0\}.$$

Thanks to the isomorphism $n \to n + 1$, we could also define $\omega$ as $\mathbb{N}$ (without zero):

$$\omega \equiv \mathbb{N}.$$

Thanks to this new definition, the isomorphism $n \to n + 1$ can now be written as a new arithmetic operation on ordinals:

$$1 + \omega \cong \mathbb{N} \cup \{0\} \cong \mathbb{N} = \omega.$$

Be careful: there is no commutativity here! This is discussed next.

## 2.8 Infinity Point

### 2.8.1 No Commutativity

What is $\omega$? It is a kind of infinity point: a new ordinal, greater than all natural numbers.

Still, be careful: adding ordinals is not commutative. On one hand, $1 + \omega$ is the same as $\omega$: introducing zero at the beginning changes nothing in terms of order. On the other hand, $\omega + 1$, is different: it adds a new maximal element, missing in both $\omega$ and $1 + \omega$. This is the new infinity point, denoted by '$\omega$' as well:

$$\omega + 1 \cong \omega \cup \{\omega\} \ncong \omega \cong 1 + \omega.$$

Thus, $\omega + 1$ is the next ordinal that follows $\omega$: it "closes" $\mathbb{N}$ at infinity, and adds a full stop on the right.

This is indeed a suitable way to continue the original Von-Neumann process. Indeed, to "close" $\mathbb{N}$ at infinity, $\omega + 1$ adds just one new element, preserving the inclusion order:

$$\omega \subset \omega \cup \{\omega\}.$$

In terms of inclusion, these ordinals follow one another: there is no room for any other ordinal in between.

### 2.8.2 What's Beyond the Infinity Point?

This process can now continue yet further, preserving the inclusion order:

$$0 \subset 1 \subset 2 \subset 3 \subset \cdots \subset n \subset \cdots \subset \omega \subset \omega + 1 \subset \omega + 2 \subset \cdots.$$

This way, the next ordinal not only includes but also *contains* all previous ones:

$$n \equiv \{0, 1, 2, \ldots, \text{"}n-1\text{"}\}$$
$$\omega \equiv \{0, 1, 2, \ldots\}$$
$$\omega + 1 \equiv \{0, 1, 2, \ldots, \omega\}$$
$$\omega + 2 \equiv \{0, 1, 2, \ldots, \omega, \omega + 1\}$$
$$\omega + 3 \equiv \{0, 1, 2, \ldots, \omega, \omega + 1, \omega + 2\}$$
$$\cdots \quad \cdots$$
$$\omega + \omega \equiv \{0, 1, 2, \ldots, \omega, \omega + 1, \omega + 2, \ldots\}$$
$$\omega + \omega + 1 \equiv \{0, 1, 2, \ldots, \omega, \omega + 1, \omega + 2, \ldots, \omega + \omega\}$$
$$\cdots \quad \cdots$$

and so on.

### 2.8.3 Associativity

This way, the associative law does hold:

$$(\omega + 1) + \omega \cong \omega + (1 + \omega) \cong \omega + \omega.$$

Indeed, all three support the same (anti-lexicographic) order: an infinite row, followed by yet another infinite row, with no maximal element at all. We can now define the new product

$$\omega \times 2 \equiv \omega + \omega.$$

After all, $\omega \times 2$ looks the same: two (infinite) horizontal lines, as in Figure 2.2.

## 2.9 Multiplying Ordinals

### 2.9.1 Multiplying Well-Ordered Sets

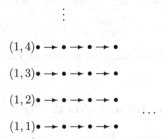

**Fig. 2.3.** $\omega^2 = \omega \times \omega$ contains $\aleph_0$ duplicate copies of $\omega$, ordered line by line, bottom to top. Within each horizontal line, the inner order is left to right.

The above can now be extended to multiply two well-ordered sets, $A$ and $B$:

$$A \times B \equiv \{(a, b) \mid a \in A, \ b \in B\}.$$

So far, this is just a product of two sets, as in Chapter 1, Section 1.5.1. Now, let's go ahead and introduce an order in it. For this purpose, use the anti-lexicographic order, as in a Hebrew dictionary: $(a, b) < (a', b')$ if

- either $b < b'$
- or $b = b'$ and $a < a'$.

Geometrically, the elements in $A$ are listed one by one in a long horizontal line (Figure 2.3). On top of this, place an isomorphic line. On top of this, place yet another isomorphic line, and so on ($|B|$ times). This makes a two-dimensional grid of pairs of the form $(a, b)$.

In geometrical terms, how is the grid ordered? Each horizontal line uses some fixed element $b \in B$:

$$(A, b) \equiv \{(a, b) \mid a \in A\} \subset A \times B.$$

Now, to move to the next line on top, just increase $b$ a little, and switch to the next element in $B$. This process goes on, producing more and more lines on top. From line to line, $b$ increases monotonically, bottom to top. This way, the lines are indeed ordered bottom to top, as required. Within each individual line, the order is left to right, as in the original set $A$.

### 2.9.2 Is the Product Well-Ordered?

Is $A \times B$ well-ordered? To see this, let's use some geometrical intuition. For this purpose, let

$$S \subset A \times B$$

be a nonempty subset. In terms of the anti-lexicographic order, what is its first element?

Geometrically, the answer is apparent from Figure 2.4. We need to find the horizontal line tangent to $S$ from below. The tangent point is what we want: the minimum (or the first element) of $S$, as required.

Let's write this in set-theory language. Look at those horizontal lines that pass through (or intersect with) $S$. Each line uses some fixed $b \in B$. Together, these $b$'s make a new subset:

$$T \equiv \{b \in B \mid (A, b) \cap S \neq \emptyset\} \subset B.$$

Geometrically, $T$ marks the shadow of $S$ on the $b$-axis on the left (Figure 2.4).

Now, $T$ is a nonempty subset of the well-ordered set $B$. As such, it has a first element $b_1 \in T$.

So far, we worked vertically. Now, it's time to work horizontally. For this purpose, look at $(A, b_1)$: the horizontal line tangent to $S$ from below. Where is the tangent point?

Fortunately, we already know its second coordinate: $b_1$. Or do we? Wait a moment: there is a subtle issue here: we may never find $b_1$ explicitly. For example,

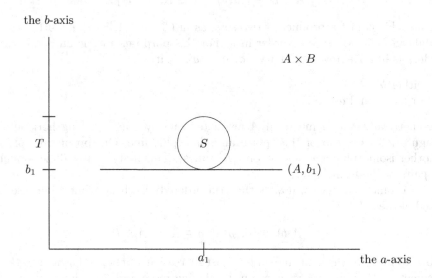

**Fig. 2.4.** The product $A \times B$ makes the $a$-$b$ plane. In it, $S$ is a nonempty subset. Where is its minimum? To find out, project $S$ leftwards, onto the $b$-axis. This produces the shadow $T$. Let $b_1$ be the minimum of $T$ in $B$. Draw the horizontal line $(A, b_1)$. It is tangent to $S$ at its minimum.

assume that $B$ contains a smaller element $b_0 < b_1$. To make sure that $b_0 \notin T$, we must make sure that the lower line $(A, b_0)$ never intersects $S$. To check on this, we must scan $(A, b_0)$, pair by pair. Unfortunately, if $A$ is infinite, then this could take forever. This is the infinity paradox.

Fortunately, we may still assume that this information is already available to us from some external source. Furthermore, we don't need to specify $b_1$ explicitly: it is good enough to know that it does exist. After all, to design $T$, we only used legitimate operations: intersection and union.

Now, what is $(A, b_1)$? This is the first horizontal line that shares a common element with $S$:

$$S \cap (A, b_1) \neq \emptyset.$$

Still, there could be many tangent points. Which one to pick? The leftmost one! How to specify it? Project onto the $a$-axis below. This produces the shadow

$$\{a \mid (a, b_1) \in S\} \neq \emptyset.$$

This is a nonempty subset of the well-ordered set $A$. As such, it has a first element $a_1 \in A$. This is what we wanted: $(a_1, b_1)$ is the first element in $S$ (in terms of the anti-lexicographic order), as required.

### 2.9.3 Multiplying Ordinals

Things get even simpler when $A$ and $B$ are interpreted as ordinals, defined up to isomorphism only. The process in Section 2.8.2 can now continue: $\omega \times 3$ contains three isomorphic copies of $\omega$ on top of each other, $\omega \times 4$ contains one more copy on top, and so on, until $\omega^2 \equiv \omega \times \omega$, which contains $\aleph_0$ copies of $\omega$, making an infinite two-dimensional grid (Figures 2.3 and 1.6).

The process can now continue in the third dimension as well: place $\aleph_0$ isomorphic copies of $\omega^2$ on top of each other to form $\omega^3$: an infinite three-dimensional grid, with the anti-lexicographic order.

Furthermore, the same could also be done in the fourth dimension as well: place $\aleph_0$ isomorphic copies of $\omega^3$ on top of one another to form $\omega^4$: an infinite four-dimensional grid, and so on.

Once interpreted in terms of the inclusion order (Sections 2.6.1–2.8.2), each such ordinal is just the union of all the previous ones:

$$\omega^2 \equiv \cup_{\{\omega \times n \mid n \in \mathbb{N}\}} \equiv \cup_{n=1}^{\infty} \omega \times n$$
$$\omega^3 \equiv \cup_{\{\omega^2 \times n \mid n \in \mathbb{N}\}} \equiv \cup_{n=1}^{\infty} \omega^2 \times n$$
$$\omega^4 \equiv \cup_{\{\omega^3 \times n \mid n \in \mathbb{N}\}} \equiv \cup_{n=1}^{\infty} \omega^3 \times n$$
$$\cdots \qquad \cdots$$
$$\omega^\omega \equiv \cup_{\{\omega^n \mid n \in \mathbb{N}\}} \equiv \cup_{n=1}^{\infty} \omega^n.$$

The latter ordinal is greater than all previous ones: it contains all finite (arbitrarily long) sequences of natural numbers.

In this great ordinal, what is the order? Well, given two finite sequences of natural numbers, we must first compare their lengths. The rule is that a shorter sequence comes before a longer one.

Now, what if both have the same length? To decide which one is prior, use the anti-lexicographic order, as before.

Alternatively, given two sequences of just any length, force them to have the same length: take the shorter one, and add tailing zeroes to it on the right, until it gets to have the same length as the other sequence. Then, use the anti-lexicographic order, as before. Fortunately, zero is smaller than any natural number. This way, the sequence that was originally shorter is prior to the longer one, as required.

This ordinal, although big, is still countable:

$$|\omega^\omega| = \aleph_0.$$

(See exercises at the end of Chapter 1.) Later on, we'll see that there are yet bigger ordinals. Still, we'll never get to see how they look like explicitly.

## 2.10 Transfinite Induction

### 2.10.1 Sequential vs. Parallel Process

Why are ordinals so important? Because they are ready for transfinite induction: a general form of mathematical induction.

So far, we often used mathematical induction: march in $\omega$, number by number. This is a sequential process: scan the natural numbers, one by one in a row. But what if the ordinal is a little bigger? For example, look at $\omega \times 2$: two horizontal rows, as in Figure 2.2. Scanning the bottom row could take forever, getting no chance to start scanning the top row! To fix this, the process must be more parallel and concurrent: start working on the top row even before finishing the bottom row. Let's extend this to bigger ordinals as well.

### 2.10.2 Proof by Transfinite Induction

Let $A$ be a well-ordered set. As a matter of fact, $A$ could even be a container that is not a set, provided that the elements in it are sets in their own right, ordered by inclusion. In this case, the trick in Section 2.4.3 could be used to make sure that every initial segment of $A$ is indeed a legitimate set.

How to march in $A$? Instead of the standard mathematical induction, we need a more general technique: transfinite induction. After all, $A$ might be bigger than $\omega$, and even as big as $\omega^\omega$.

What is transfinite induction? In principle, it works like standard mathematical induction: it has its own induction hypothesis and step.

To introduce transfinite induction, consider some property that the elements in $A$ could satisfy. Assume that the induction step can always be carried out: for every $x \in A$, if all the elements in the initial segment $A_x$ satisfy the property (this is the induction hypothesis), then $x$ satisfies it as well.

That's it! Thanks to the induction step, we can now prove (by contradiction) that all elements in $A$ indeed satisfy the property. Otherwise, pick some element $z \in A$ that doesn't. Consider those elements that

- are not larger than $z$
- and don't satisfy the property.

Together, they make a new subset.

This subset is nonempty: it contains $z$ itself. Let $y$ be its first element. This way, all the elements in $A_y$ must satisfy the property, so the induction step tells us that $y$ must satisfy it as well. in violation of the very definition of $y$.

Thus, the above could never happen: no such $z$ could ever be found, because all elements in $A$ do satisfy the property, as asserted. This gives us a powerful tool: a new method to prove that a certain property always holds. Thanks to this generalization, mathematical induction is no longer an axiom, but rather a theorem: a robust method of proof. In fact, we've just proved that it always works, for every property that supports an induction step. Moreover, we proved it in its strongest form: not only when $A = \omega$ but also when $A = \omega^\omega$, or even bigger.

### 2.10.3 Definition by Transfinite Induction

As a matter of fact, transfinite induction could be used not only to prove a certain property but also to define a set or a function in the first place. (For example, the ordinals in Section 2.9.3 are actually defined by transfinite induction.) Let's use this method for yet another purpose: to prove that every two well-ordered sets $A$ and $B$ could be embedded in one another as an initial segment. More precisely, there is a unique isomorphism

- either from $A$ onto $B$,
- or from $A$ onto a unique initial segment of $B$,
- or from $B$ onto a unique initial segment of $A$.

To design the unique isomorphism, use transfinite induction. For this purpose, consider a particular element $x \in A$, and assume that the induction hypothesis indeed holds for $A_x$: there is a unique isomorphism

- either from $A_x$ onto $B$,
- or from $A_x$ onto a unique initial segment $B_y \subset B$ (for a unique $y \in B$),
- or from $B$ onto a unique initial segment of $A_x$.

In the former or latter option, we're done. In the middle option, on the other hand, there is still some work to do: to extend the isomorphism. How to do this uniquely? Easy: just map $x$ to $y$. This maps $A_x \cup \{x\}$ uniquely onto

- $B$ (if $y$ is maximal)
- or $B_{\hat{y}}$ (if $y$ is followed by $\hat{y}$ in $B$).

This completes the induction step, as required.

## 2.11 Ordinals and Their Order

### 2.11.1 How to Order the Ordinals?

We are now ready to define a strict order between ordinals. Let $A$ and $B$ be two ordinals. We say that $A$ is smaller than $B$ if $A$ is isomorphic to an initial segment of $B$:

$$A < B \quad \text{if} \quad A \cong B_y \text{ for some } y \in B.$$

Is this a proper strict order? In other words, is it transitive, anti-reflexive, and anti-symmetric, as required in Section 2.3.1? Thanks to Section 2.5.3, it indeed is. (Check!)

Furthermore, thanks to Section 2.10.3, this new order is not only strict but also complete: every two ordinals can be compared to one another. Later on, we'll see that this order is also well.

Moreover, from Section 2.5.3, an ordinal $A$ is isomorphic to the set of its own initial segments:

$$A \cong \{A_x \mid x \in A\}.$$

This set can now be interpreted as the set of those ordinals smaller than $A$. After all, each and every ordinal smaller than $A$ is isomorphic to some initial segment of $A$.

This motivates the definitions in Sections 2.6.1–2.8.2, in which an ordinal is not only isomorphic but also *identical* to the set of smaller ordinals, strictly included in it.

### 2.11.2 Are the Ordinals Well-Ordered?

Let $A$ be the container that contains all ordinals. We've just defined a complete order in $A$.

Later on, we'll see that $A$ is just a container, but not a legitimate set. Still, one could ask: is $A$ well-ordered? After all, the well-order property is relevant not only to a set but also to a container like $A$ (Section 2.4.2).

To answer this, let $S \subset A$ be a nonempty subset of ordinals. Does $S$ contain a first ordinal? Well, $S$ could be too big to study. Let's go ahead and intersect it with an initial segment of $A$. This will give us a smaller subset, easier to study.

Let $x \in S$ be some ordinal. Let's design an initial segment of $A$ that still contains $x$. For this purpose, let's define the next ordinal in $A$:

$$\hat{x} \cong x + 1.$$

This way, the initial segment $A_{\hat{x}}$ is big enough: it contains $x$, as required. Furthermore, from Section 2.11.1,

$$A_{\hat{x}} \cong \hat{x}.$$

To use this isomorphism in practice, let's give it a name:

$$J : A_{\hat{x}} \to \hat{x}.$$

In particular, $J$ is onto $\hat{x}$, so its image is

$$J(A_{\hat{x}}) = \hat{x}.$$

Now, look at the intersection

$$S \cap A_{\hat{x}} \subset A_{\hat{x}}.$$

It is nonempty: it contains $x$. Therefore, its image

$$J(S \cap A_{\hat{x}}) \subset \hat{x}$$

is nonempty as well: it contains $J(x)$. Therefore, it must contain a first element in the ordinal $\hat{x}$, say $y \in \hat{x}$. As a result, $J^{-1}(y)$ is the first ordinal in $S$, as required.

### 2.11.3 The Ordinal Paradox

By now, $A$ is the container that contains all ordinals. Still, is $A$ a legitimate set? No, it is too big! Indeed, by contradiction: if it were, then it would be a well-ordered set, or an ordinal in $A$:

$$A \in A.$$

Is this possible? Well, from Section 2.11.1, each ordinal $x \in A$ is isomorphic to the set of smaller ordinals:

$$x \cong A_x.$$

In particular, this would apply to the special case $x \equiv A \in A$:

$$A \cong A_A.$$

But this is impossible: no ordinal could ever be isomorphic to its own initial segment (Section 2.5.3).

This proves the (so-called) ordinal paradox: $A$ can never be a legitimate set, but only a container.

## 2.12 Zorn's Lemma in Quantum Mechanics

### 2.12.1 The Uncertainty Principle

How does an atom look like? In a naive model, the nucleus is at the center, and the electrons orbit it, round and round. Quantum mechanics, on the other hand, tells us a different story: more stochastic, and less deterministic. In fact, the principle of uncertainty says that one can never know all. At a given time, one could measure one quantity only: either the position of the electron, or its momentum, but not both. Thus, the electron has no definite place: it only has some probability (or likelihood) to occupy a certain place at a given time.

### 2.12.2 Tree of Universes

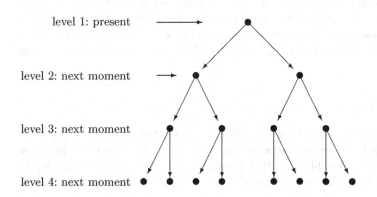

level 1: present

level 2: next moment

level 3: next moment

level 4: next moment

**Fig. 2.5.** Tree of potential universes: at the head, there is our familiar universe, at the present. At the next moment, it splits into two (or more) potential universes. Each has its own probability to be picked as the correct (real) universe. At the next moment, it will also split into two (or more) new universes, and so on.

To model this, one could design a tree of potential universes (Figure 2.5). In each moment, the universe splits into many different universes, each assumes that the electron is at a different place in the atom. Together, these new universes make a new level in the tree. From each universe in this level, many more universes could emerge in the next moment, each at a certain probability. Together, they form the next level, and so on, moment by moment, producing more and more potential universes, each with its own (small) probability. Together, all these levels make an infinite tree of many potential universes that may appear later.

As a matter of fact, the same could be done not only for one particular electron but also for each and every electron in the world. Fortunately, the total number of electrons is still finite (Chapter 6). For simplicity, we consider one electron only.

This is particularly easy to illustrate in our tree. Later on, we'll extend this to a more continuous model (Chapter 12).

### 2.12.3 Entropy: Partial Order

In our tree, define a new (strict) partial order: a particular universe is "smaller" than those new universes that may emerge from it later. Is this a legitimate order? Well, let's check:

- It is clearly transitive.
- It is anti-symmetric: there is no going back to an old universe. After all, entropy must increase, as in the second law of thermodynamics (Chapter 6, Section 6.3.5).
- It is anti-reflexive: no universe is static.

### 2.12.4 Experiment or Observation

At each individual moment, only one particular universe appears in reality. To pick it, there is a need to do an experiment, and observe the exact location of the electron in the atom. At each level, this picks just one (correct and real) universe. By repeating this moment by moment, we get an infinite path, level by level, down the tree, along those universes that really appear. All others remain science fiction.

### 2.12.5 Parallel Universes

This models the theory of parallel universes [9]. In it, you must use your imagination: you only *imagine* that you are in this world, here and now. As a matter of fact, you also have many duplicate copies in other potential universes (at the same level), who also think that they are the only ones!

### 2.12.6 Zorn's Lemma

But time is unlimited: there are infinitely many moments, each with its own level in the tree. How to form the infinite path down the tree? At each moment, which universe should be picked from the level?

Fortunately, the tree must contain no dead-end: from each universe, at least one new universe emerges. In terms of our partial order, this means no maximal universe.

Thanks to this property, such a path indeed exists. This is indeed Zorn's lemma: there is a chain with no upper bound. This way, we can always advance along the path, down the tree. This can be done for arbitrarily long time. This may help "solve" Zeno's paradox: time never stops, but always restarts.

## 2.13 Three Equivalent Theorems

### 2.13.1 The Well-Ordering Theorem

Se far, we've seen a few examples of ordinals: $\omega$, $\omega^2$, $\omega^3$, etc. All these are countable: they have cardinality $\aleph_0$. As a matter of fact, even the great ordinal $\omega^\omega$ is still countable:

$$|\omega^\omega| = \aleph_0.$$

(See exercises at the end of Chapter 1.) Still, is there any uncountable ordinal?

The first candidate that comes to mind is the real axis $\mathbb{R}$. After all, it is indeed uncountable:

$$|\mathbb{R}| = \aleph > \aleph_0.$$

Still, is $\mathbb{R}$ a good candidate? In other words, is it well-ordered? Not with the standard order. After all, the standard order, although complete, is not well: every open interval has no first element at all. Fortunately, there is a nonstandard order, with which $\mathbb{R}$ is indeed well-ordered. We may never know how it looks like. Still, we do know that it does exist. This follows from the well-ordering theorem.

From its name, it sounds like the well-ordering theorem can be proved. After all, it is just a theorem. Still, to prove it, we must use a heavy tool: either the axiom of choice, or Zorn's lemma.

This also works the other way around: if the well-ordering theorem was accepted as an axiom with no proof at all, then we could use it to prove both the axiom of choice and Zorn's lemma. As we'll see below, all three have equal status: they are equivalent to each other, and follow from one another:

$$\text{well-ordering theorem} \quad \Leftrightarrow \quad \text{axiom of choice} \quad \Leftrightarrow \quad \text{Zorn's lemma.}$$

Thus, all three have the same status: accept just one of them as an axiom with no proof at all, and you can easily prove the other two. The terms "lemma," "theorem," and "axiom" mean nothing: they are just traditional names that stuck.

### 2.13.2 Naive "Proof"

To help prove the well-ordering theorem, let's look at a naive model. Let $B$ be a star, made of white particles. This is not a physical star: it could be infinite, and even uncountable:

$$|B| \geq \aleph_0.$$

Could the particles be listed in any way?

To do this, let's apply the axiom of choice to $P(B)$. This is the power set of $B$: the set that contains all subsets of $B$. In other words, each element in $P(B)$ is a subset of $B$. From each such subset, the axiom of choice helps pick just one element: the chosen particle.

The axiom of choice is rather implicit: it doesn't tell us which particle is picked. For example, the particle picked from one subset may differ from the particle picked from another subset, or not. If both subsets have a nonempty intersection, then the same particle could be picked.

To list the particles one by one, we can't use transfinite induction. After all, $B$ is not yet well-ordered. Still, we could use the same idea: look at those particles that are still white. They make a subset of $B$. Thanks to the axiom of choice applied above, pick one white particle, list it, and paint it red.

This could be done for every subset of white particles. In the "end," no particle remains white: all are painted red and listed, as required. Or are they?

Is this a good proof? Well, it may work in a sequential process, as in standard mathematical induction, in which $|B| = \aleph_0$. In such an elementary case, we know what subset to use to pick from: the subset that contains those particles that are still white, not red. In a more general case, in which $|B| > \aleph_0$, on the other hand, the process is no longer sequential, but much more parallel and concurrent. In such a case, what subset to use to pick from? How could we tell whether it is made of red or white particles? After all, this depends on the very order that is still unspecified! To fix this, we must use yet another important tool: the ordinal paradox.

### 2.13.3 The Axiom of Choice and the Well-Ordering Theorem

Here we show that the axiom of choice and the well-ordering theorem are equivalent to each other, and follow from one another:

$$\text{axiom of choice} \quad \Leftrightarrow \quad \text{well-ordering theorem?}$$

For this purpose, we must first state them in their precise mathematical form.

Let $D$ be a set of sets: each element in $D$ is a (nonempty) set in its own right. The axiom of choice says that, from each element of $D$, it is possible to pick one inner element.

This may seem trivial, but is not. After all, if $D$ is infinite, then who could design the rule (or mechanism) to decide which inner element to pick? Fortunately, the axiom of choice guarantees that such a mechanism indeed exists, even though it may remain implicit and unspecified.

What's so good about the axiom of choice? Well, it could employ this mechanism once and for all in advance, to get ready for any (tedious) infinite process. This way, the act of choosing shouldn't "slow" the process down: it is complete even before the process starts, and will take care of every puzzling situation that may arise later.

More formally, the axiom of choice tells us that there is a choice function

$$f : D \to \cup D$$

that picks some inner element

$$f(d) \in d, \quad d \in D.$$

Let's show that this is indeed equivalent to the well-ordering theorem that says that every set can be ordered well.

Let's start with the easy bit:

$$\text{axiom of choice} \quad \Leftarrow \quad \text{well-ordering theorem.}$$

In other words, assume that the well-ordering theorem holds: every set can be ordered well. This is now considered as an axiom, which needs no proof. Traditionally, however, it is still referred to as a theorem.

Let's use it to prove the axiom of choice. This will degrade the axiom of choice to the lower status of a mere theorem. Traditionally, however, it is still referred to as an axiom.

Indeed, the well-ordering theorem can be used to order $\cup_D$ well. This way, for each element $d \in D$, $d \subset \cup_D$ contains a first element, which may now serve as $f(d)$. This defines the new choice function $f$, as required.

This was easy. Next, let's prove the difficult bit — the other way around:

$$\text{axiom of choice} \quad \Rightarrow \quad \text{well-ordering theorem.}$$

In other words, assume now that the axiom of choice holds. Let's use it to prove the well-ordering theorem. For this purpose, let $B$ be some given set. Our task is to order $B$ well.

Let $A$ be the container that contains all ordinals. Let's show that there is an ordinal $a \in A$ that orders $B$ well, or is isomorphic to $B$:

$$a \cong B.$$

For this purpose, let's add a fictitious element $e$ to $B$. This way, $B$ is extended to $B \cup \{e\}$.

To order the elements of $B$, let's design a new mapping:

$$g : A \to B \cup \{e\}.$$

How to define $g$? For this purpose, apply the axiom of choice to $P(B)$ (the power set of $B$). This way, from each subset of $B$, one element is picked. Most of them will never be used: we'll only need those subsets of a special kind.

Now, for each ordinal $o \in A$, $g(o)$ should be an element in $B \setminus g(A_o)$: a new element in $B$ that is never used to define $g$ for any smaller ordinal. Thanks to the axiom of choice applied above, such an element is already available. Only if

$$B \setminus g(A_o) = \emptyset$$

must we step outside $B$, and define

$$g(o) \equiv e \notin B.$$

Is this familiar? This is nothing but transfinite induction on $A$. Fortunately, transfinite induction works not only in a set but also in a container like $A$ (Section 2.10.2).

In the above process, $e$ must have been used at least once. Otherwise, the entire container $A$ would have been embedded in the set $B$, in violation of the ordinal paradox.

Let's identify the first ordinal mapped to $e$. For this purpose, we can't consider all those ordinals mapped to $e$: there may be too many. We must restrict to some initial segment of $A$.

To this end, let's use our good old trick. Let $x \in A$ be some ordinal mapped to $e$:

$$g(x) = e.$$

Look at the next ordinal:

$$\hat{x} \in A, \qquad \hat{x} \cong x + 1.$$

Now, in the initial segment $A_{\hat{x}}$, look at those ordinals mapped to $e$. Together, they make a legitimate subset of ordinals.

This subset is nonempty: it contains $x$. Let $a$ be its first element: the smallest ordinal mapped to $e$:

$$g(a) = e.$$

The one-to-one function $g$ from the initial segment $A_a$ onto $B$ can now be used to order $B$:

$$b_1 < b_2 \ \text{ if } \ g^{-1}(b_1) < g^{-1}(b_2) \ \text{ in } A_a.$$

From Section 2.11.1, we then have

$$a \cong A_a \cong B,$$

so $a$ indeed orders $B$ well, as required.

### 2.13.4 Zorn's Lemma

In Section 2.12.2, we've already used Zorn's lemma in the context of quantum mechanics. Let's introduce it more formally, in the context of set theory.

In Section 2.1.1, we've introduced a partial order for the first time. Let $B$ be such a partially-ordered (nonempty) set.

In Section 2.2.1, we've also introduced the concept of a chain. Zorn's lemma says that, if every chain has an upper bound in $B$, then $B$ must also contain a maximal element. Below, we'll see that Zorn's lemma is equivalent to the axiom of choice:

$$\text{axiom of choice} \quad \Leftrightarrow \quad \text{Zorn's lemma.}$$

Let's start with the difficult bit:

$$\text{axiom of choice} \quad \Rightarrow \quad \text{Zorn's lemma.}$$

Why is this difficult? Because, as before, the axiom of choice mustn't be used on its own. It should better be combined with two other powerful tools: transfinite induction and the ordinal paradox.

Indeed, let's use the axiom of choice to prove Zorn's lemma by contradiction. For this purpose, assume momentarily that $B$ had no maximal element at all.

As before, let $A$ be the container that contains all ordinals. Recall that transfinite induction works not only in a set but also in a (well-ordered) container like $A$ (Section 2.10.2). Let's use it to define an isomorphism $J$ from $A$ onto a chain in $B$.

Let $o \in A$ be some ordinal, and assume that the induction hypothesis holds: the isomorphism $J$ has already been defined on $A_o$ in such a way that $J(A_o)$ is indeed a chain in $B$. This chain has an upper bound in $B$. As a matter of fact, it may have many. Use the axiom of choice to pick one.

This upper bound must be nonmaximal: it must have a greater element in $B$. As a matter of fact, it may have many. As before, use the axiom of choice to pick one, and define it as $J(o)$. This way, the original chain $J(A_o)$ has extended by one more element, and $J$ has extended to map $A_o \cup \{o\}$ onto this longer chain. This completes the induction step.

By transfinite induction, this can be done for every ordinal $o \in A$. Thus, the entire container $A$ has been embedded in the set $B$, in violation of the ordinal paradox. Thus, our momentary assumption must have been false: $B$ must contain a maximal element, as Zorn's lemma indeed says.

In the above proof, $A_o$ was well-ordered, so $J(A_o)$ was a well-ordered chain in $B$. Thus, in Zorn's lemma, there is actually no need to assume that *every* chain has an upper bound, but only *well-ordered* chains. Thus, we've actually proved a slightly stronger version of Zorn's lemma, with a slightly weaker assumption:

If every *well-ordered* chain has an upper bound in $B$, then $B$ must contain a maximal element.

Chains that are not well-ordered, on the other hand, may have an upper bound or not — we don't care.

Thanks to this new version, we have a new result: if there is no maximal element, then there is a *well-ordered* chain with no upper bound. In the context of quantum mechanics (Section 2.12.2), this is quite useful: it implies that reality is indeed well-ordered in the time axis. This may help "solve" Zeno's paradox: time never stops, but always restarts. Still, this version is rather nonstandard, and rarely used.

Finally, let's work the other way around, and prove the easy bit:

$$\text{axiom of choice} \quad \Leftarrow \quad \text{Zorn's lemma.}$$

In other words, use Zorn's lemma to prove the axiom of choice. This way, Zorn's lemma is now accepted as an axiom, which needs no proof. Traditionally, however, it is still referred to as a lemma. The axiom of choice, on the other hand, is now degraded to the lower status of a mere theorem, which needs a proof. Traditionally, however, it is still referred to as an axiom.

Let $D$ be a set whose elements are (nonempty) sets in their own right. From the discussion in Section 2.2.5, every chain of choice functions has an upper bound. From Zorn's lemma, there is a maximal choice function $f$, defined on some subset $S \subset D$:

$$f(s) \in s, \quad s \in S.$$

Because $f$ can no longer be extended, we must have $S = D$. This guarantees that $f$ is indeed a proper choice function on $D$, as required. This proves the axiom of choice, as required.

### 2.13.5 Cardinalities: A Complete Order

In Chapter 1, we've studied sets and their cardinality. Fortunately, the well-ordering theorem and Zorn's lemma can now help in this study.

We've already met a few important cardinalities:

$$\aleph_0 < \aleph < 2^\aleph \cdots.$$

This comparison uses the strict order '$<$'. This also induces the weak order '$\leq$': either $<$, or $=$.

Is this a legitimate partial order? Well, let's check the requirements in Section 2.1.2. Clearly, the present '$\leq$' is reflexive and transitive. Furthermore, thanks to Cantor-Bernstein's theorem, it is also "anti-symmetric" (Chapter 1, Section 1.6.2).

So far, we only know that this is a partial order. Is it also complete? In other words, given two cardinalities, could they always be compared to one another? To check on this, pick two sets: $O$ (of the former cardinality), and $Q$ (of the latter cardinality). Use the well-ordering theorem to order them well.

Now, as well-ordered sets, they must embed in one another:

- either $O$ embeds in $Q$ as an initial segment, so

$$|O| \leq |Q|,$$

- or $Q$ embeds in $O$ as an initial segment, so

$$|Q| \leq |O|,$$

- or $O \cong Q$, so

$$|O| = |Q|.$$

In summary, $O$ and $Q$ are indeed comparable in terms of cardinality:

$$\text{either} \quad |O| \leq |Q| \quad \text{or} \quad |Q| \leq |O|.$$

Thus, cardinalities are ordered completely: every two cardinalities could be compared to one another.

## 2.14 Zorn's Lemma and Its Applications

### 2.14.1 How to Add Cardinalities?

How to add two cardinalities to each other? Pick two (disjoint) sets: $O$ (of the former cardinality) and $Q$, (of the latter cardinality), and define

$$|O| + |Q| \equiv |O \cup Q|.$$

For example, we already know that

$$\aleph_0 + \aleph_0 = \aleph_0,$$

and

$$\aleph + \aleph = \aleph.$$

Is this true for every infinite cardinality? Yes, it is!

To see this, let $A$ be some infinite set. We want to show that

$$|A| + |A| = |A|.$$

For this purpose, let's write $A$ in a new form: as a union of disjoint $B_i$'s:

$$A = \cup_{\{B_i \mid i \in I\}} \equiv \cup_{i \in I} B_i.$$

Here, $I$ is an index set: it contains those indices $i$ used to index the $B_i$'s. There may be really many $i$'s: in fact, $I$ may even be uncountable. In this case, we may need not only $i = 1, 2, 3, \ldots$ but also $i = 1/2$ or any other fraction. Fortunately, $I$ can't be too big:

$$|I| \leq |A|.$$

The $B_i$'s, on the other hand, should be countable:

$$|B_i| = \aleph_0, \quad i \in I.$$

Must $A$ have such a form? Yes, it must! To see this, let's split $A$ into disjoint countable pieces, and define a new function on them.

As an infinite set, $A$ must contain at least $\aleph_0$ elements (Chapter 1, Section 1.4.1). Together, they make a new subset $B_1 \subset A$. Let

$$f_1 : B_1 \to \mathbb{N} \times \{1\}$$

be one-to-one and *onto* (not just *into*). This means that its image is

$$f_1(B_1) = \mathbb{N} \times \{1\}.$$

Now, if there are still $\aleph_0$ elements left in $A \setminus B_1$, then they make a new (disjoint) subset $B_2 \subset A$. Extend $f_1$ into a new function $f_2$, defined in $B_2$ as well:

$$f_2 : B_1 \cup B_2 \to \mathbb{N} \times \{1, 2\},$$

which agrees with $f_1$ on $B_1$:

$$f_2 \mid_{B_1} \equiv f_1,$$

and is still one-to-one and onto:

$$f_2(B_2) = \mathbb{N} \times \{2\}.$$

This way, $I$ increases monotonically in terms of inclusion. In $f_1$, $I = \{1\}$. In $f_2$, on the other hand, $I = \{1, 2\}$.

What are $f_1$ and $f_2$? Well, they are special functions. Of what kind? To understand this, let's put both $f_1$ and $f_2$ in a new set $F$, containing those functions defined on a union of countable subsets as above, which are also one-to one and onto:

$$f_1, f_2 \in F \equiv \{u : \cup_{i \in I} B_i \to \mathbb{N} \times I \mid I \neq \emptyset, \; |I| \leq |A|,$$
$$\cup_{i \in I} B_i \subset A, \; B_i \cap B_j = \emptyset \; (i, j \in I, \; i \neq j),$$
$$u \text{ is one-to-one and onto: } u(B_i) = \mathbb{N} \times \{i\}, \; i \in I\}.$$

Is $F$ a legitimate set? Yes, it is:

$$|F| \leq \left| \cup_{S \subset A} (\mathbb{N} \times A)^S \right|.$$

Furthermore, $F$ is nonempty: it contains at least one function: $f_1$.

Moreover, $F$ has a partial order, with every chain having an upper bound in $F$ (Section 2.2.3). In fact, in such a chain, the $I$'s increase monotonically in terms of inclusion. In the upper bound, on the other hand, $I$ is just their union.

Thanks to Zorn's lemma, $F$ must contain a maximal element: a one-to-one function

$$g : \cup_{i \in I} B_i \to \mathbb{N} \times I$$

(for some $I$) that has the above properties and can never be extended. Thus, there may be only a few last elements left in $A$ unaccounted for:

$$|A \setminus \cup_{i \in I} B_i| < \aleph_0.$$

Let's go ahead and add them to $B_1$:

$$B_1 \leftarrow B_1 \cup (A \setminus \cup_{i \in I} B_i).$$

Once this substitution is made,

$$A = \cup_{i \in I} B_i.$$

In its up-to-date form, $B_1$ is still countable. So, we can now redefine $g$ on the new $B_1$, to become again one-to-one and onto:

$$g(B_1) = \mathbb{N} \times \{1\}.$$

In its final form,

$$g : A \to \mathbb{N} \times I$$

is one-to-one and onto, as required. Thus,

$$|A| = \aleph_0 |I|,$$

so

$$|A| + |A| = \aleph_0 |I| + \aleph_0 |I| = (\aleph_0 + \aleph_0) |I| = \aleph_0 |I| = |A|,$$

as asserted.

## 2.14.2 Multiplying Cardinalities: Geometrical Approach

Let's use the above result to multiply cardinalities. In Chapter 1, we've already seen that

$$\aleph_0^2 = \aleph_0,$$

and

$$\aleph^2 = \aleph.$$

Is this true for every infinite cardinality? Yes, it is!

To see this, let $A$ be some infinite set. We need to show that

$$|A|^2 = |A|.$$

To see this, let $G$ be the set of one-to-one functions from a subset $S \subset A$ onto $S^2$:

$$G \equiv \left\{ u : S \to S^2 \mid S \subset A, \ u \text{ is one-to-one and onto: } u(S) = S^2 \right\}.$$

Is $G$ a legitimate set? Yes, it is:

$$|G| \leq \left| \cup_{S \subset A} \left( S^2 \right)^S \right|.$$

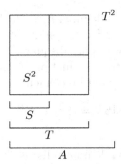

**Fig. 2.6.** Proof by contradiction: if $|A \setminus S| > |S|$, then one could define an intermediate subset $T \equiv 2S \subset A$. $T^2$ would then contain three more subsquares of size $S^2$ each, which could be covered by elements from $T \setminus S$. This way, $T$ would cover $T^2$, in violation of Zorn's lemma.

Clearly, $G$ is nonempty. Indeed, from the infinite set $A$, pick $\aleph_0$ elements, place them in a new subset $S \subset A$, and map them by a one-to-one function onto $S^2$ (Chapter 1, Sections 1.4.1 and 1.5.2).

Furthermore, $G$ has a partial order, with every chain having an upper bound in $G$ (Sections 2.2.3–2.2.4). Thanks to Zorn's lemma, $G$ also contains a maximal element: a one-to-one function $h$ (from some subset $S \subset A$ onto $S^2$) that can never be extended to any one-to-one function from $T$ onto $T^2$, for any intermediate subset $S \subset T \subset A$, $T \neq S$.

Now, how big is this $S$? Is $S$ as big as $A$? In other words, is

$$|S| = |A|?$$

If it is, then we're done. After all, thanks to $h$,

$$|A|^2 = |S|^2 = |S| = |A|,$$

as required. If, on the other hand,

$$|S| < |A|,$$

then we'd have

$$|A \setminus S| + |S| = |A| > |S|.$$

Thanks to the previous section, we must then have

$$|A \setminus S| > |S|.$$

We could then easily design an intermediate subset $S \subset T \subset A$ that is twice as big as $S$. For instance, $T$ could contain all the original elements in $S$, plus $|S|$ more elements from $A \setminus S$.

This would contradict Zorn's lemma: $h$ could then extend to a one-to-one function from $T$ onto $T^2$. Indeed, $T^2$ could split into four disjoint subsquares of size $|S|^2$ each (Figure 2.6). The lower-left subsquare, $S^2$, is already covered by elements from $S$, as in the original one-to-one function $h$. The three other subsquares, on the other hand, could be easily covered by elements from $T \setminus S$:

$$|T \setminus S| = |S| = |S^2| = |S^2| + |S^2| = |S^2| + |S^2| + |S^2|$$

(Section 2.14.1).

But we know better. Zorn's lemma tells us that $h$ can never extend in such a way. Thus, we must have

$$|A \setminus S| \leq |S|,$$

so

$$|A| = |S| + |A \setminus S| \leq |S| + |S| = |S|.$$

From Cantor-Bernstein's theorem, we therefore have

$$|A| = |S|.$$

Thanks to $h$, we therefore have

$$|A| = |S| = |S|^2 = |A|^2,$$

as asserted.

## 2.15 Exercises: No Commutativity!

### 2.15.1 Ordinals: No Commutativity

1. If God is all-mighty and is everywhere, could he create a new place where he is missing?
2. Could God create a stone so heavy that even he could never lift?
3. What's this got to do with set theory?
4. What's this got to do with Russell's paradox? Hint: the container $A$ is so big that neither $A \in A$ nor $A \notin A$.
5. What's this got to do with the ordinal paradox? Hint: again, the container $A$ is too big: neither $A \in A$ nor $A \notin A$.
6. Consider two real numbers:

$$0 < a < b \leq \infty.$$

Consider the half open interval
$$[-a, b).$$

In it, order the real numbers by absolute value:

$$x_1 \text{ is prior to } x_2 \quad \text{if} \quad |x_1| < |x_2|.$$

7. Is this transitive?
8. Is this anti-reflexive?
9. Is this anti-symmetric?
10. Conclude that this is indeed a legitimate (strict) partial order.
11. Is it complete? Hint: $x$ is not comparable to $-x$.
12. Use this order to design a chain.
13. Is the chain completely-ordered? Hint: by definition, it must be.
14. Is it well-ordered? Hint: not necessarily. For example, the chain

$$\left\{ c_{1/i} \equiv \frac{1}{i} \right\}_{i \in \mathbb{N}}$$

has no first element at all.
15. Does it have an upper bound?
16. Design a chain with no upper bound. Hint: let it approach $b$.
17. Is there a maximal element in $[-a, b)$?
18. Does this contradict Zorn's lemma in any way?
19. Assume now that

$$0 < b \leq a < \infty.$$

20. Consider the new half open interval

$$[-a, b).$$

21. Order it by absolute value, as before.
22. Design a chain in it.
23. Does it have an upper bound? Hint: $-a$.
24. Is there a maximal element? Hint: $-a$.
25. Is this as expected from Zorn's lemma?
26. Show that the inclusion order (Section 2.1.3) is indeed a legitimate weak order: reflexive, transitive, and "anti-symmetric."
27. Why the quotation marks?
28. From it, deduce a strict order: included, but not equal.
29. Show that it is indeed transitive, anti-reflexive, and anti-symmetric, as required.
30. Why there are now no quotation marks?
31. Is this order complete? Give an example of two elements that can never be compared to each other. Hint: design two sets that don't include one another: each has an element not in the other.
32. Show that the partial orders defined in Sections 2.2.3–2.2.5 are indeed legitimate weak orders: reflexive, transitive, and "anti-symmetric."
33. Give an example of two elements that can never be compared in terms of this order.
34. Conclude that this order is not complete.
35. In a well-ordered set, does every element has a next element? Hint: only a nonmaximal element.
36. Is it unique?
37. Show that, in a well-ordered set, there is no intermediate element between $x$ and its next element $\hat{x}$, defined in Section 2.4.3.
38. Show that a subset of a well-ordered set is well-ordered as well (in terms of the same order).

39. Conclude that an initial segment of a well-ordered set is well-ordered as well.
40. What is an isomorphism? Hint: one-to-one, onto, and order preserving.
41. Could it be defined on a set that is only partially-ordered, but not completely-ordered, let alone well-ordered? Hint: yes, provided that it is one-to-one, onto, and order preserving.
42. In what sense does an isomorphism preserve the structure of the original partially-ordered set?
43. Show that isomorphism between partially-ordered sets (denoted by '$\cong$') is a mathematical equivalence relation: reflexive, symmetric, and transitive.
44. Show that isomorphism preserves the next element, if exists.
45. Show that isomorphism preserves a maximal element, if exists.
46. Show that isomorphism preserves a minimal element, if exists.
47. Show that isomorphism preserves the greatest element, if exists.
48. Show that isomorphism preserves the smallest (or first) element, if exists.
49. Show that isomorphism preserves completeness.
50. Show that isomorphism preserves the well-order property.
51. Could a well-ordered set be isomorphic to its own initial segment? Hint: see Section 2.5.3.
52. Could two distinct initial segments of the same well-ordered set be isomorphic to each other? Hint: see Section 2.5.3.
53. Show that, in the induction step in Section 2.10.3, the isomorphism is extended uniquely. Hint: use the induction hypothesis.
54. Show that the new (extended) isomorphism indeed proves the induction step.
55. Use this to prove the assertion.
56. Interpret an ordinal as a typical well-ordered set that mirrors or represents all those that are isomorphic to it.
57. Conclude that an ordinal is not defined uniquely, but only up to isomorphism.
58. Show that a well-ordered set is isomorphic to the set of its own initial segments. Hint: see Section 2.5.3.
59. Conclude that an ordinal is isomorphic to the set of smaller ordinals. Hint: here, "smaller" means isomorphic to an initial segment of the original ordinal (Section 2.11.1).
60. Show that, if $A$ and $B$ are well-ordered sets, then their sum $A + B$ (defined in Section 2.7.1) is well-ordered as well.
61. Show that adding ordinals is associative.
62. Is it also commutative? Hint: give a counter example.
63. Show that Von-Neumann's natural numbers (defined in Section 2.6.2) are ordered one by one in terms of inclusion. Hint: use mathematical induction.
64. Is this a complete order? Hint: each natural number is greater than all the previous ones, and smaller than all the following ones.
65. Is this a well order? Hint: pick some (infinite) subset of natural numbers. Does it contain a minimal number? Well, pick some number from it. There are only a few smaller numbers. Pick their minimum.
66. Conclude that their union $\omega$ (defined in Section 2.7.2) is well-ordered as well. Hint: use the same technique.
67. Show that Von-Neumann's ordinals sum as in standard arithmetic:

$$3 + 4 \cong 7.$$

68. What are the individual elements in $\omega^k$? Hint: finite sequences of $k$ natural numbers in a row.
69. What is the order between them? Hint: anti-lexicographic.
70. Is it complete?
71. Is it well? Hint: see below.
72. Show that, if $A$ and $B$ are well-ordered sets, then their product $A \times B$ is well-ordered as well. Hint: see Section 2.9.2.
73. Conclude that, for every natural number $k$, $\omega^k$ is well-ordered as well. Hint: use mathematical induction on $k \geq 1$.
74. Look at their union $\omega^\omega$. What are the individual elements in it? Hint: arbitrarily long finite sequences of natural numbers.
75. What is the order between them? Hint: see Section 2.9.3.
76. Is it complete?
77. Is it well? Hint: pick some (infinite) subset of (arbitrarily long) finite sequences. Does it contain a first sequence? Well, pick some sequence from it. It must have a finite length $k$. We already know that $\omega^k$ is well-ordered.
78. Show that $\omega^\omega$ is countable:
$$|\omega^\omega| = \aleph_0.$$

Hint: see exercises at the end of Chapter 1.
79. Show that multiplying ordinals is associative.
80. Is it also commutative? Hint: give a counter example.
81. Show that Von-Neumann's ordinals (Section 2.6.2) multiply as in standard arithmetic:
$$3 \cdot 4 \cong 12.$$

82. Is
$$3 \cdot 4 \cong 4 \cdot 3?$$

Why?
83. Is
$$2 \cdot \omega \cong \omega \cdot 2?$$

Why? Hint: see below.
84. Do they have a maximal element? Hint: let one coordinate approach infinity.
85. What would Zorn's lemma say about this?
86. In view of the above, must they have a chain with no upper bound?
87. What would Zorn's lemma say about this?
88. Design such a chain.
89. On the other hand, could they have a chain with an upper bound? Hint: Zorn's lemma never said they couldn't.
90. Furthermore, do they have an infinite chain with an upper bound? Hint: only $\omega \cdot 2$ does, but $2 \cdot \omega$ doesn't.
91. Do they contain an element with infinitely many smaller elements? Hint: only $\omega \cdot 2$ does, but $2 \cdot \omega$ doesn't.
92. Could they possibly be isomorphic to one another? Hint: the previous exercise shows they couldn't.
93. How are ordinals compared to one another? Hint: one is embedded in the other as an initial segment.

94. Show that this is indeed a legitimate strict order: transitive, anti-reflexive, and anti-symmetric. Hint: no ordinal could ever be isomorphic to its own initial segment (Section 2.5.3).

95. Show that this is also a complete order. Hint: see Section 2.10.3.

96. Show that this is also a well order. Hint: see Section 2.11.2.

97. Prove the ordinal paradox: the container of all ordinals is too big — it could never be a legitimate set. Hint: if it were, then it would be isomorphic to its own initial segment, which is forbidden (Section 2.11.3).

98. Fortunately, it is not a set but a mere container. This way, there is no problem any more. Why? Hint: in a container that is not a set, the monotonically decreasing chain designed in Section 2.5.3 is not a set either, so it may have no first element at all, with no problem.

99. Show that Zorn's lemma could have been stronger: it is sufficient to assume that the *well-ordered* chains have upper bounds.

100. How could this help "solve" Zeno's paradox? Hint: see Section 2.13.4.

101. Let $A$ and $B$ be nonempty sets. Assume that at least one of them is infinite. Show that

$$|A| + |B| = \max(|A|, |B|).$$

Hint: see Section 2.14.1.

102. Furthermore, show also that

$$|A| \cdot |B| = \max(|A|, |B|).$$

Hint: see Section 2.14.2.

# Part II

## Applications
## in Functional Analysis

# Applications in Functional Analysis

So far, we introduced Zorn's lemma, and saw that it is equivalent to two other fundamental theorems: the well-ordering theorem, and the axiom of choice. Furthermore, we also used Zorn's lemma in a nice application in set theory. Let's see how useful it is in yet another field: functional analysis. In fact, it helps prove two important Han-Banach theorems.

This material is written in a new style: exercise by exercise. This way, you are welcome to try and solve each exercise, before looking at the solution. This way, you get to see how the theory develops linearly, step by step.

# 3

# Zorn's Lemma
# in Han-Banach Theory

In this chapter, we use set theory in yet another important field: functional analysis. More specifically, we prove two Han-Banach theorems: how to extend a functional linearly and continuously, and how to separate two convex sets from each other by a hyperplane. Both have a lot of applications in physics, detailed elsewhere. This is a good exercise in using Zorn's lemma in practice.

This chapter is written in a new style: list of exercises. In each exercise, you are welcome to spend some time trying to solve it. Then, proceed to the hint that actually provides the solution. This way, you get to see how the theory develops linearly, step by step.

## 3.1 The Han-Banach Extension Theorem

1. Consider a linear space (or vector space) $S$.
2. What is the dimension of $S$? Hint: it could be infinite, and even uncountable. For example, $S$ could be a space of functions.
3. What algebraic operations are defined in $S$? Hint: addition and scalar multiplication.
4. Is $S$ closed under addition?
5. What does this mean? Hint: for every two vectors in $S$, their sum is in $S$ as well.
6. Is $S$ closed under scalar multiplication?
7. What does this mean? Hint: for every vector in $S$, all its scalar multiples are in $S$ as well.
8. What laws does addition obey? Hint: associativity and commutativity.
9. What laws does scalar multiplication obey? Hint: associativity and distributivity.
10. Specify two kinds of distributivity. Hint: open parentheses around the sum of two scalars, and also around the sum of two vectors.
11. Consider a subspace $V \subset S$. What are its properties? Hint: $V$ is a linear space in its own right.

12. Consider a real function on $S$:

$$P : S \to \mathbb{R},$$

with two important properties:
- A positive scalar could be pulled out:

$$P(\alpha v) = \alpha P(v), \quad v \in S, \ \alpha > 0.$$

- The triangle inequality:

$$P(u + v) \leq P(u) + P(v), \quad u, v \in S.$$

13. $P$ is also called a sublinear functional on $S$.
14. Consider a chain, as in Chapter 2, Sections 2.2.1–2.2.2. More precisely, assume that the $c_i$'s are subspaces of $S$ that include each other in a row.
15. On the $c_i$'s, define real functions $f_i$ that extend each other in a row (as in Chapter 2, Section 2.2.3), and have two important properties:
- They are linear (it is allowed to open parentheses):

$$f_i(\alpha u + \beta v) = \alpha f_i(u) + \beta f_i(v), \quad u, v \in c_i, \ \alpha, \beta \in \mathbb{R}.$$

- They are bounded by $P$:

$$f_i(v) \leq P(v), \quad v \in c_i.$$

16. Do they have an upper bound? Hint: as in Chapter 2, Sections 2.2.1–2.2.3, consider the union of subspaces

$$\cup_C \equiv c_1 \cup c_2 \cup c_3 \cup \cdots \cup c_i \cup \cdots.$$

(Here, $i$ is not necessarily an integer, but could also be a fraction, so the number of subspaces could be uncountable.) This union contains all the vectors in the $c_i$'s. On this union, define the new function $f$ that extends all the $f_i$'s. After all, each and every vector $v \in \cup_C$ belongs to some $c_i$, so $f$ could be defined just like $f_i$:

$$f(v) \equiv f_i(v), \quad v \in c_i.$$

17. Is $f$ a legitimate upper bound? Hint: we only allow functions that are linear and bounded by $P$. Fortunately, $f$ is. To show this, use the same technique as in Chapter 2, Section 2.2.3.
18. Consider a real function on $V$:

$$F : V \to \mathbb{R},$$

with our two properties: linear, and bounded by $P$.
19. $F$ is also called a linear functional on $V$.
20. Consider the set of those functions that extend $F$, and still enjoy our two properties: linear, and bounded by $P$.
21. Is it nonempty? Hint: yes. After all, it contains $F$ itself.
22. Does every chain in it have an upper bound?

23. How does this upper bound look like? Hint: we've already designed it above: $f$.

24. What does Zorn's lemma (Chapter 2, Section 2.13.4) say about this? Hint: $F$ could be extended to a maximal function (still called $F$, and still linear and bounded by $P$) that can't be extended any more. After being extended, $F$ is defined not only on $V$ but also on a bigger subspace $W$:

$$V \subset W \subset S.$$

Moreover, $F$ can never be extended to any subspace bigger than $W$.

25. Is $W = S$? Hint: prove this by contradiction, step by step, as done below.

26. Assume momentarily that

$$W \neq S.$$

27. In this case, there were some vector

$$z \in S \setminus W.$$

28. Extend $F$ to $z$ as well. For this purpose, we must define $F(z)$ properly. How?

29. Assume that we've already defined $F(z)$ somehow. (Later on, we'll specify how.) To remain linear, $F$ must satisfy

$$F(x + \alpha z) = F(x) + \alpha F(z), \quad x \in W, \ \alpha \in \mathbb{R}.$$

30. Moreover, to remain bounded by $P$, $F$ must also satisfy

$$F(x + \alpha z) \leq P(x + \alpha z) \quad x \in W, \ \alpha \in \mathbb{R}.$$

31. For $\alpha = 0$, this is obvious. How to guarantee this for $\alpha \neq 0$ as well? Hint: split the problem into two cases: $\alpha > 0$, and $\alpha < 0$, as done below.

32. For $\alpha > 0$, divide the above requirement by $\alpha$:

$$F\left(\frac{x}{\alpha} + z\right) \leq P\left(\frac{x}{\alpha} + z\right).$$

33. For $\alpha < 0$, on the other hand, divide the original requirement by $-\alpha$:

$$F\left(-\frac{x}{\alpha} - z\right) \leq P\left(-\frac{x}{\alpha} - z\right).$$

34. In other words, for every vector $x \in W$, for every positive scalar $\alpha > 0$, we must have

$$\frac{1}{\alpha} F(x) + F(z) = F\left(\frac{x}{\alpha} + z\right)$$
$$\leq P\left(\frac{x}{\alpha} + z\right).$$

Hint: $F$ should be linear, and bounded by $P$.

35. Likewise, for every vector $x \in W$, for every negative scalar $\alpha < 0$, we must also have

$$\frac{1}{\alpha} F(x) + F(z) = F\left(\frac{x}{\alpha} + z\right)$$
$$= -F\left(-\frac{x}{\alpha} - z\right)$$
$$\geq -P\left(-\frac{x}{\alpha} - z\right).$$

36. Rearrange these two requirements a little, and rewrite them as new requirements on $F(z)$ (the former with $\alpha > 0$, and the latter with $\alpha < 0$):

$$F(z) \le P\left(\frac{x}{\alpha} + z\right) - \frac{1}{\alpha}F(x)$$
$$F(z) \ge -P\left(-\frac{x}{\alpha} - z\right) - \frac{1}{\alpha}F(x).$$

37. Could $F(z)$ be defined in such a way? Hint: yes! See below.
38. Let $p, q \in W$ be two vectors. Apply $F$ to both of them, and look at the difference:

$$\begin{aligned} F(p) - F(q) &= F(p - q) \\ &\le P(p - q) \\ &= P(p + z + (-q - z)) \\ &\le P(p + z) + P(-q - z). \end{aligned}$$

39. Rearrange this a little, so that $p$ is on the right, and $q$ is on the left:

$$-F(q) - P(-q - z) \le P(p + z) - F(p).$$

40. Take the maximum (or supremum) on the left-hand side, and the minimum (or infimum) on the right-hand side:

$$\max_{q \in W}\left(-F(q) - P(-q - z)\right) \le \min_{p \in W}\left(P(p + z) - F(p)\right).$$

41. Define $F(z)$ cleverly: plug it in between the left-hand side and the right-hand side:

$$\max_{q \in W}\left(-F(q) - P(-q - z)\right) \le F(z) \le \min_{p \in W}\left(P(p + z) - F(p)\right).$$

42. Use this to make sure that the original requirements are indeed satisfied. For this purpose, just pick $p \equiv x/\alpha$ (with $\alpha > 0$), or $q \equiv x/\alpha$ (with $\alpha < 0$).
43. Conclude that $F$ could indeed be extended to $z$, remaining linear and bounded by $P$.
44. Is this a contradiction?
45. Conclude that
$$W = S.$$

46. This is indeed the Han-Banach extension theorem in functional analysis: $F$ could indeed be extended from the original subspace $V$ to the entire linear space $S$, remaining linear and bounded by $P$. This theorem has important applications in physics.

## 3.2 Vector Norm and Continuity

1. Let $0 \le t \le 1$ be a given number. Let $u, v \in S$ be two vectors. The sum

$$tu + (1 - t)v$$

is called a convex combination of $u$ and $v$.

2. Show that our functional $P$ is convex in the sense that

$$P(tu + (1 - t)v) \leq tP(u) + (1 - t)P(v), \quad u, v \in S, \ 0 \leq t \leq 1.$$

Hint: for $t = 1$ (or $t = 0$), this is obvious: both sides are equal to $P(u)$ (or $P(v)$). For $0 < t < 1$, on the other hand, use the triangle inequality:

$$P(tu + (1 - t)v) \leq P(tu) + P((1 - t)v) = tP(u) + (1 - t)P(v).$$

3. Assume now that $P$ has another attractive property — it is insensitive to sign:

$$P(-v) = P(v), \quad v \in S.$$

4. Show that, in this case, our linear functional $F$ is bounded not only in the usual sense but also in absolute value:

$$|F(v)| \leq P(v), \quad v \in S.$$

Hint: we already know that
$$F(v) \leq P(v).$$

As a result,
$$-F(v) = F(-v) \leq P(-v) = P(v).$$

5. Assume now that $P$ has yet another useful property — it vanishes at the zero vector:
$$P(\mathbf{0}) = 0.$$

6. Show that, in this case, $P$ is nonnegative. Hint: thanks to the triangle inequality,

$$0 = P(\mathbf{0}) = P(v + (-v)) \leq P(v) + P(-v) = 2P(v), \quad v \in S.$$

7. Assume now that $P$ has yet another attractive property — it vanishes *only* at the zero vector:
$$P(v) = 0 \iff v = \mathbf{0}.$$

8. Show that, in this case, for every nonzero vector, $P$ must be positive. Hint: we've already seen that $P$ is nonnegative. Now, we also know that it can't vanish.

9. In this case, $P$ is also called a norm, and is also denoted by

$$\|v\| \equiv P(v), \quad v \in S.$$

10. In this case, our linear functional $F$ satisfies

$$|F(v)| \leq \|v\|, \quad v \in S.$$

11. Show that, in this case, every two vectors that are "close" to each other in $S$ have nearly the same value under $F$:

$$|F(u) - F(v)| = |F(u - v)| \leq \|u - v\|, \quad u, v \in S.$$

12. In this case, we say that $F$ is not only linear but also continuous in $S$.

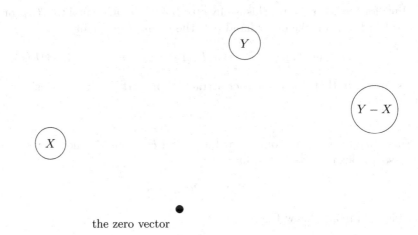

**Fig. 3.1.** $Y$ is not only convex but also open (contains no boundary point). $X$, on the other hand, is convex, but not necessarily open (may contain some boundary points). The difference $Y - X$ is obtained by "shifting" $Y$ by $-X$. Therefore, it is convex and open as well.

## 3.3 Open Convex Set

1. Let $r > 0$ be a given number. The ball $B_r$ contains those vectors whose norm is smaller than $r$:
$$B_r \equiv \{v \in S \mid \|v\| < r\}.$$

2. Let
$$Y \subset S$$
be some nonempty subset (not necessarily a subspace). Assume that $Y$ is open: each vector in $Y$ is surrounded by a small ball in $Y$:
$$y \in Y \Rightarrow y + B_r \subset Y,$$
where $r \equiv r(y) > 0$ depends on $y$. Here, the sum means
$$y + B_r \equiv \{y + b \mid b \in B_r\}.$$

3. Pick some fixed $r > 0$. Could $B_r$ contain any boundary point?
4. Is $B_r$ open? Hint: pick some $p \in B_r$. This way, $\|p\| < r$. Now, in $B_r$, is $p$ surrounded by a yet smaller ball? Well, around $p$, draw a ball of radius $r - \|p\| > 0$. Now, for every vector $q$ with norm as small as $\|q\| < r - \|p\|$, thanks to the triangle inequality,

$$\|p + q\| \le \|p\| + \|q\| < \|p\| + r - \|p\| = r.$$

Thus, in $B_r$, $p$ is indeed surrounded by a yet smaller ball, of radius $r - \|p\| > 0$:

$$p + B_{r-\|p\|} \subset B_r.$$

5. Assume that $Y$ is not only open but also convex: for every two vectors in $Y$, their convex combination is in $Y$ as well:

$$tu + (1 - t)v \in Y, \quad u, v \in Y, \ 0 \le t \le 1.$$

6. Let $X \subset S$ be another (nonempty) convex set (not necessarily open). Define the difference of these sets:

$$Y - X \equiv \{y - x \mid y \in Y, \ x \in X\}.$$

Is $Y - X$ still open? Hint: see Figure 3.1. $Y$ is just "shifted" by $-X$, remaining open. Indeed, pick some $y \in Y$. We already have $y + B_r \subset Y$. Therefore, for each $x \in X$,

$$y - x + B_r = y + B_r - x \subset Y - X.$$

7. Is $Y - X$ still convex? Hint: pick $y, \tilde{y} \in Y$ and $x, \tilde{x} \in X$. For each $0 \le t \le 1$,

$$t(y - x) + (1 - t)(\tilde{y} - \tilde{x}) \stackrel{*}{=} ty + (1 - t)\tilde{y} - (tx + (1 - t)\tilde{x}) \in Y - X.$$

## 3.4 Zorn's Lemma: Maximal Cone

1. Assume also that $Y$ and $X$ are disjoint (share no joint vector):

$$Y \cap X = \emptyset.$$

2. Conclude that $Y - X$ doesn't contain the zero vector:

$$\mathbf{0} \notin Y - X.$$

3. In our linear space $S$, look at all those subsets that share these four properties:
   - open,
   - convex,
   - include $Y - X$,
   - exclude the zero vector.
4. Give an example of such a set. Hint: $Y - X$ itself.
5. Consider a chain of such sets, ordered by inclusion: they include each other in a row. Hint: see Chapter 2, Section 2.2.1.
6. Does it have an upper bound? Hint: take its union.
7. Is the union a legitimate upper bound?
8. Does it still include $Y - X$?
9. Does it still exclude the zero vector?
10. Is it still open and convex? Hint: to prove this, use the same technique as in Chapter 2, Section 2.2.3: pick two vectors in the union. They must already belong to some set in the chain, etc.

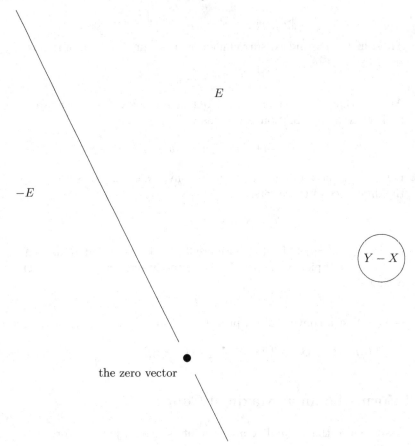

**Fig. 3.2.** $E$ is the maximal cone that includes $Y - X$. $E$ has good properties: open, convex, closed under positive multiplication and addition, and doesn't contain the zero vector.

11. What would Zorn's lemma (Chapter 2, Section 2.13.4) say about this?
12. Conclude that $Y - X$ could be extended to a maximal open convex set $E \subset S$ that doesn't contain the zero vector. Hint: see Figure 3.2.
13. $E$ is also called a maximal cone. Let's see why.
14. Define a yet "bigger" set that contains all the positive multiples of vectors in $E$:

$$\hat{E} \equiv \{ae \mid a > 0,\ e \in E\}.$$

15. Does it include $E$? Hint: pick $a = 1$.
16. Does it exclude the zero vector? Hint: $e \neq \mathbf{0}$.
17. Is it still open? Hint: pick some $e \in E$. We already know that, for some $r \equiv r(e) > 0$, $e + B_r \subset E$. Therefore, for each $a > 0$,

$$ae + B_{ar} = ae + aB_r = a\,(e + B_r) \subset \hat{E}.$$

18. Is it still convex? Hint: pick some $e, \tilde{e} \in E$ $a, \tilde{a} > 0$, and $0 \leq t \leq 1$. Define

$$b \equiv ta + (1 - t)\tilde{a} > 0.$$

Then,

$$tae + (1-t)\tilde{a}\tilde{e} = b\left(\frac{ta}{b}e + \frac{(1-t)\tilde{a}}{b}\tilde{e}\right) \in \hat{E}.$$

19. Could $\hat{E}$ be really bigger than $E$? Hint: no! $E$ is maximal.
20. Conclude that

$$\hat{E} = E.$$

21. Is $E$ closed under positive multiplication? Hint: for every $e \in E$ and $a > 0$, $ae \in \hat{E} = E$.
22. Is $E$ closed under addition? Hint: pick some $e, \tilde{e} \in E$. Since $E$ is convex, $(e + \tilde{e})/2 \in E$. Therefore, $e + \tilde{e} \in \hat{E} = E$.
23. This is why $E$ is also called the maximal cone that includes $Y - X$.


## 3.5 The Han-Banach Separation Theorem

1. Is $E$ a subspace? Hint: no! It is not closed under the minus sign.
2. Define the negative of $E$:

$$-E \equiv \{-e \mid e \in E\}.$$

3. Look at those vectors that are neither in $E$ nor in $-E$:

$$U \equiv S \setminus (E \cup -E).$$

4. Is $U$ closed under positive multiplication? Hint: pick some $a > 0$ and $u \in U$. If $au \notin U$, then there are two possibilities: either $au \in E$, so

$$u = \frac{1}{a}au \in \hat{E} = E$$

(a contradiction), or $au \in -E$, so

$$u = \frac{1}{a}au \in -E$$

(a contradiction as well).
5. Is $U$ closed under the minus sign? Hint: pick some $u \in U$. If $-u \notin U$, then $-u$ must be in $E$ (or $-E$), so $u$ must be in $-E$ (or $E$), in violation of our assumption that $u \in U$.
6. Is $U$ closed under addition? Hint: see below.
7. Prove this by contradiction (Figure 3.3). For this purpose, pick some $u, v \in U$. Assume momentarily that $u + v \notin U$.
8. This means that $u + v \in E$ or $u + v \in -E$.
9. Without loss of generality, assume that $u + v \in E$. (If, on the other hand, $u + v \in -E$, then work with $-u, -v \in U$.)
10. Now, where is $u - v$? Could $u - v \in E$? Hint: if so, then $2u = u + v + u - v \in E$, in violation of our assumption that $u \in U$.
11. On the other hand, could $u - v \in -E$? Hint: if so, then $v - u \in E$, and $2v = u + v + v - u \in E$, in violation of our assumption that $v \in U$.

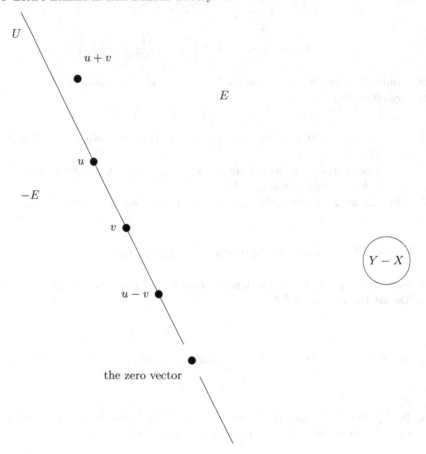

**Fig. 3.3.** The subspace $U$ (which separates $E$ from $-E$) is closed under addition. Indeed, $u + v$ can never be in $E$. If it were, then $u$ would be in $E$ as well, which is not true.

12. Conclude that $u - v \in U$.
13. Define the "bigger" cone
$$\hat{E} \equiv \{a(u - v) + e \mid a \geq 0, \ e \in E\}.$$
14. Does it include $E$? Hint: pick $a = 0$.
15. Does it still exclude $\mathbf{0}$? Hint: recall that $e \neq \mathbf{0}$. Now, could $a(u - v) + e = \mathbf{0}$? If it were, then we'd have $u - v = -e/a \in -E$ (a contradiction).
16. Is it still open? Hint: since $e + B_r \subset E$,
$$a(u - v) + e + B_r \subset \hat{E}.$$
17. Is it still convex? Hint: pick some $a, \tilde{a} \geq 0$, $e, \tilde{e} \in E$, and $0 \leq t \leq 1$. Then,
$$t(a(u - v) + e) + (1 - t)(\tilde{a}(u - v) + \tilde{e}) = (ta + (1 - t)\tilde{a})(u - v) + te + (1 - t)\tilde{e} \in \hat{E}.$$
18. Conclude that
$$\hat{E} = E.$$

Hint: $E$ is maximal.

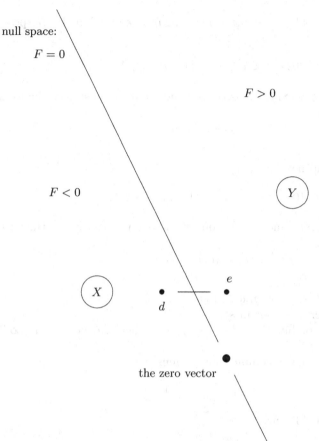

null space:

$F = 0$

$F > 0$

$F < 0$

$Y$

$e$

$X$

$d$

the zero vector

**Fig. 3.4.** The linear functional $F$ that separates $Y$ from $X$. $F(e) \equiv 1$ is used to define $F(d)$ as well. This way, $F$ is positive on $E$ on the upper right, negative on $-E$ on the lower left, and zero on $U$ in between.

19. Is this a contradiction? Hint: in $\hat{E}$, pick $a = 1$ and $e = u + v$:

$$2u = u - v + u + v = a(u - v) + e \in \hat{E} = E,$$

in violation of our assumption that $u \in U$.

20. What's wrong? Hint: our momentary assumption must have been wrong. The truth is that $u + v$ could never be in $E$. On the contrary: it must have been in $U$ all along, as required.

21. Conclude that $U$ is indeed closed under addition.

22. Is $U$ a legitimate subspace? Hint: yes! It is closed under addition and scalar multiplication.

23. Define a new functional $F$ that vanishes on $U$. This way, $U$ is the null space of $F$.

24. How to extend $F$ linearly? Hint: see below.

25. Pick some fixed $e \in E$ (Figure 3.4). Define

$$F(e) \equiv 1.$$

26. Now, for each $d \in -E$, look at the interval leading from $e$ to $d$. For this purpose, define two real sets:

$$s_0 \equiv \{0 \le t \le 1 \mid td + (1-t)e \in E\}.$$

27. Clearly, $0 \in s_0$, but $1 \notin s_0$. Since $E$ is convex and open, $s_0$ must take the form of a half-open interval:

$$s_0 = [0, a)$$

(for some fixed $0 < a < 1$).

28. Likewise, define yet another (disjoint) real set:

$$s_1 \equiv \{0 \le t \le 1 \mid td + (1-t)e \in -E\}.$$

29. Clearly, $1 \in s_1$, but $0 \notin s_1$. Since $-E$ is convex and open, $s_1$ must take the form of a half-open interval:

$$s_1 = (b, 1]$$

(for some fixed $a \le b < 1$).

30. Now, pick some fixed $t$ in between $a$ and $b$: $a \le t \le b$.
31. Clearly, this $t$ is neither in $s_0$ nor in $s_1$.
32. Therefore, the convex combination $td + (1-t)e$ is neither in $E$ nor in $-E$, so it must be in $U$.
33. Conclude that, at this convex combination, $F$ must vanish:

$$F(td + (1-t)e) = 0.$$

34. Use this to define $F$ at $d$ linearly:

$$tF(d) + (1-t)F(e) = F(td + (1-t)e) = 0.$$

Hint: divide this by $t \ge a > 0$, and use the original definition $F(e) \equiv 1$:

$$F(d) \equiv -\frac{1-t}{t}F(e) = -\frac{1-t}{t}.$$

35. Do this for every $d \in -E$.
36. Conclude that $F$ is negative throughout $-E$. Hint: $t \le b < 1$.
37. Use this to define $F$ throughout $E$ as well. Hint: $F(-d) \equiv -F(d)$.
38. Conclude that $F$ is positive throughout $E$.
39. Conclude that $F$ is positive throughout $Y - X$. Hint: $Y - X \subset E$.
40. Conclude that

$$F(y) > F(x), \quad y \in Y, \ x \in X.$$

41. Take the maximum (or supremum) on the right-hand side, and the minimum (or infimum) on the left-hand side:

$$\min_{y \in Y} F(y) \ge \max_{x \in X} F(x).$$

42. This is indeed the Han-Banach separation theorem in functional analysis, with a lot of applications in physics.

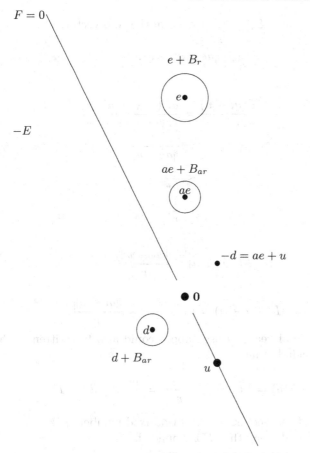

**Fig. 3.5.** $e$ is the fixed vector at which $F(e) \equiv 1$. Since $E$ is open, $e$ is surrounded by a ball of radius $r$. The vector $d = -ae - u$, on the other hand, is a general vector in $-E$. Since $-E$ is open, $d$ is surrounded by a ball of radius $ar$ that doesn't contain the zero vector.

## 3.6 Continuous Bounded Functional

1. Is $F$ continuous? Hint: see below.
2. Recall that $E$ is open. Therefore, the fixed vector $e$ defined above is surrounded by a little ball in $E$. Define $r \equiv r(e) > 0$ such that

$$e + B_r \subset E.$$

3. Pick some $a > 0$. Show that $ae$ is surrounded by a little ball as well. Hint:

$$ae + B_{ar} = ae + aB_r = a\left(e + B_r\right) \subset E.$$

4. Pick some $u \in U$ (Figure 3.5). Show that $ae + u$ is surrounded by a little ball as well. Hint: for every $b \in B_{ar}$,

$$F(ae + u + b) = F(ae + b) > 0.$$

5. Conclude that $ae + u + B_{ar}$ doesn't contain the zero vector.
6. Conclude that

$$\|ae + u\| \ge ar.$$

7. Conclude that

$$\frac{F(ae + u)}{\|ae + u\|} = \frac{aF(e) + F(u)}{\|ae + u\|}$$

$$= \frac{a}{\|ae + u\|}$$

$$\le \frac{a}{ar}$$

$$= \frac{1}{r}.$$

8. Rewrite this as

$$F(ae + u) \le \frac{\|ae + u\|}{r}.$$

9. Conclude that

$$-F(-ae - u) = F(ae + u) \le \frac{\|ae + u\|}{r}.$$

10. Note that the general vector $d$ used above could also be written in this form: $d = -ae - u$. Conclude that

$$-F(d) = F(-d) \le \frac{\| - d\|}{r} = \frac{\|d\|}{r}, \quad d \in -E.$$

11. Recall that $e$ is a fixed vector, so $r > 0$ is a fixed number.
12. For this reason, we also say that $F$ is bounded.
13. Conclude that, for every two vectors $v, w \in S$,

$$|F(v) - F(w))| = |F(v - w)| \le \frac{\|v - w\|}{r}.$$

14. Conclude that $F$ is indeed continuous: every two vectors that are "close" to each other have nearly the same value under $F$.

## 3.7 Boundary Points

1. Use the above to generalize the Han-Banach separation theorem yet more.
2. For this purpose, assume now that $Y \subset S$ is a general set of vectors, not necessarily open any more.
3. An interior point of $Y$ is a point $y \in Y$ surrounded by a ball $B_r$:

$$y + B_r \subset Y$$

(for some $r \equiv r(y) > 0$ that depends on $y$).
4. Let $O$ be the interior of $Y$ — the set of all interior points:

$$O \equiv \{y \in Y \mid y + B_{r(y)} \subset Y\}.$$

$O$

**Fig. 3.6.** If $Y$ is convex, then its interior $O$ is convex as well. Indeed, pick two points $o, q \in O$. Both are surrounded by a little ball in $Y$. Therefore, their convex combination is surrounded by a little ball as well.

5. Is $O$ open? Hint: to prove this, we must show that each such ball is included not only in $Y$ but also in $O$. This is done below.
6. Let $\hat{O}$ be an open set included in $Y$:

$$\hat{O} \subset Y.$$

Is $\hat{O} \subset O$? In other words, is $\hat{O}$ made of interior points only? Hint: pick some vector $o \in \hat{O}$. Since $\hat{O}$ is open, there is some $r \equiv r(o) > 0$ for which

$$o + B_r \subset \hat{O} \subset Y.$$

In other words, $o$ is indeed an interior point of $Y$. As such, $o \in O$. In summary, $\hat{O} \subset O$, as asserted.

7. Conclude that every open set is indeed included in $O$.
8. Note that, in the above definition of $O$, each ball of the form $y + B_{r(y)}$ is open. Hint: every ball is in fact open (Section 3.3).
9. Conclude that each such ball is included in $O$:

$$y + B_{r(y)} \subset O.$$

10. Conclude that each such ball is made of interior points only.
11. Conclude that $O$ contains not only the interior points but also the balls surrounding them:

$$O = \cup_{y \in Y} \left( y + B_{r(y)} \right).$$

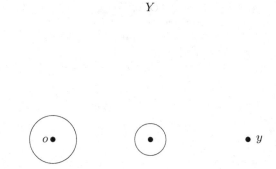

**Fig. 3.7.** The boundary point $y \in \partial Y$ could be approached by interior points in $O$. Indeed, pick some interior point $o \in O$. It is surrounded by a little ball in $Y$. Therefore, each convex combination of $y$ and $o$ is surrounded by a little ball as well, so it is indeed an interior point as well.

12. Conclude that $O$ is indeed open.
13. Show that $O$ is the maximal open set included in $Y$. Hint: let $\hat{O}$ be an open set included in $Y$. We've just seen that $\hat{O} \subset O$.
14. Could $O$ be characterized in yet another way?
15. Show that $O$ is also the union of all open sets included in $Y$. Hint: this union is still open. After all, each point in it must belong to some open set in the union, along with a little ball around it. Thus, the union is indeed the maximal open set included in $Y$.
16. Here, although $O$ is maximal in some sense, there is no need to use Zorn's lemma. Why? Hint: here, the union still has the desired property: it is still open. In Section 3.4, on the other hand, we required yet another property: convexity. Unfortunately, even the union of two convex sets is not necessarily convex any more. This is why we had to use Zorn's lemma there.
17. Assume now that $Y$ is convex. Is $O$ convex as well? Hint: pick some $0 < t < 1$, and some $o, q \in O$ (Figure 3.6). Since $o \in O$, $o$ has a little ball around it:

$$o + B_r \subset Y$$

(for some $r \equiv r(o) > 0$). Therefore, the convex combination of $o$ and $q$ has a little ball around it as well:

$$to + (1-t)q + B_{tr} = t(o + B_r) + (1-t)q \subset Y$$

(because $Y$ is convex). Thus, $to+(1-t)q \in O$, so $O$ is convex as well, as asserted.

18. In $Y$, what is the boundary? Hint: this is a special kind of set, containing those points that are not interior:
$$\partial Y \equiv Y \setminus O.$$

19. This is also called the boundary of $O$ in $Y$:
$$\partial O \equiv \partial Y.$$

20. Each point in it is called a boundary point.

21. As in Section 3.3, assume now that $X \subset S$ is nonempty and convex as well, and that $\mathbf{0} \notin Y - X$.

22. Assume also that $Y$ contains at least one interior point. In other words, $O$ is nonempty:
$$O \neq \emptyset.$$

23. Design a linear functional $F$ to separate $O$ from $X$:
$$F(o) > F(x), \quad o \in O, \ x \in X.$$

Hint: use the Han-Banach separation theorem (Section 3.5).

24. Let $y \in \partial Y$ be a boundary point in $Y$. Is $y$ separated from $X$ as well? In other words, is
$$F(y) \geq F(x), \quad x \in X?$$

Hint: see below.

25. Pick some $0 < t < 1$, and some interior point $o \in O$. As discussed above, $o$ has a little ball around it:
$$o + B_r \subset Y$$

(for some $r \equiv r(o) > 0$). Does the convex combination of $o$ and $y$ have a little ball around it as well? Hint: in Figure 3.7,
$$to + (1-t)y + B_{tr} = t(o + B_r) + (1-t)y \subset Y$$

(because $Y$ is convex).

26. Could the boundary point $y \in \partial Y$ be approached by interior points in $O$? Hint: in the above, pick $t$ as small as you like.

27. Could $F(y)$ be obtained as a limit of the value of $F$ at these interior points? Hint: recall that $F$ is continuous (Section 3.6). Therefore,
$$F(to + (1-t)y) \to_{t \to 0+} F(y).$$

28. Conclude that $y$ is indeed separated from $X$ as well:
$$F(y) \geq F(x), \quad x \in X.$$

29. Do the same for every boundary point $y \in \partial Y$.

30. Conclude that
$$F(y) \geq F(x), \quad y \in Y, \ x \in X.$$

31. Take the maximum (or supremum) on the right-hand side, and the minimum (or infimum) on the left-hand side:
$$\min_{y \in Y} F(y) \geq \max_{x \in X} F(x).$$

32. This is the Han-Banach separation theorem in its stronger version: $Y$ is still convex, but not necessarily open any more.

## 3.8 Open Set and Its Compact Subsets

1. Consider some property. Call it property-C. Assume that each set may satisfy property-C or not.
2. Consider compact sets (Chapter 4, Sections 4.5.1–4.5.5). In particular, each compact set may satisfy property-C or not.
3. We say that a set satisfies property-O if every compact subset of it satisfies property-C.
4. Assume that there is at least one open set with property-O.
5. Is there a maximal open set with property-O? Hint: see below.
6. Consider a chain of open sets with property-O, ordered by inclusion.
7. Does it have an upper bound? Hint: take their union.
8. Is the union open? Hint: a union of open sets is open as well.
9. Does the union have property-O as well? Hint: see below.
10. Consider a subset of the union. Call it $S$. Is $S$ a subset of some set in the original chain as well? Hint: see below.
11. Pick some element $x \in S$. Must $x$ be in the union as well?
12. Must

$$x \in B(x)$$

for some set $B(x)$ in the original chain?
13. Conclude that

$$S \subset \cup_x B(x).$$

14. What is the geometrical meaning of this? Hint: $S$ is covered by a union of (infinitely many) open sets.
15. Still, thanks to another definition of compactness, $S$ is already covered by a finite number of open sets:

$$S \subset B(x_1) \cup B(x_2) \cup B(x_3) \cup \cdots \cup B(x_m)$$

for some natural number $m$.
16. What is the right-hand side? Hint: it is a set in the original chain in its own right.
17. Conclude that $S$ indeed has property-C.
18. Conclude that the union of the entire chain indeed has property-O.
19. Conclude that the union of the entire chain is indeed a legitimate upper bound.
20. What would Zorn's lemma say about this?
21. Conclude that there is a maximal open set with property-O: every compact subset of it has property-C.
22. In what sense is it maximal? Hint: if you add even one element to it, then it is no longer open, or no longer of property-O: it would have a compact subset without property-C.
23. Is it compact? Hint: no, it is open and noncompact (Chapter 4, Sections 4.5.1–4.5.5).

# Part III

## Cantor Set
## and Stability

# Cantor Set and Stability

In this part, we introduce some more material in set theory and discrete math, and use it in physics. In discrete math, there is no continuum: the mathematical objects are discrete, and well-separated from each other. In a mathematical graph, for example, two nodes may be connected to each other by an edge. Still, this doesn't tell us how "close" they are to each other. In fact, there is no notion of distance at all.

In a general graph, the inner structure is not obvious. A tree, on the other hand, is easier to understand: it develops hierarchically, level by level. In fact, it starts from a unique root: the head, at the very top. From the root, a few branches may issue downwards. At the end of each branch, there is a new node. Together, these nodes form the next lower level. This process may then continue recursively. This way, the tree may grow and develop, level by level, until the bottom level, which contains the leaves. This is indeed a multilevel hierarchy.

A binary tree, for example, grows exponentially: each level contains twice as many nodes as the previous level. This kind of growth also appears in stability analysis in classical mechanics.

Furthermore, a binary tree is closely related to Cantor set. Together, they help form fractals, and introduce chaos theory from a geometrical point of view.

Chaos theory studies instability in dynamical processes. It teaches us that the initial data are not always good enough to predict the future. Indeed, however accurate they are, they may still lead to a substantial inaccuracy in the long run. This teaches us to be more humble, and not always trust our own eyes. Better have a little bit of doubt, and never take things as an absolute truth. This will be emphasized even more in quantum mechanics and relativity.

# 4

# The Pigeonhole Principle
# and Its Applications
# in Calculus
# and Classical Mechanics

For ages, people used the pigeonhole principle without noticing. Here, we give it the honor it deserves.

In its original version, the pigeonhole principle talks about a finite set. Even so, it can still help study stability in classical mechanics. Furthermore, it also has an infinite version, useful to prove theorems in calculus. What's so good about this proof? It will help design the Cantor set, and study stability and instability in classical mechanics.

Thus, the pigeonhole principle lies on the interface between finite and infinite. Is infinity physical? No! Indeed, later on, we'll use Cantor set to show that the universe cannot possibly be infinite.

## 4.1 Geometrical Preliminaries

### 4.1.1 The Unit Interval

0                1

**Fig. 4.1.** The closed unit interval contains both endpoints: 0 and 1. For this reason, it is denoted by $[0, 1]$.

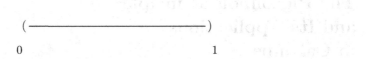

0                                                1

**Fig. 4.2.** The open unit interval, on the other hand, doesn't contain its endpoints. For this reason, it is denoted by $(0, 1)$.

We start with some geometrical background. The closed unit interval contains all real numbers between 0 and 1, including the endpoints (Figure 4.1). This is denoted by square brackets:

$$[0, 1] \equiv \{x \text{ is a real number} \mid 0 \le x \le 1\}.$$

Why is this a closed interval? Because it contains both endpoints: 0 and 1. Indeed, in the above formula, one could legitimately pick $x = 0$ or $x = 1$. For this reason, both 0 and 1 are included:

$$0 \in [0, 1] \quad \text{and} \quad 1 \in [0, 1].$$

In fact, we could form a small set, containing two numbers only: 0 and 1 alone. This set is denoted by braces: $\{0, 1\}$. It is then *included* in the closed unit interval:

$$\{0, 1\} \subset [0, 1].$$

However, it is not included in the *open* unit interval. Indeed, the open unit interval doesn't contain its endpoints (Figure 4.2). This is why it is denoted by round parentheses: $(0, 1)$. In fact, it could be obtained from the closed unit interval by dropping (or subtracting) both endpoints:

$$(0, 1) \equiv \{x \text{ is a real number} \mid 0 < x < 1\} = [0, 1] \setminus \{0, 1\}.$$

Don't get confused! In the Cartesian plane, $(0, 1)$ stands for just one point: the $x$-coordinate is 0, and the $y$-coordinate is 1. In the context of the real axis, on the other hand, there is no $y$-coordinate at all. This is why $(0, 1)$ stands for the open unit interval, as discussed above.

Finally, the closed unit interval could be obtained from the open unit interval by adding the endpoints:

$$[0, 1] = (0, 1) \cup \{0, 1\}.$$

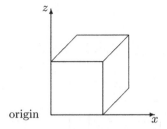

**Fig. 4.3.** The unit cube in the three-dimensional Cartesian space.

### 4.1.2 The Unit Cube

Consider now the three-dimensional Cartesian space, spanned by the standard $x$-, $y$-, and $z$-axes. This way, each individual point could be written as $(x, y, z)$, where $x$, $y$, and $z$ are its spatial coordinates in our coordinate system.

There is nothing special about these axes: you could pick them as you like, provided that they are perpendicular to each other. In particular, pick one point to serve as the origin: $(0, 0, 0)$. What's so special about the origin? It is the unique point where the axes meet (and cross) each other.

Next to the origin, we have the unit cube of length 1, width 1, and height 1 (Figure 4.3). It contains those points with coordinates between 0 and 1: $0 \le x, y, z \le 1$. For this reason, it is often written in terms of the closed unit interval:

$$[0, 1]^3 \equiv [0, 1] \times [0, 1] \times [0, 1] \equiv \{(x, y, z) \mid 0 \le x, y, z \le 1\}.$$

Let's convert it into a more complicated domain: a three-dimensional torus.

### 4.1.3 Three-Dimensional Torus

Now, let's "stretch" the above cube, and "bend" it: the left side loops clockwise, until it meets the right side and sticks to it (Figure 4.4).

For example, look at some fixed $y$ and $z$. On the left side of the cube, where $x = 0$, we have the point $(0, y, z)$. Across from it, on the right side, where $x = 1$, we have the point $(1, y, z)$. Now, let $(0, y, z)$ "travel," and make a complete loop clockwise in the $x$-$y$ plane, until meeting $(1, y, z)$ from the right (Figure 4.4).

Once these twin points are stuck together, these are just two different names for the same geometrical point:

$$(0, y, z) = (1, y, z), \quad 0 \le y, z \le 1.$$

Likewise, do the same in the $z$-coordinate as well — identify the top with the bottom:

$$(x, y, 0) = (x, y, 1), \quad 0 \le x, y \le 1.$$

Finally, do the same in the $y$-coordinate as well:

$$(x, 0, z) = (x, 1, z), \quad 0 \le x, z \le 1.$$

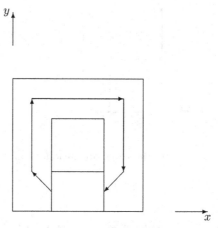

**Fig. 4.4.** Stretching and bending the unit cube (a view from above): the left side makes a complete loop clockwise, to meet the right side from the right and stick to it. The result is a thick square torus.

This is a bit tricky to visualize. Indeed, to make the latter identification, three dimensions are not enough. A new leap of the imagination is needed: go outside the physical world, and loop through a fourth dimension. Alternatively, keep working in our world: stick to the original cube in Figure 4.3, but keep the above identifications in mind.

So, what have we got? A three-dimensional "torus." It has no boundary any more: each point in it is an inner point, surrounded by other points from it. To imagine this, better stick to Figure 4.3, but keep in mind that sides that lie across from each other are now considered as one and the same.

### 4.1.4 Vector and Its Direction

A point $(x, y, z)$ can also be viewed as a vector: an arrow issuing from the origin, with its head pointing at the point $(x, y, z)$. From Pythagoras theorem, its length (or magnitude, or norm) is

$$\|(x, y, z)\| = \sqrt{x^2 + y^2 + z^2}.$$

Still, this is not so important. What is more important is its direction. Fortunately, the direction is rather stable: it doesn't change if all coordinates are multiplied by the same nonzero scalar:

$$\alpha(x, y, z) = (\alpha x, \alpha y, \alpha z), \quad \alpha \neq 0.$$

After all, to specify the direction uniquely, the important things are the ratios (proportions) between the coordinates, which remain the same for all $\alpha \neq 0$.

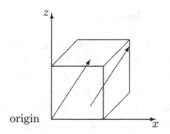

**Fig. 4.5.** In the three-dimensional torus, the ceiling coincides with the floor. For this reason, the light ray issuing from the origin in direction $(1,1,2)$ will make two complete loops. Indeed, upon hitting the ceiling at the middle, it will reappear at the bottom, and reenter from below. Then, upon reaching the ceiling for the second time, it will hit the corner $(1,1,1)$, which coincides with the origin.

## 4.2 Integrable System and Its Stability

### 4.2.1 Rational Proportion

Consider now a light ray issuing from the origin at some direction. In our three-dimensional torus, the light ray will loop. For example, upon hitting the ceiling of the original cube, it will reappear at the bottom and reenter, and so on (Figure 4.5). This way, it may go on making a lot of loops, never exiting the three-dimensional torus.

Will the light ray hit the origin ever again? Only if the direction vector could be written with integer coordinates:

$$(x, y, z) = \alpha(l, m, n),$$

where $l$, $m$, and $n$ are some integer numbers, at least one of which is nonzero. We then say that $x$, $y$, and $z$ have a common length unit: $\alpha$. In this case, the ratios between the coordinates could be written as rational numbers. For example, if $n \neq 0$, then

$$\frac{x}{z} = \frac{l}{n} \quad \text{and} \quad \frac{y}{z} = \frac{m}{n}.$$

For this reason, once the light ray has coordinate $z = n$, it must also have coordinates $x = l$ and $y = m$, so it must hit the point $(l, m, n)$, which coincides with the origin.

### 4.2.2 Irrational Proportion

But what if the ratio between two coordinates was irrational, and could never be written in such a way? In this case, the coordinates of the direction vector have no common length unit. As a result, the light ray will never hit the origin any more, however many times it may loop. Fortunately, it will still get arbitrarily close to the origin. This is indeed stability.

### 4.2.3 Poincare Recurrence Theorem

To see this, pick an arbitrarily small number $\varepsilon > 0$. Consider a small $\varepsilon \times \varepsilon \times \varepsilon$ cube: a small cell of length $\varepsilon$, width $\varepsilon$, and height $\varepsilon$. Let's use many such cells to cover the entire unit cube. For this purpose, we may need as many as $\varepsilon^{-3}$ cells. For simplicity, assume that $\varepsilon^{-3}$ is an integer number. Otherwise, just pick a slightly smaller $\varepsilon$.

This number, however big, is still finite. For this reason, the light ray can't go on visiting new cells only. After sufficiently many loops, it will have to step once again in an old cell, where it has already been before. Since $\varepsilon$ is very small, the ray will eventually get ever so close to a point where it has already been before.

Now, there is nothing special about this point: it could be easily shifted to the origin. After all, in the three-dimensional torus, the origin is just an ordinary inner point. Thus, after sufficiently many loops, the ray will get ever so close to the origin. In other words, since $\varepsilon$ is arbitrarily small, the ray will eventually get arbitrarily close to the origin. This is Poincare recurrence theorem.

### 4.2.4 Approaching the Infinity

In the above proof, infinity was never mentioned! After all, however small, $\varepsilon$ is still nonzero. Likewise, however big, $\varepsilon^{-3}$ is still finite.

Only at the end of the above proof was the infinity used, implicitly and indirectly. Indeed, by saying "arbitrarily close," we actually used the implicit limit process

$$\varepsilon \to 0.$$

Since $\varepsilon$ must remain positive, this is often written as

$$\varepsilon \to 0^+.$$

In this process, as $\varepsilon$ gets smaller and smaller, the total number of cells gets bigger and bigger, with no bound:

$$\varepsilon^{-3} \to_{\varepsilon \to 0^+} \infty.$$

Still, for each fixed $\varepsilon > 0$, $\varepsilon^{-3}$ is fixed as well. However big, this is still finite, and can be used in the above proof. It approaches infinity, but never reaches it!

## 4.3 The Pigeonhole Principle

### 4.3.1 The Finite Version

So, the above proof is purely discrete and algebraic: it uses finite sets only, and never mentions infinity. As a matter of fact, it is based on a very simple principle in discrete math and set theory: the pigeonhole principle.

Assume that ten pigeons need to occupy nine holes. There is not enough room for all! As a result, at least one hole must host two pigeons or more. This seems trivial, yet it is a fundamental principle. Who said math wasn't simple?

### 4.3.2 The Pigeonhole Principle in Stability

How was this principle used in the above proof? Well, instead of nine holes, we had $\varepsilon^{-3}$ cells. Instead of ten pigeons, we had $\varepsilon^{-3}$ loops (or even more). For this reason, once the light ray loops $\varepsilon^{-3}$ times, it must revisit an old cell, where it has already been before.

### 4.3.3 The Infinite Version

So far, we've stated the pigeonhole principle in its finite version. Now, let's see an infinite version as well. For this purpose, assume that there are infinitely many pigeons, but only a finite number of holes. In this case, at least one hole must host infinitely many pigeons. Let's use this extended version as well.

## 4.4 Application: the Bolzano–Weierstrass Theorem

### 4.4.1 Infinite Sequence

Let's use the infinite version in calculus and chaos theory. For this purpose, consider an infinite sequence:
$$x_1, x_2, x_3, x_4, \ldots, x_n, \ldots.$$
Assume that the entire sequence is contained in the closed unit interval. This means that, for all $n$, $0 \leq x_n \leq 1$.

### 4.4.2 Splitting the Interval

**Fig. 4.6.** The first step: split the closed unit interval into two nonoverlapping subintervals. Pick one subinterval (say, the left one). Initialize $\tilde{x}$ to be its left endpoint (say, $\tilde{x} \equiv 0$).

Now, let's split the unit interval into two equal subintervals (Figure 4.6). In other words, let's write it as the union

$$[0,1] = \left[0, \frac{1}{2}\right] \cup \left[\frac{1}{2}, 1\right].$$

Here, '$\cup$' stands for the union of two sets.

Now, where are the $x_n$'s located? Well, they must lie somewhere: either in the left subinterval, or in the right one. Thanks to the pigeonhole principle (in its infinite version), at least one subinterval must contain infinitely many $x_n$'s. Let's pick it, and drop the other one.

More precisely, if the left subinterval contains infinitely many $x_n$'s, then let's pick it, and drop all the other $x_n$'s that are not in it. In this case, let's mark its left endpoint by

$$\tilde{x} \equiv 0.0 \equiv 0 \cdot 1 + 0 \cdot \frac{1}{2}.$$

This is a binary representation: it tells us that $\tilde{x}$ indeed lies in the left subinterval. This is just an initialization: later on, we may add more binary digits on the right.

If, on the other hand, the left subinterval contains just a finite number of $x_n$'s, then drop them, and pick the right subinterval. In this case, $\tilde{x}$ is initialized as its left endpoint:

$$\tilde{x} \equiv 0 \cdot 1 + 1 \cdot \frac{1}{2} \equiv 0.1.$$

Here, the binary digit 1 behind the point stands for 1 times 1/2, as required. Later on, we may add more binary digits behind it, to update $\tilde{x}$ further.

### 4.4.3 The Induction Step

$$\begin{array}{ccccc} | & | & | & & | \\ 0 & \frac{1}{4} & \frac{1}{2} & & 1 \end{array}$$

**Fig. 4.7.** The second step: split one subinterval (say, the left one) into two nonoverlapping subsubintervals. Pick one subsubinterval. Update $\tilde{x}$ to be its left endpoint.

Next, let's look at our subinterval, and do the same to it: split it into two subsubintervals, and pick one of them (Figure 4.7). If the left subsubinterval is picked, then add 0 to the binary representation of $\tilde{x}$, on the far right. If, on the other hand, the right subsubinterval is picked, then add 1 instead. In either case, our subsubinterval contains infinitely many $x_n$'s, and the up-to-date $\tilde{x}$ indeed marks its left endpoint, as required.

### 4.4.4 Accumulation Point

This is actually the induction step. By mathematical induction, it can be repeated on and on forever, producing a hierarchy of nested intervals, contained in one another. The final (possibly irrational) $\tilde{x}$ will be an accumulation point of our original sequence: every open interval that contains $\tilde{x}$, however short, must also contain infinitely many $x_n$'s.

This is indeed Bolzano–Weierstrass theorem: every infinite bounded sequence must have an accumulation point.

## 4.5 Compact Domain in 3-D

### 4.5.1 Extension to 3-D

Let's extend the above to three dimensions as well. For this purpose, let $D$ be a bounded three-dimensional domain. This way, $D$ can be confined in a box:

$$D \subset [x_{\min}, x_{\max}] \times [y_{\min}, y_{\max}] \times [z_{\min}, z_{\max}].$$

Here, $x_{\min}$ stands for the minimal possible $x$-coordinate in any point in $D$, and so on.

Consider now an infinite sequence of points in $D$:

$$(x_1, y_1, z_1), \ (x_2, y_2, z_2), \ \ldots, (x_n, y_n, z_n), \ \ldots.$$

Does it have an accumulation point?

### 4.5.2 Splitting the Interval

To answer this, let's apply one step as in Figure 4.6: split $[x_{\min}, x_{\max}]$ into two subintervals, pick one of them, and initialize $\tilde{x}$ to mark its left endpoint. While doing this, drop some $(x_n, y_n, z_n)$'s from the original sequence, yet leave infinitely many.

### 4.5.3 The Induction Step

Then, do the same in the $y$-direction: split $[y_{\min}, y_{\max}]$ into two subintervals, pick one of them, and initialize $\tilde{y}$ to mark its left endpoint. While doing this, drop some more $(x_n, y_n, z_n)$'s from the up-to-date sequence, yet leave infinitely many.

Finally, do the same in the $z$-coordinate as well: split $[z_{\min}, z_{\max}]$ into two subintervals, pick one of them, and initialize $\tilde{z}$ to mark its left endpoint. While doing this, drop some more $(x_n, y_n, z_n)$'s, yet leave infinitely many.

### 4.5.4 Accumulation Point

This completes the induction step. What have we done in it? We've actually split the original box into eight subboxes, and picked one of them, which still contains infinitely many points from the original sequence. All other points that are not in this subbox have been dropped, never to be regarded any more.

The above step can now be repeated time and again in a mathematical induction, updating $\tilde{x}$, $\tilde{y}$, and $\tilde{z}$ by adding a suitable binary digit to each of them on the far right. This way, these binary representations are in $[0, 1]$. In the end, they may need to shift, to make the required accumulation point:

$$(x_{\min} + (x_{\max} - x_{\min})\tilde{x}, \ y_{\min} + (y_{\max} - y_{\min})\tilde{y}, \ z_{\min} + (z_{\max} - z_{\min})\tilde{z}).$$

Indeed, every open cell that contains this point, however small, must also contain infinitely many points from the original sequence. This extends the original theorem to three dimensions as well.

### 4.5.5 Compact Domain

Assume now that $D$ is not only bounded but also closed: it contains all its accumulation points. In this case, we say that $D$ is compact. So, thanks to the pigeonhole principle (in its infinite version), we've just proved that, in a compact domain, every infinite sequence must have an accumulation point.

## 4.6 Nonintegrable System: Poincare Stability

### 4.6.1 Domain and Its Mapping

So far, we've considered a three-dimensional torus. This is a linear model. In classical mechanics, it is called an integrable system. In it, a light ray issuing from some point could shift, and produce a new (parallel) light ray, issuing from a new point. This is why Poincare recurrence theorem is so easy to prove: the pigeonhole principle (in its finite version) is good enough (Section 4.2.3).

But what about a nonintegrable system? In this case, instead of the three-dimensional torus, we have now a more complicated domain. Furthermore, instead of a straight light ray, we have now a curved trajectory, describing the flow of a particle from the initial point $v_0$. After one second, it arrives at the new point $T(v_0)$. After two seconds, on the other hand, it arrives at $T^2(v_0)$, and so on.

Thus, $T$ actually maps the entire domain. It describes the entire dynamics in terms of discrete math and set theory. The system is now ready to benefit from the pigeonhole principle.

### 4.6.2 Mapping and Its Origin

In a flow as above, is $T$ one-to-one? In other words, could two particles occupy the same point? If not, then $T$ is invertible: one could time-travel, from the position after one second, back to the initial position. In this case, the inverse mapping is denoted by $T^{-1}$.

Here, however, we drop this assumption. Indeed, we never assume that $T$ is invertible. On the contrary: two different trajectories may meet each other. Thus, $T^{-1}$ is *not* a mapping any more. What is $T^{-1}$? Well, if $A$ is a set in the domain $D$, then $T^{-1}(A)$ contains all those points that are mapped into $A$:

$$T^{-1}(A) \equiv \{v \in D \mid T(v) \in A\}.$$

We then say that $T^{-1}(A)$ is the origin of $A$ under $T$.

### 4.6.3 Intersection and Its Origin

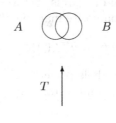

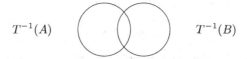

**Fig. 4.8.** The lower-left circle is mapped to the upper-left circle. The lower-right circle is mapped to the upper-right circle. As a result, the lower overlap is mapped to the upper overlap.

Consider now two subsets of $D$:

$$A, B \subset D.$$

Their intersection (or overlap) contains those common points that are in both $A$ and $B$:

$$A \cap B \equiv \{v \in D \mid v \in A \text{ and } v \in B\}.$$

In other words, $v$ is in the intersection if (and only if) $v$ is in both $A$ and $B$:

$$v \in A \cap B \Leftrightarrow v \in A \text{ and } v \in B.$$

In other words, $A \cap B$ is the maximal set included in both $A$ and $B$.

For example, if $A$ and $B$ are disjoint from each other, then they have no overlap at all:

$$A \cap B = \emptyset$$

(the empty set). Still, this is a rather rare case. Usually, $A$ and $B$ may overlap, and have common points.

What is the origin of $A \cap B$ under $T$? Well, as in Figure 4.8,

$$
\begin{aligned}
T^{-1}(A \cap B) &\equiv \{v \in D \mid T(v) \in A \cap B\} \\
&= \{v \in D \mid T(v) \in A \text{ and } T(v) \in B\} \\
&= \{v \in D \mid T(v) \in A\} \cap \{v \in D \mid T(v) \in B\} \\
&= T^{-1}(A) \cap T^{-1}(B).
\end{aligned}
$$

Thus, we have a sort of "distributive" law, teaching us how to open parentheses. It tells us that order doesn't matter: instead of intersecting and then applying $T^{-1}$, we can first apply $T^{-1}$, and then intersect.

In summary, the origin of the overlap is the overlap of the origins. This works not only for $T$ but also for the composite mapping $T^n$ $(n \geq 0)$:

$$T^{-n}(A \cap B) = T^{-n}(A) \cap T^{-n}(B).$$

We're now ready to use the pigeonhole principle (in a new finite version) to prove Poincare recurrence theorem in a nonintegrable system as well.

### 4.6.4 Stationarity: Volume Preservation

Assume now that our domain has a finite volume:

$$V(D) < \infty.$$

Furthermore, assume also that our mapping is stationary: it preserves volume. In other words, the origin has the same volume:

$$V\left(T^{-1}(A)\right) = V(A)$$

for any subset $A \subset D$.

In theory, $A$ could be as thin as a sheet of paper. In this case, it would have zero volume:

$$V(A) = 0.$$

Here, on the other hand, we assume that $A$ is thick and substantial, so it has a positive volume:

$$V(A) > 0.$$

Now, use $A$ to form an infinite sequence of sets: the origin of $A$, the origin of the origin, and so on:

$$A, \ T^{-1}(A), \ T^{-2}(A), \ \ldots, \ T^{-n}(A), \ \ldots.$$

Thanks to stationarity, all these sets have the same volume: $V(A) > 0$. Thanks to the pigeonhole principle, they can't be completely disjoint from each other: there is just not enough room for all. Thus, there must be two sets that overlap substantially with each other. In other words, there must be two indices $j > i \geq 0$ for which

$$V\left(T^{-i}(A) \cap T^{-j}(A)\right) > 0.$$

Thanks to our "distributive" law in Section 4.6.3, we can now factor $T^{-i}$ out:

$$T^{-i}(A) \cap T^{-j}(A) = T^{-i}\left(A \cap T^{-(j-i)}(A)\right).$$

In summary, thanks to volume preservation, we now have

$$V\left(A \cap T^{-(j-i)}(A)\right) = V\left(T^{-i}\left(A \cap T^{-(j-i)}(A)\right)\right)$$
$$= V\left(T^{-i}(A) \cap T^{-j}(A)\right)$$
$$> 0.$$

Is this familiar? The same process has already been used on the light ray in Section 4.2.3, including the shift in the end. Here, however, $T$ is not necessarily one-to-one: this is why there is a need to use the origins.

Finally, define

$$N \equiv j - i \geq 1.$$

In summary, we've used the pigeonhole principle to prove that, for every $A$ of positive volume, there exists a natural number $N \geq 1$ for which $T^{-N}(A)$ overlaps with $A$ substantially:

$$V\left(A \cap T^{-N}(A)\right) > 0.$$

Note that each $A$ may have its own $N$, associated with it. In other words, two different $A$'s may have two different $N$'s. Thus, $N$ depends on $A$:

$$N \equiv N(A).$$

Still, so long as it is clear what $A$ we're talking about, we often drop the '$(A)$,' and write $N$ on its own for short.

Let's use the above result to show that a nonintegrable system is in principle the same as our original integrable system: it enjoys stability as well.

### 4.6.5 Poincare Stability

So far, infinity was never mentioned. Furthermore, we've talked about sets only, not individual points. This is a good sign: it indicates that we're on the right track. After all, sets are more fundamental than points. Now, it is time to talk about individual points as well.

In an integrable system (our three-dimensional torus), we've already established stability: for every cell, however small, every light ray issuing from it will eventually return to it (Section 4.2.3). Let's see that this is also the case for almost every point in a nonintegrable system. For this purpose, we need the infinity.

Let $A \subset D$ be a substantial set of positive volume: $V(A) > 0$. (Still, this volume could be as small as you like.) Let $B \subset A$ contain those points that never return to $A$:

$$B \equiv \{v \in A \mid T^n(v) \notin A, \ n = 1, 2, 3, \ldots\}.$$

The rest of the points in $A$, on the other hand, do return to $A$ eventually. Now, what is stability? Well, stability means that almost all points in $A$ do return to $A$ eventually. In other words, $B$ is unsubstantial: as thin as a sheet of paper.

Let's prove stability by contradiction. Assume momentarily that $B$ was substantial:

$$V(B) > 0.$$

In this case, thanks to Section 4.6.4, there must be a natural number

$$N \equiv N(B) \geq 1$$

for which

$$V\left(B \cap T^{-N}(B)\right) > 0.$$

But these sets are disjoint:

$$B \cap T^{-N}(B) = \emptyset.$$

Indeed, consider a particular point $v \in T^{-N}(B)$. After $N$ applications of $T$, $v$ returns to $B \subset A$. Therefore, $v \notin B$.

Thus, our momentary assumption must have been false: $B$ could never have been substantial. On the contrary:

$$V(B) = 0.$$

This is indeed Poincare stability: almost all $v \in A$ must eventually return to $A$.

Where was the infinity used? Only implicitly, in the very definition of $B$. Indeed, to define $B$, one must have a mechanism to check whether a given point $v \in A$ ever returns to $A$. If it does ($v \notin B$), then this is easy enough to verify: just apply $T$ time and again, until getting back into $A$. This way, you get your answer in a finite time. If, on the other hand, $v \in B$, then you may apply $T$ time and again forever, and got no answer in any finite time. This is the infinity paradox. To believe in infinity, you must trust your imagination. This is indeed an axiom in set theory: there exists an infinite set.

## 4.7 Exercises

### 4.7.1 Three-Dimensional Torus in $\mathbb{R}^4$

1. Define the real axis:
$$\mathbb{R} \equiv \{x \mid -\infty < x < \infty\}.$$

2. Define the three-dimensional Cartesian space:
$$\mathbb{R}^3 \equiv \{(x, y, z) \mid -\infty < x, y, z < \infty\}.$$

3. To the three-dimensional Cartesian space, add also a fourth dimension: the $w$-coordinate:
$$\mathbb{R}^4 \equiv \{(x, y, z, w) \mid -\infty < x, y, z, w < \infty\}.$$

4. State the pigeonhole principle in its finite version.
5. Prove it.
6. Use it to prove Poincare recurrence theorem in the three-dimensional torus. Hint: see Section 4.2.3.

7. State the pigeonhole principle in its infinite version.
8. Prove it.
9. Use it to prove the Bolzano–Weierstrass theorem. Hint: see Sections 4.4.1–4.4.4.
10. Extend it to a three-dimensional compact domain. Hint: see Sections 4.5.1–4.5.5.
11. Extend it to a four-dimensional compact domain as well.
12. Consider a three-dimensional compact domain $D \subset \mathbb{R}^3$. Does it have a finite volume? Hint: since $D$ is bounded, its volume could be defined as

$$V(D) \equiv \int \int \int_D dxdydz < \infty.$$

13. Establish Poincare stability in a nonintegrable system as well: a compact domain $D \subset \mathbb{R}^3$. Hint: see Sections 4.6.4–4.6.5.
14. Extend this to a four-dimensional compact domain $D \subset \mathbb{R}^4$ as well.
15. Conclude once again that the original three-dimensional torus indeed enjoys Poincare stability. Hint: it is indeed a compact domain in $\mathbb{R}^4$.
16. What is its volume? Hint: it is still 1, just like the unit cube. Indeed, the integration could still take place in 3-D: the sides are only two-dimensional, so sticking them together (as in Figure 4.4) has no effect.
17. To define $B$ in Section 4.6.5, how to "solve" the infinity paradox? Hint: see Section 4.6.5.

# 5

## Cantor Set
## and Its Applications

In its infinite version, the pigeonhole principle helped design an accumulation point. Let's use the same process to design a fundamental set: the Cantor set.

Later on, we'll use the Cantor set in chaos theory. Before that, let's use it to approximate a real number by a rational number. For this purpose, we design the Cantor set not only in one but also in two dimensions.

Is nature infinite? No! Indeed, later on, we'll use Cantor set to show that the universe cannot possibly be infinite. As a matter of fact, nature is finite in both the macroscale and microscale, as we'll see in quantum mechanics and relativity later on.

## 5.1 Probability

### 5.1.1 Throwing a Dice

Let's play a game: throw a dice. If you get 1 or 2 or 3 or 4 or 5, then you win. If, on the other hand, you get 6, then I win.

How likely are you to win? Well, in five of six cases, you win. So, the probability that you win is as high as 5/6. The probability that I win, on the other hand, is as low as 1/6.

Now, let's play the game twice. How likely are you to win twice? Clearly, the probability for this is lower:

$$\frac{5}{6} \cdot \frac{5}{6} = \frac{25}{36}.$$

Now, let's play $n$ times, for some large $n$. How likely are you to win $n$ times? Unfortunately for you, the probability for this is already very low. In fact, as $n$ increases, it gets closer and closer to zero. In other words, it approaches zero:

$$\left(\frac{5}{6}\right)^n \to_{n\to\infty} 0.$$

In summary, if we play sufficiently many times, then I'm highly likely to win at least once.

Where is this useful? Pick an arbitrarily long decimal number at random. How likely is it that the digit 7 never appears in it? The probability for this is as low as zero! We'll discuss this later from a geometrical point of view.

### 5.1.2 Throwing Two Dices

Now, let's move on to a yet more interesting game: throw two dices together at the same time. If you get $(6, 6)$, then I win. Otherwise, you win.

The probability that I win is now as low as $1/36$. The probability that you win, on the other hand, is now as high as $35/36$. Still, how likely are you to win $n$ times in a row? Well, as $n$ increases, this approaches zero again:

$$\left(\frac{35}{36}\right)^n \to_{n \to \infty} 0.$$

So, if we play many times, then I'm highly likely to win at least once.

Later on, we'll use this to model two decimal numbers, and their individual digits. For this purpose, let's look at things from a geometrical point of view.

## 5.2 Geometrical Setting

### 5.2.1 Splitting the Interval

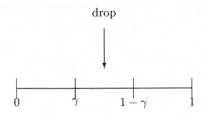

**Fig. 5.1.** In the first step, drop the middle subinterval: from $\gamma$ to $1 - \gamma$.

In set theory, we get to see a few kinds of infinity. For instance, look at a rather strange kind of infinity: the Cantor set. In what way is it strange? Well, it seems to contain nothing, yet contains infinitely many points!

To design the Cantor set, start from the closed unit interval $[0, 1]$: the real numbers between 0 and 1 (including the endpoints). (See Chapter 4, Section 4.1.1.) Let $\gamma$ be a fixed number (parameter) between 0 and $1/2$:

$$0 < \gamma < \frac{1}{2}.$$

(For example, $\gamma = 1/3$.) This way, $2\gamma < 1$, so $1 - 2\gamma > 0$, and

$$1 - \gamma > \gamma.$$

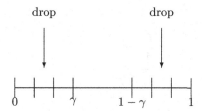

**Fig. 5.2.** In the second step, apply the above procedure twice: look at the left and right subintervals. From each of them, drop its own middle subsubinterval.

This is good: we can now start our geometrical process (Figure 5.1). Split the original interval into three disjoint subintervals: the first one is from 0 to $\gamma$, the second from $\gamma$ to $1 - \gamma$, and the third from $1 - \gamma$ to 1:

$$[0,1] = [0,\gamma] \cup (\gamma, 1 - \gamma) \cup [1 - \gamma, 1].$$

Here, the left and right subintervals are closed, whereas the middle one is open: it doesn't contain its own endpoints.

This is indeed a symmetric splitting: the left and right subintervals have the same length: $\gamma$. The middle subinterval, on the other hand, may have a different length: $1 - 2\gamma$. Let's go ahead and drop it!

We are now left with two subintervals only: the right one, and the left one. To each of them, apply the same procedure recursively. This way, we are now left with four subsubintervals, of length $\gamma^2$ each (Figure 5.2).

This is the induction step. The same procedure of splitting and dropping may now go on and on (recursively) time and again, infinitely many times.

What is left in the end? Well, the points that are never dropped make a new set: the Cantor set. How does it look like? Well, to illustrate it, better look at a new version, in a more familiar setting: decimal. After all, who says we must drop the middle subinterval? Why not drop a new subinterval, of length $1/10$?

### 5.2.2 Decimal Base

A real number $0 \le x \le 1$ can be written as an infinite decimal fraction:

$$x = 0.a_1 a_2 a_3 \ldots,$$

where each $a_i$ is some digit between 0 and 9. For example, $1/10$ could be written in two different styles: either a finite style:

$$\frac{1}{10} \equiv 0.1 \equiv 0.1000000000000000000\ldots,$$

or an infinite style:

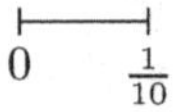

**Fig. 5.3.** In the first step, one subinterval drops from the unit interval.

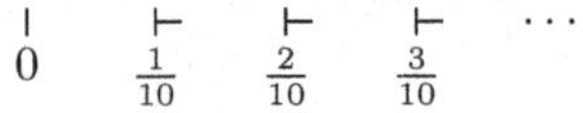

**Fig. 5.4.** The second step: from each subinterval, drop one subsubinterval. In total, nine subsubintervals are dropped.

$$\frac{1}{10} \equiv 0.0999999999999999\ldots.$$

In this case, pick the former style. Only for the number 1 must we pick the infinite style:

$$1 \equiv 0.999999999999999\ldots.$$

In set theory, one could ask an interesting question: pick a random $0 \leq x \leq 1$. How likely is $x$ to have no zero digit at all? In other words: how many $x$'s have nonzero digits only?

The surprising answer is: next to none! Indeed, let's consider the half-open unit interval:

$$[0,1) \equiv \{x \text{ is a real number} \mid 0 \leq x < 1\}.$$

Why is it called half-open? Because it contains only its left endpoint, but not its right one (Chapter 4, Section 4.1.1).

What is the size (or length) of this interval? Just take the right endpoint, and subtract the left one:

$$1 - 0 = 1.$$

As a matter of fact, the closed (or open) unit interval has the same length as well. After all, in terms of length, excluding an endpoint has no effect whatsoever. Still, a half-open interval is easier to work with.

Now, drop those $x$'s with some zero digit. This could be done step by step. In the first step, drop the first (half-open) subinterval:

$$\left[0, \frac{1}{10}\right) = \left\{x \text{ is a real number} \mid 0 \leq x < \frac{1}{10}\right\}$$

(Figure 5.3). This drops those $x$'s with $a_1 = 0$. Clearly, the length (or size) of this subinterval is:

$$\frac{1}{10} - 0 = \frac{1}{10}.$$

So, we are left with 90% only: the remaining (half-open) subinterval

$$\left[\frac{1}{10}, 1\right) = \left\{x \text{ is a real number } \mid \frac{1}{10} \leq x < 1\right\},$$

whose length is

$$1 - \frac{1}{10} = \frac{9}{10}.$$

Now, split this into nine disjoint (half-open) subintervals, of size 1/10 each:

$$\left[\frac{1}{10}, 1\right) = \left[\frac{1}{10}, \frac{2}{10}\right) \cup \left[\frac{2}{10}, \frac{3}{10}\right) \cup \cdots \cup \left[\frac{9}{10}, 1\right).$$

These subintervals are now forwarded to the next step. In this step, from each subinterval of length 1/10, drop the first subsubinterval of length 1/100 (Figure 5.4). This drops those $x$'s with $a_1 > 0$ and $a_2 = 0$. Again, we are left with 90% only, or 81% of the original interval.

This is the induction step. The process may now continue step by step in the same way. In each step, the total remaining size gets 0.9 times as small.

Thanks to mathematical induction, we can now go ahead and carry out infinitely many steps. Indeed, assuming that $n$ steps have already been made, go ahead and do the next step as well. In the "end," we are left with a size as small as

$$\lim_{n \to \infty} 0.9^n = 0.$$

This is Cantor's set (in base 10): a sparse "sieve" of those $x$'s that never use 0 in their decimal representation. Although it is infinite, this set is as thin as air: zero total length, or zero measure.

Thus, real numbers with no zero digit are very rare. If you pick a random $0 \leq x < 1$, then it is highly likely to have at least one zero behind the decimal point. Let's extend this result to a higher dimension as well.

### 5.2.3 Two Dimensions

Let us extend the above to two dimensions as well. For this purpose, consider the $x$-$y$ Cartesian plane. In it, consider the (half-open) unit square, which contains points with coordinates between 0 and 1:

$$[0, 1)^2 \equiv \{(x, y) \mid 0 \leq x, y < 1\}.$$

Clearly, the area of the unit square is 1. Now, let $(x, y)$ be some point in it. This way, both $x$ and $y$ have their own decimal representations:

$$x = 0.a_1 a_2 a_3 \ldots$$
$$y = 0.b_1 b_2 b_3 \ldots,$$

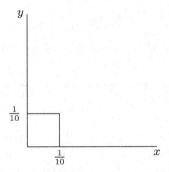

**Fig. 5.5.** In the first step, the lower-left subsquare drops from the unit square.

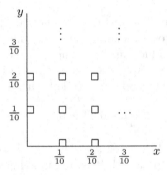

**Fig. 5.6.** The second step: from each subsquare, drop its own lower-left subsubsquare. In total, 99 subsubsquares are dropped.

where the $a_i$'s and the $b_i$'s are some digits between 0 and 9. As we've seen in Section 5.2.2, $x$ is most likely to have some index $k(x)$ for which $a_{k(x)} = 0$. Likewise, $y$ is most likely to have some other index $k(y)$ for which $b_{k(y)} = 0$.

Still, $k(x)$ and $k(y)$ may be different from each other. Is there yet another index $k \equiv k(x, y)$, for which both $a_k = b_k = 0$?

Most likely, there is! To see this, drop one (half-open) subsquare: the lower-left subsquare, of size $(1/10) \times (1/10)$:

$$\left[0, \frac{1}{10}\right)^2 = \left\{(x, y) \mid 0 \le x, y < \frac{1}{10}\right\}$$

(Figure 5.5). This drops those $(x, y)$'s with $a_1 = b_1 = 0$. The area of this subsquare is as small as

$$\frac{1}{10} \cdot \frac{1}{10} = \frac{1}{100}.$$

So, we are left with 99% only. This completes the first step.

The remaining 99 subsquares are now forwarded to the second step: from each of them, drop the lower-left subsubsquare (Figure 5.6). This drops those $(x, y)$'s with $a_2 = b_2 = 0$ (but $a_1 > 0$ or $b_1 > 0$). Again, we are left with 99% only, or 98.01% of the original square.

This is the induction step. The process may now continue step by step in the same way. After each step, the total remaining area gets 0.99 times as small.

Thanks to mathematical induction, we can now go ahead and carry out infinitely many steps. Indeed, assuming that $n$ steps have already been made, go ahead and do the next step as well. Eventually, we are left with area as small as

$$\lim_{n \to \infty} 0.99^n = 0,$$

or with a set as thin as air: it has zero area or size or measure.

Thus, if you pick a random point $(x, y)$ from the unit square, then both $x$ and $y$ are highly likely to have at least one joint zero digit $a_k = b_k = 0$ (for some $k \geq 1$). Let's use this observation to approximate both $x$ and $y$ by rational numbers.

## 5.3 Application: How to Approximate a Real Number?

### 5.3.1 Rational Approximation

Let's pick a random point $(x, y)$ from the unit square. Let $k \geq 1$ be the first index for which both

$$a_k = b_k = 0.$$

As discussed above, at probability 1, $k$ indeed exists. Furthermore, in our dropping process, in the $k$th step, $(x, y)$ must drop from the unit square. This way, both $x$ and $y$ have a decimal approximation of length $k - 1$:

$$|x - 0.a_1 a_2 a_3 \ldots a_{k-1}| \leq 10^{-k}$$
$$|y - 0.b_1 b_2 b_3 \ldots b_{k-1}| \leq 10^{-k}.$$

In other words, both $x$ and $y$ have rational approximations that share the same denominator $m \equiv 10^{k-1}$, but may have different numerators $l \equiv a_1 a_2 a_3 \ldots a_{k-1}$ and $n \equiv b_1 b_2 b_3 \ldots b_{k-1}$:

$$\left| x - \frac{l}{m} \right| \leq \frac{1}{10m}$$
$$\left| y - \frac{n}{m} \right| \leq \frac{1}{10m}.$$

For instance, if $k = 1$, then $l = n = 0$.

As a matter of fact, one could even improve on this: pick a yet bigger $k$, and construct many more rational approximations, with $m$ as large as you like. Alternatively, one could pick a bigger base.

### 5.3.2 Nondecimal Base

So far, we've used decimal digits only. After all, we humans have ten fingers to count with, so base 10 suits us best. The computer, on the other hand, has just two "fingers:" 0 and 1. This is why it uses base 2 rather than 10: not decimal but binary digits.

Base 2 is the smallest base possible. In our application, however, better use a much larger base. To design such a base, let $\varepsilon > 0$ be an arbitrarily small parameter. Without loss of generality, assume that $\varepsilon^{-1}$ is an integer number. (Otherwise, just make $\varepsilon$ a little smaller.)

Now, in the discussion in Sections 5.2.2–5.3.1, replace base 10 by base $\varepsilon^{-1}$. This way, the now integer numbers $a_i$ and $b_i$ are different from before: they are no longer between 0 and 9, but between 0 and $\varepsilon^{-1} - 1$.

With these changes, the new $k$, $l$, $n$, and $m$ are also different from before. This leads to a new (and better) rational approximation:

$$\left| x - \frac{l}{m} \right| \leq \frac{\varepsilon}{m}$$

$$\left| y - \frac{n}{m} \right| \leq \frac{\varepsilon}{m}.$$

This result will be used in the next chapter.

## 5.4 Exercises: Cantor Null Set

### 5.4.1 Cantor Set — Big or Small?

1. After the first step in Figure 5.1, how many subintervals remain? Hint: 2.
2. What is their total length? Hint: $2\gamma$.
3. After the second step in Figure 5.2, how many subsubintervals remain? Hint: 4.
4. What is their total length? Hint: $4\gamma^2$.
5. Show that, after each step, the total remaining length gets $2\gamma$ times as short.
6. After the $n$th step, how many intervals remain? Hint: $2^n$.
7. What is their total length? Hint: $(2\gamma)^n$.
8. Use the assumption that $\gamma < 1/2$ to conclude that Cantor's set has zero length (or size, or measure). Hint: since $2\gamma < 1$,

$$(2\gamma)^n \to_{n \to \infty} 0.$$

9. For instance, set $\gamma \equiv 1/3$. In this case, what is the measure of Cantor's set? Hint:

$$\left( \frac{2}{3} \right)^n \to_{n \to \infty} 0.$$

10. Pick some $0 \leq x \leq 1$. In base 3, write $x$ as a new trinary fraction:

$$x = 0.a_1 a_2 a_3 \ldots,$$

where the new $a_i$'s are now different from before: each new $a_i$ is now a trinary digit: either 0 or 1 or 2.

11. For example, to write 1/3, use the infinite style:

$$\frac{1}{3} \equiv 0.02222222222222222222222222222\ldots.$$

12. To write 2/3, on the other hand, use the finite style:

$$\frac{2}{3} \equiv 0.2 \equiv 0.2000000000000000000000000000\ldots.$$

13. Note that, in either of these examples, the digit '1' is never used!

14. Assume that $x$ belongs to Cantor's set, constructed with $\gamma \equiv 1/3$. How does the trinary representation of $x$ look like? Hint: for each $i$, it may use $a_i \equiv 0$ or $a_i \equiv 2$, but never $a_i \equiv 1$.

15. Conversely, show that every trinary fraction that looks like this indeed belongs to this Cantor set.

16. Map this Cantor set as follows: in the above trinary fraction, change each '2' into '1'. Then, interpret the result as a binary fraction, not trinary.

17. Conclude that the above mapping covers the entire unit interval.

18. Conclude that Cantor's set is a rather strange kind of infinity: it has zero measure, yet contains as many numbers as the entire unit interval!

19. Explain this paradox. Hint: an infinite set may still be sparse, and have zero size.

20. Extend the construction in Section 5.2.3 to three dimensions as well.

21. What is the volume of the resulting Cantor set? Hint: zero.

22. Could this Cantor set serve as the set $B$ in Chapter 4, Section 4.6.5? Hint: yes! Indeed, $B$ should have zero volume. As proved above, this Cantor set indeed has zero volume.

# 6

## Is the Universe Infinite?

In ancient philosophy, it was often assumed that the universe was infinite and unbounded. Nowadays, on the other hand, we know better: the universe contains only a finite number of galaxies, each containing only a finite number of stars. Could this be proved with elementary tools only, with no background in physics at all?

In modern physics, the universe is clearly bounded. As a matter of fact, one can even tell the "size" of the observable universe. This follows from Einstein's principle: the speed of light is maximal, and can never be exceeded. Furthermore, the speed of light is the same in every coordinate system, moving or not. In your own (private) system, you'd also measure the same speed of light: $300,000$ kilometer per second (in terms of your own seconds and kilometers). We'll discuss this widely later, in the chapter about relativity.

Thanks to Einstein's principle, the observable universe is actually a three-dimensional "hypersphere." $13.7$ billion year ago, at the big bang, it started as a singular point. Since then, it has expanded as fast as light, until today.

What is the "radius" of this hypersphere? It is the long line made by a light ray issuing from this singularity and traveling since the big bang until today. During this time, the ray made quite a long journey: $13.7$ billion light year: $13.7$ billion year, times $365 \cdot 24 \cdot 3600$ second per year, times $300,000$ kilometer per second.

However big, this radius is still finite. Still, can this also be proved using simple math only, assuming no background in physics at all? In this chapter, we prove this from scratch, using elementary math only.

For this purpose, we need just one simple observation. This is indeed the power of experimental physics. This is also the difference between physics and math. Math is easy: it requires no experiment at all. Physics, on the other hand, is more advanced: it must be supported by some experiment, and contradicted by no other experiment.

The universe is finite not only in the macroscale but also in the microscale: matter is discrete, not continuous. In fact, each piece of matter contains only a finite (albeit huge) number of molecules. Each molecule contains only a few atoms. Each atom contains only a few subatomic particles: protons, neutrons, and electrons. Still, in this chapter, we focus on the macroscale only: the universe is bounded, not infinite.

## 6.1 Flux of Light

### 6.1.1 The Geometrical Model

**Fig. 6.1.** The light coming from a star to the Earth makes a complete cone, full of light.

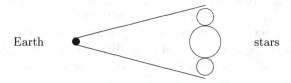

**Fig. 6.2.** The star is now twice as far. From it, the Earth looks now four times as small, and receives four times less light. To compensate for this, there are now more stars nearby, to make a complete "layer" of stars. Together, they shed on the Earth the same light flux as before.

How much light do we receive here on the Earth? To help estimate this, let's draw a geometrical model. For this purpose, consider a star shining at the Earth (Figure 6.1). Assume that the star is (nearly) as big as the sun: much bigger than the Earth. Assume also that the star is far away from the Earth: nearly as far as the sun. This way, the light rays make a complete cone (three-dimensional sector), full of light.

The total amount of light that the Earth receives from the star is called the light flux. How does it change with distance? To see this, let's imagine a thought experiment.

Suddenly, the star jumps away, and gets twice as far. For an observer who sits on the star, the Earth looks now four times as small: it occupies now four times less area in the sky. Since light shines uniformly in all directions, the Earth receives now four times less light.

To compensate for this, place more stars nearby (Figure 6.2). This makes a vertical "layer" of stars, filling the cone with as much light as before. In fact, the vertical cross-section area of the entire layer is about four times as before. This

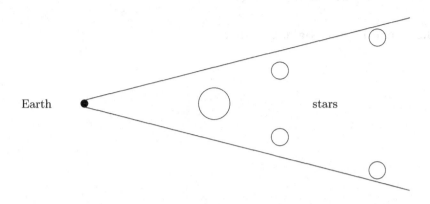

**Fig. 6.3.** The star still sheds four times less light onto the Earth. To compensate for this, there are now more stars behind it, to fill the entire cone with light, and give the Earth the same light flux as before.

indeed compensates: we have the same cone as before, full of light. So, the light flux remains unchanged: people here on the Earth would see nothing unusual.

Better yet, the layer could get nonuniform: some stars may go ahead and "jump" farther, provided that more stars are added nearby or behind, to make sure that there are no dark "holes" in the sky (Figure 6.3). This way, each individual star that jumped away and gives now less light to the Earth also triggered an immediate compensation: new stars nearby or behind, to make sure that no flux is lost: people on the Earth still see the same cone, full of light.

## 6.2 Olbers' Paradox

### 6.2.1 A Wrong Model

In ancient days, it was often assumed that the universe was infinite and unbounded, with infinitely many stars in it. Let's prove this wrong.

The proof is by contradiction. Assume momentarily that the universe contained infinitely many stars. They must have been distributed rather uniformly throughout the outer space. After all, there is no reason why one place should be denser than another, and contain more stars per unit volume.

For example, consider a three-dimensional infinite grid. Each point of the form $(l, n, m)$ (where $l$, $n$, and $m$ are some integer numbers) has a small star centered at it. Only at the origin $(0, 0, 0)$ is there no star: the viewer sits there, staring at the sky.

Each individual star is rather small compared to the distance between stars. Still, a star is not infinitesimal: it must still confine a little cube of size $(2\varepsilon) \times (2\varepsilon) \times (2\varepsilon)$, for some small $\varepsilon > 0$.

But this model is wrong! Indeed, if it were true, then we'd have a cone full of light, as in Figure 6.3, even at night!

### 6.2.2 The Bright-Night Paradox

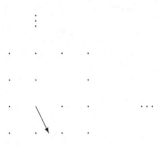

**Fig. 6.4.** Olbers' bright-night paradox: if the universe were infinite, then every ray issuing from the Earth would eventually hit some star.

Indeed, in the above infinite grid, consider a viewer who sits on the Earth at $(0,0,0)$, staring at the sky in some direction (Figure 6.4). Let's prove that the ray issuing from his/her eye must eventually hit some star, and get some light from it, even at night!

Indeed, let $(x, y, z)$ be some point on this ray. For simplicity, assume that all three coordinates are nonnegative. Furthermore, assume also that $z$ is positive and maximal:

$$z \geq \max(x, y).$$

(Otherwise, just interchange the role of $z$ with that of $x$ or $y$.) Consider now the point

$$\left(\frac{x}{z}, \frac{y}{z}, 1\right)$$

on the ray. (This trick is often used in projective geometry.) From Chapter 5, Section 5.3.2, the pair $(x/z, y/z)$ is highly likely to have a rational approximation of the form

$$\left|\frac{x}{z} - \frac{l}{m}\right| \leq \frac{\varepsilon}{m}$$

$$\left|\frac{y}{z} - \frac{n}{m}\right| \leq \frac{\varepsilon}{m},$$

or

$$\left|m\frac{x}{z} - l\right| \leq \varepsilon$$

$$\left|m\frac{y}{z} - n\right| \leq \varepsilon.$$

Thus, the point

$$m\left(\frac{x}{z}, \frac{y}{z}, 1\right) = \left(m\frac{x}{z}, m\frac{y}{z}, m\right)$$

must lie in the star centered at $(l, n, m)$ in the grid. As a matter of fact, the existence of such a star also follows from the Poincare recurrence theorem in a three-dimensional torus (Chapter 4, Section 4.2.3).

Once we've established this, we can now work the other way around: the light from this star could now travel all the way back, along the same line, but in the opposite direction, to meet the viewer's eye. Thus, people here on the Earth would see light everywhere. As in Figure 6.3, there would be no dark hole, even at night!

### 6.2.3 The Observable Universe

Evidently, night is dark, not bright. Why? Because the universe could never be modeled as an infinite grid as above.

Here one may ask: why not? Perhaps there are infinitely many stars, only they are as far as 13.7 billion light year away from us, so we can never see them! After all, no light could make such a long journey in time to reach our eyes here today!

The answer is simple: such stars, if existed, could never be considered as part of our own observable universe. After all, they are completely detached from us, and have no effect on us whatsoever. As such, they actually belong to a hidden part of the universe, completely irrelevant to us. This part also came from the big bang, so it must be finite as well. Still, this is no longer a scientific question: we'll never know anything about this part, because we'll never get any information from it.

## 6.3 Energy and Entropy: Laws of Thermodynamics

### 6.3.1 Stars and Galaxies

By now, we've seen that the universe could never be both infinite and uniform at the same time. Still, could it be infinite and nonuniform? For example, could it be denser near the origin, and sparser elsewhere? Fortunately not. After all, in our infinite grid, the origin is picked arbitrarily. Physically, all places have the same opportunity to attract matter. Why should one place be denser than another?

In a small scale, there could be some nonuniformity. In a large scale, on the other hand, this could be averaged off, and wiped out. For example, in the above model, instead of individual stars, work with galaxies. This way, thanks to the same paradox, the universe could contain only a finite number of galaxies. Furthermore, each galaxy could contain only a finite number of stars. Thus, even in a hierarchical setting, the universe must still be finite and bounded.

### 6.3.2 Big Bang and Inflation

In fact, the universe must be not only finite but also nonuniform. We've already seen two levels: in the global level, the universe contains galaxies. Within each galaxy, there is a smaller level: stars. Why do we have this hierarchy and nonuniformity?

One theory is like this. Right after the big bang, the universe expanded exponentially fast, in what is known as inflation. If it were completely uniform, then matter could never form. Fortunately, thanks to the uncertainty principle, there were some subtle quantum fluctuations in energy level, at which elementary particles formed.

Fortunately, these fluctuations were so tiny that inflation didn't last for too long. (There could be parallel universes with bigger fluctuations and longer inflation, but no stars or galaxies could possibly form or survive in them for too long.) Later on, the universe cooled down, and gravity helped form clouds of particles and gas, galaxies, and stars.

### 6.3.3 Symmetry Had Broken

Ideally, the universe should actually have been completely uniform. Right after the big bang, no matter should actually have formed at all. After all, why should a particle form here, not there? Why is one place better than another? Everywhere should have been just the same, in a complete symmetry and uniformity. This could be quite boring.

Fortunately, thanks to the principle of uncertainty, some quantum fluctuations occurred. This way, right after the big bang, some nonsymmetry has also developed. In the inflation that followed, uniformity had broken down. Around small islands of different energy levels, particles formed.

### 6.3.4 The First Law of Thermodynamics

Thanks to gravity, when the universe cooled down, clouds of particles and gas formed as well, and then galaxies and stars. Thanks to the first law of thermodynamics, energy is never lost: all the original energy that was at the big bang remains with us until today. Still, energy may change form: from a well-ordered (useful) energy in the beginning, to a poorly-ordered (used) energy later on.

### 6.3.5 The Second Law of Thermodynamics

Indeed, the second law of thermodynamics tells us that energy may change form, but only in one direction: on average, good useful energy (with low entropy) could get more and more random and useless (high entropy). For example, the original pure energy, concentrated at the singularity, was used to produce elementary particles and atoms, containing a lot of potential nuclear energy. This energy is still of high quality: it still has low entropy. Still, in the process, a lot of heat was also generated. This is random energy, with high entropy. Thus, the overall entropy always increases: on average, energy can only get less and less useful. This is the second law of thermodynamics.

The first atoms were rather light: they contained only a few protons, neutrons, and electrons. Billions of years later, in a process of nuclear fusion, they merged into heavier atoms, with yet more potential nuclear energy of yet higher quality (lower entropy). Still, this generated a lot of heat, increasing entropy on average. This way, in the entire system, the total entropy increased, as in the second law of thermodynamics. Later on, we'll use the binomial coefficients to discuss entropy from an algebraic-geometrical point of view.

### 6.3.6 Nature is Finite and Discrete

Nature is finite in two ways. In terms of astronomic macroscale, the universe contains only a finite number of galaxies, each containing only a finite number of stars. To prove this, we've used mathematical tools only: geometry and set theory. No background in physics was used.

Still, nature is finite not only in the macroscale but also in the microscale and nanoscale: each piece of matter contains only a finite (albeit large) number of molecules, each containing only a few atoms, each containing only a few subatomic particles.

Thus, matter is discrete rather than continuous. As a matter of fact, the distance between two atoms is far larger than the size of an individual atom. In other words, most of matter is just vacuum or empty space. In between the atoms, there is nothing.

Still, to help model nature, the continuum is vital. It gives us the mathematical tools needed to model mechanics, electromagnetics, and relativity.

### 6.3.7 Continuous Models

Although matter is discrete, it is better described in a continuous geometrical model, with infinitely many points: the real axis, the two-dimensional plane, or the three-dimensional space. Each point could be interpreted as a vector: this makes a linear vector space.

To describe the elementary laws of nature, we better use infinitely many numbers: the real numbers, with their attractive (algebraic and analytic) properties. In this context, differentiation and integration are often used to describe motion, flow, and dynamics in general.

## 6.4 Exercises: What is Time?

### 6.4.1 Entropy and the Time Arrow

1. Is the universe infinite? Why?
2. Use Olbers' bright-night paradox to prove that it is not.
3. In this proof, use the rational approximation in Chapter 5, Section 5.3.2. Hint: see Section 6.2.2.
4. Alternatively, use Poincare recurrence theorem. Hint: see Chapter 4, Section 4.2.3.
5. What is time? Hint: some agreed process that, in our eyes, ticks at a constant (unchanging) rate.
6. What is time in special relativity? Hint: in a given system, time is a process that, in the eyes of an observer who lives in this system, seems to tick at a constant rate (say, a radioactive decay.)
7. Draw the arrow of time.
8. Does it point to the future? Why not to the past?
9. What is the direction of time?
10. Who says time flows (or progresses) forward, not backward?

11. What is entropy? Hint: in a specific kind of energy, entropy is the amount of randomness and disorder that make the energy hard to use.
12. Give an example of a kind of energy with low entropy. Hint: potential nuclear energy in the atom.
13. Is this energy of high quality? Hint: yes, it could convert to electricity.
14. Give an example of a kind of energy with high entropy. Hint: heat.
15. Is this energy of high quality? Hint: no, it is too random, disordered, and hard to use.
16. What happens to entropy in time? Hint: in a closed system, the total entropy must increase, not decrease. In other words, although the energy remains constant, it gets less and less useful.
17. What is the correct direction of time? Hint: towards the future! After all, this is the direction of increasing entropy.
18. What was the amount of entropy at the singularity? Hint: at its absolute minimum.
19. What happens in nuclear fusion? Hint: two small atoms, with only a few protons, neutrons, and electrons, merge into one heavy atom.
20. Look at the original small atoms. What energy do they have? Hint: potential nuclear energy.
21. Is this energy of high quality? Hint: yes, this is useful energy, of low entropy.
22. Now, look at the new heavy atom. What energy does it have? Hint: potential nuclear energy.
23. More or less? Hint: more than the original atoms together.
24. It seems that entropy has decreased. After all, there is now more good energy, with low entropy. How come? Hint: in the process, a lot of heat is also generated. This is useless energy, with high entropy. Thus, in a closed system, the total entropy has increased, as in the second law of thermodynamics.

# 7

## Binary Trees
## and Chaos Theory

To study stability in classical mechanics, we use a simple tool from discrete math and set theory: a multilevel hierarchy, or a tree. This may help not only to solve practical problems but also to develop a suitable theory in the first place. To make things clearer, we take a geometrical point of view.

From the dawn of civilization, people used to ask: how to measure weight, distance, or time? This is indeed a most practical problem. For example, you could ask your grocer to have 1700 grams of apples, and pay 400 cents. But it would make more sense to ask 1.7 kilogram, and pay 4 dollars. By grouping 1000 grams in one kilogram, and 100 cents in one dollar, one could move from a too fine scale to a coarser (and more practical) unit of measurement.

The above example shows how often we use multilevel in day-to-day life to group small elements in one composite unit. This way, we "climb" up the hierarchy, from an elementary object to a more complex object, containing many small objects.

In this chapter, we look at a few useful multilevel hierarchies. In particular, we focus on binary trees, with their applications in chaos theory. In this vein, the Cantor set will play a major role: it will help design fractals, and look at chaos theory from a geometrical point of view.

Thanks to trees, we can also look at stability and instability in dynamical systems. This is indeed the start of modern physics: initial data are not always good enough to predict the future. In fact, due to instability, even a tiny error may grow exponentially, and contaminate the results in the long run. We must therefore be more humble, and always have a little doubt. Things are not always as they seem: we can never understand everything. This will be emphasized even more in quantum mechanics and relativity, later on in the book.

## 7.1 Induction and Deduction

### 7.1.1 Induction

To show how useful multilevel could be, we start with an important example from the philosophy of science. Thanks to multilevel, this will take a geometrical face, easy to comprehend.

Induction and deduction are probably the most fundamental analytical tools introduced by the ancient Greeks. To help comprehend the deep insight behind them, let's put them in a new two-level scheme.

Assume that you are given a concrete engineering problem: say, to build a road between two particular cities. For this purpose, you are also given the relevant data about resources, costs, topography, etc.

Unfortunately, the specific details may be too tedious and technical. They may cloud the issue, and obscure the entire problem area. To see the light, better drop them.

Furthermore, tomorrow you may be given a slightly different problem: say, to build yet another road between two other cities. Should you then start all over again from scratch? Why not extend the original problem to a more general one, and introduce more general guidelines, to help solve every problem of the kind?

This is indeed induction: extend the concrete problem into a more general one, by introducing new concepts and terminology. The extended problem may seem harder, but is not. On the contrary: the new general terms may help identify the main ingredients, and lead to a new theory, useful in other fields as well.

This way, induction contributes not only to solving the practical problem but also to a deeper understanding of the theoretical components in it. Furthermore, induction may enrich our language and vocabulary, introducing new terms and concepts not only for the present problem but also for many others to come.

### 7.1.2 Deduction

Once the general problem has been solved efficiently, deduce a specific solution to your concrete problem. For this purpose, just feed the original data to your solver, to specify the relevant circumstances explicitly. This will adapt your well-written code to the original special case, as required. This is indeed deduction.

To help understand the entire process as a whole, let's put it in a two-level scheme.

### 7.1.3 Two-Level Tree

To help visualize the induction–deduction process geometrically, let's model it in a two-level tree (Figure 7.1). The original problem is placed at the bottom-left. Induction is then used to extend it into a more general problem at the top level. Once the required theory has been developed to help solve the problem in most general terms, deduction is used to help solve practical instances at the bottom-right.

### 7.1.4 V-Cycle

To visualize this in yet another way, reverse the picture upside-down. This gives the so-called V-cycle (Figure 7.2).

Where did this name come from? Well, the picture is like the Latin letter 'V,' which has two legs: induction at the left leg, to "slide" to the general case at the bottom, and deduction at the right leg, to "climb" up to the top-right, where the general model is applied to the original case.

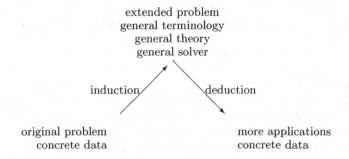

**Fig. 7.1.** The induction–deduction tree: the original problem at the bottom-left is extended into a more general problem. To help solve it, a general theory is developed at the top level. From this, concrete applications are then deduced at the bottom-right.

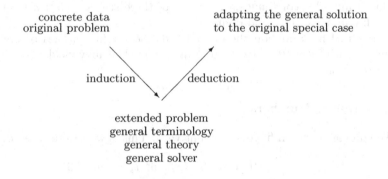

**Fig. 7.2.** The V-cycle — the upside-down tree: the original problem at the top-left is rewritten in most general terms at the bottom. General theory is then developed to form the general solution, which is then adapted to the original data at the top-right.

So far, we've only used a two-level tree. Let's go ahead and introduce more levels, to have a more complicated structure: multilevel. To help motivate this, let's see how common this is in practice.

## 7.2 Multiscale

### 7.2.1 How Did Multiscale Develop?

From the dawn of history, people faced a common practical question: how to measure? The answer was simple: multiscale! In the big (coarse) scale, use big units: meter, kilogram, or hour. This helps measure most of the amount. To measure the rest, on the other hand, we can no longer use big units. We must turn to smaller units: centimeter, gram, or second. This is indeed the fine scale. By combining (or adding) both scales, we can now measure the entire amount accurately and efficiently.

This is quite practical: one can now say that the distance between two points is 17.63 meters (17 meters plus 63 centimeters). Likewise, one could also say that the weight of some object is 8.130 kilograms (8 kilograms plus 130 grams). This combines large-scale units (meter or kilogram) with small-scale units (centimeter or gram). Finally, one could also say that the time is 5:31:20 (5 hours, 31 minutes, and 20 seconds). This combines three scales: coarse, intermediate, and fine.

Moreover, multiscale is vital not only in physics but also in elementary math.

### 7.2.2 Decimal Numbers

To write a natural number, we must use multiscale. For example,

$$178 = 100 + 70 + 8 = 1 \cdot 100 + 7 \cdot 10 + 8 \cdot 1 = 1 \cdot 10^2 + 7 \cdot 10^1 + 8 \cdot 10^0.$$

This is indeed a combination (or sum) of three scales — large, intermediate, and small: hundreds, tens, and digits.

Decimal fractions, on the other hand, use yet finer scales behind the decimal point: multiples of $10^{-1}$, $10^{-2}$, $10^{-3}$, and so on. This may yield a yet higher accuracy or resolution.

### 7.2.3 Binary Numbers

We humans have ten fingers to count with. This is why we use ten digits:

$$0, \ 1, \ 2, \ 3, \ 4, \ 5, \ 6, \ 7, \ 8, \ \text{and} \ 9$$

to help write a decimal number in base 10, as above. The computer, on the other hand, has just two "fingers:" 0 and 1. This is why it uses binary numbers, in base 2.

A binary number, written in base 2, uses powers of 2 rather than 10. A binary number like 1101.01 is interpreted as

$$1 \cdot 2^3 + 1 \cdot 2^2 + 0 \cdot 2^1 + 1 \cdot 2^0 + 0 \cdot 2^{-1} + 1 \cdot 2^{-2}.$$

This representation combines (or sums) six different scales: from $2^3$ (the largest or coarsest scale) to $2^{-2}$ (the smallest or finest scale). The coefficients (either 0 or 1) are stored in the original binary number: 1101.01.

## 7.3 Dimension

### 7.3.1 Dimension: a New Definition

Let's use multiscale to define dimension. So far, we've interpreted the concept of dimension geometrically: an interval has dimension 1, a square has dimension 2, and so on. But what is the deeper meaning of dimension?

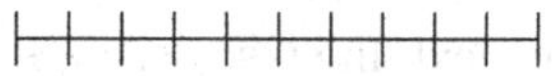

**Fig. 7.3.** To cover the unit interval, use ten intervals, of length 1/10 each.

Well, consider the (closed) unit interval

$$[0,1] \equiv \{x \text{ is a real number} \mid 0 \le x \le 1\}.$$

Let's cover it with smaller intervals, of length 1/10 each. How many are needed? Clearly, ten (Figure 7.3).

Now, let's move on to the next finer scale, and use yet smaller intervals, of length 1/100 each. How many are needed? Clearly, one hundred. In general, we need $10^n$ intervals of length $10^{-n}$ each ($n \ge 0$). In this case, we say that our original interval has dimension 1.

### 7.3.2 Two Dimensions

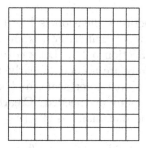

**Fig. 7.4.** To cover the unit square, on the other hand, use 100 cells, of size $(1/10) \times (1/10)$ each.

Consider now the (closed) unit square:

$$[0,1]^2 \equiv \{(x,y) \mid 0 \le x, y \le 1\}.$$

Let's cover it with small cells, of size $(1/10) \times (1/10)$ each. How many are required? Clearly, one hundred (Figure 7.4).

Now, let's move on to the next finer scale, and use yet smaller cells, of size $(1/100) \times (1/100)$ each. How many are needed? Clearly, $10,000$. In general, to cover the original square with cells of size $10^{-n} \times 10^{-n}$, we need as many as $10^{2n}$. This is much more than in Section 7.3.1. We therefore say that the unit square has dimension 2.

What's so good about this new interpretation? Well, it may help define a new kind of geometry, with a strange dimension: no longer an integer number, but a fraction.

## 7.4 Fractals and Hausdorff Dimension

### 7.4.1 Fractal and Its Dimension

Look at Cantor's set (in base 10) (Chapter 5, Section 5.2.2). Let's cover it with small intervals, of length $1/10$ each. How many are needed? Clearly, nine (Figure 5.3).

Now, let's move on to the next finer scale, and use yet smaller intervals, of length $1/100$ each. How many are needed? Clearly, $9 \cdot 9 = 81$ (Figure 5.4). In general, we need $9^n$ intervals of length $10^{-n}$ each ($n \ge 0$). This is substantially less than in Section 7.3.1:

$$9^n < 10^n.$$

For this reason, Cantor's set is a fractal: it has a fractional dimension, smaller than 1. This is called Hausdorff dimension.

### 7.4.2 Hausdorff Dimension

In the above example, Hausdorff dimension is less than 1. But this is now the only option: it could also be more than 1. For example, look at the Cantor set in two dimensions (Chapter 5, Section 5.2.3). Let's cover it with small cells, of size $(1/10) \times (1/10)$ each. How many are needed? Clearly, 99 (Figure 5.5).

Now, let's move on to the next finer scale, and use yet smaller cells, of size $(1/100) \times (1/100)$ each. How many are needed? Clearly, $99 \cdot 99$. In general, to use cells of size $10^{-n} \times 10^{-n}$, we need as many as $99^n$ ($n \ge 0$). This number lies strictly in between

$$10^n < 99^n < 10^{2n}.$$

Therefore, this kind of Cantor set is not really two-dimensional. On the contrary: its dimension must be a fraction in between 1 and 2. This is a fractal as well.

### 7.4.3 Self Similarity

How does a fractal look like? Well, it has a nice geometrical property: it is similar to itself. For example, look at Cantor's set, restricted to just one subinterval in Figure 5.4. It looks just like the entire Cantor set in the entire interval. Likewise, look at just one subsquare in Figure 5.6. In it, Cantor's set looks just like the entire Cantor set in the entire square.

This is not just a mathematical artifact: it also appears in nature. For example, look at the coast of Norway. Now, look at a small portion of it: It could be as complicated as the entire coast. This is indeed self-similarity.

Of course, nature is finite. Therefore, at a very fine scale, self-similarity breaks down. Indeed, if you stand at the beach, then you'd see a nearly straight line. We don't talk about such a small scale. In every reasonable scale, as in a photograph taken from a satellite, the coast always looks like a fractal: similar to itself.

Later on, we'll see self-similarity in an infinite binary tree as well. Before going into this, let's see self-similarity in its simplest form: a unary tree.

## 7.5 Mathematical Induction

### 7.5.1 Unary Tree

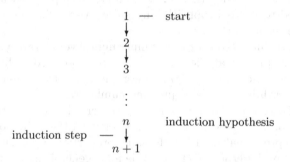

**Fig. 7.5.** Mathematical induction as a unary tree: at the top, make sure that 1 has the property. For $n = 1, 2, 3, 4, \ldots$, assume that $n$ has the property (the induction hypothesis), and prove that $n + 1$ has the same property as well (the induction step).

So far, we've seen a few examples of multiscale, used in elementary math and physics. Now, let's look at multilevel from a geometrical point of view.

In the simplest kind of multilevel, each level has just one item in it. This could be modeled in a unary tree: from each level, issue just one branch, pointing downwards, to the next lower level underneath. As a matter of fact, we've already met this model, and used it implicitly in mathematical induction.

The ancient Greeks introduced the principle of induction as a powerful thinking tool. From this wide principle, we can now obtain mathematical induction as a formal version.

Indeed, mathematical induction is similar, but much more technical. Assume that we already know that 1 has some desirable property. Assume that we also have a mechanism to prove that, if some number has the property, then its successor has the same property as well. How to show that 1000 has the same property as well?

For this purpose, we must first show that 999 has the property. For this, we must first show that 998 has the property, and so on. This is indeed recursion. It goes backward, as in reverse engineering. Why not go forward, from 1 onward, as in standard engineering? This is indeed the other face of the same process: mathematical induction.

Unfortunately, the above recursion is a long and tedious process: we must use the same mechanism over and over again, as many as 999 times! To avoid this, use a powerful thinking tool: induction. To apply it, extend the original task into a yet more general task: prove the property not only for 1000 but also for larger numbers as well. This seems more difficult, but is not. After all, what's so special about 1000? Better use the same mechanism not only up to 1000 but also beyond! For this purpose, prove one thing only: if $1, 2, \ldots, n$ have the property (the induction hypothesis), then $n + 1$ has it as well (the induction step).

That's it: we are done! We can now deduce that 1000 indeed has the property, as required. After all, in theory, we could always start from 1, and march step by step to $2, 3, 4, \ldots, 1000$, proving that all of them have the property as well. Fortunately, we don't need to do this explicitly any more. After all, we've already established that this is possible in theory.

To visualize this better, draw a simple multilevel hierarchy: a unary tree (Figure 7.5). First, place 1 at the top. For this purpose, prove that 1 indeed has the property. Then, place $2, 3, 4, \ldots, n$ in lower and lower levels, each in a separate level. This way, the $n$th level contains just one number: $n$.

To do this, you need to prove nothing. After all, the induction hypothesis says that levels $2, 3, \ldots, n$ indeed have the property. In the induction step, we use this hypothesis to prove that $n+1$ has the same property as well. This allows us to place $n + 1$ in the next lower level. This is done for a general (unspecified) $n$. In fact, the same could be done for each and every $n$. That's it: we're done!

We can now use our work, and deduce a special case: $n = 1000$. For this purpose, no more work is needed. After all, in theory, one could always start from 1, and "jump" down from level to level, until reaching the 1000th level, as required.

Mathematical induction could be used not only to establish a particular property but also to define the natural numbers in the first place. Later on, we'll use it for our own goal: to define a tree.

### 7.5.2 Example: Finite Power Series

A finite (geometric) power series can be written in two equivalent ways:

$$1 + q + q^2 + q^3 + \cdots + q^n = \sum_{i=0}^{n} q^i,$$

where $n \geq 0$ is a given integer number, and $q \neq 1$ is a given parameter. Here, $i$ is just an index that runs from 0 to $n$.

Let's use mathematical induction to prove that the sum is

$$\sum_{i=0}^{n} q^i = \frac{q^{n+1} - 1}{q - 1}.$$

Indeed, for $n = 0$, this is easy to prove. After all, in this trivial case, the series is very short: it contains one term only — the leading term $q^0 = 1$:

$$\sum_{i=0}^{n} q^i = \sum_{i=0}^{0} q^i = q^0 = 1 = \frac{q^1 - 1}{q - 1} = \frac{q^{n+1} - 1}{q - 1}.$$

This proves the above formula for $n = 0$.

Let us now consider a more interesting case: $n \geq 1$. Assume that our formula holds for a slightly shorter series:

$$\sum_{i=0}^{n-1} q^i = \frac{q^n - 1}{q - 1}.$$

This is indeed the induction hypothesis: it is obtained from the original formula by substituting $n - 1$ for $n$. Thanks to it, we can now prove the original formula. For this purpose, split the original series into two parts. The first part contains most of the terms: the $n$ leading terms. Fortunately, their sum has just been written down. The second part, on the other hand, contains one term only — the final term:

$$\sum_{i=0}^{n} q^i = \left( \sum_{i=0}^{n-1} q^i \right) + q^n$$
$$= \frac{q^n - 1}{q - 1} + q^n$$
$$= \frac{q^n - 1 + q^n(q - 1)}{q - 1}$$
$$= \frac{q^{n+1} - 1}{q - 1},$$

as required. This completes the induction step. This completes the mathematical induction, proving that

$$\sum_{i=0}^{n} q^i = \frac{q^{n+1} - 1}{q - 1}$$

for every integer number $n \geq 0$, as required.

This is just algebra: after all, we only used numbers, and arithmetic operations between them. Let's move on to calculus: the limit process. For this purpose, we need a new tool: infinity, or $\infty$.

### 7.5.3 Infinite Power Series

Consider now the infinite power series:

$$1 + q + q^2 + q^3 + \cdots = \sum_{i=0}^{\infty} q^i.$$

What is the meaning of this? We must distinguish between a few different cases. Let's start with a few strange cases. If $q = 1$, then this is just the constant series:

$$1 + 1 + 1 + \cdots = \sum_{i=0}^{\infty} 1.$$

Later on, we'll see that this series approaches infinity, or converges to $\infty$.

A more interesting case, on the other hand, is $q \neq 1$. From the previous section, we already know how to sum up the $n + 1$ leading terms, to produce the partial sum $S_n$:

$$S_n \equiv \sum_{i=0}^{n} q^i = 1 + q + q^2 + \cdots + q^n = \frac{q^{n+1} - 1}{q - 1}.$$

We can now deal with another degenerate case: $q = -1$. In this case, there are alternating signs, so each term cancels the previous one:

$$1 - 1 + 1 - 1 + 1 - \cdots = (1 - 1) + (1 - 1) + 1 - \cdots.$$

In other words, each pair of two consecutive terms sums up to 0, contributing nothing to the total sum. As a result, we must distinguish between two cases: an even $n$ and an odd $n$:

$$S_n = \begin{cases} 1 & \text{if } n \text{ is even} \\ 0 & \text{if } n \text{ is odd}. \end{cases}$$

Fortunately, this is also in agreement with our formula:

$$S_n = \frac{q^{n+1} - 1}{q - 1} = \frac{(-1)^{n+1} - 1}{-2} = \begin{cases} 1 & \text{if } n \text{ is even} \\ 0 & \text{if } n \text{ is odd}. \end{cases}$$

So, in this case, $S_n$ can never make up its mind. As $n$ grows, $S_n$ never converges: it always switches from 1 to 0 and back to 1 and so on, with no limit. We then say that the original series diverges. Likewise, if $q < -1$, then the series diverges as well.

Another strange case is $q > 1$. In this case, as $n$ grows, $q^n$ gets as large as you like:

$$q^n \to_{n \to \infty} \infty.$$

Therefore, the power series "converges" to infinity, or approaches $\infty$:

$$S_n = \frac{q^{n+1} - 1}{q - 1} \to_{n \to \infty} \infty.$$

This is often written as

$$\lim_{n \to \infty} S_n = \infty,$$

or simply

$$\sum_{i=0}^{\infty} q^i = \infty.$$

Finally, we arrive at the really interesting case: $|q| < 1$. In this case, the power series really converges: not to infinity, but to a finite limit. Indeed, in this case, as $n$ grows, $|q^n|$ gets as small as you like:

$$q^n \to_{n \to \infty} 0.$$

Thus,

$$S_n = \frac{q^{n+1} - 1}{q - 1} \rightarrow_{n \to \infty} \frac{-1}{q - 1} = \frac{1}{1 - q}.$$

This is often written as

$$\sum_{i=0}^{\infty} q^i = \lim_{n \to \infty} S_n = \frac{1}{1 - q}.$$

These results will be useful later on.

## 7.6 Trees

### 7.6.1 General Tree

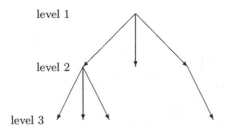

**Fig. 7.6.** A three-level tree. From the head, three branches issue. The left and right branches point to a new two-level subtree. The middle branch, on the other hand, points to a leaf.

The unary tree is rather degenerate and boring, and hardly deserves to be called a tree. Indeed, in it, each level contains just one branch, pointing to an equally boring unary subtree. A general tree, on the other hand, is much more interesting: a complex multilevel hierarchy, which develops and grows level by level.

To define a general tree, use mathematical induction. Start from a degenerate one-level tree: just one dangling node (or leaf), to serve as a head (or root).

Now, for $n = 1, 2, 3, 4, \ldots$, look at a fixed $n$. Assume that we already know how to construct any $k$-level tree ($1 \leq k \leq n$). This is indeed the induction hypothesis.

Now, how to construct a bigger $(n+1)$-level tree? First, form the head (or root), and place there a new node. From it, issue a few new edges (or branches), each pointing to a new $k$-level subtree (for some $1 \leq k \leq n$).

Thanks to the induction hypothesis, we already know how to construct these subtrees. In this construction, at least one of them must have $k = n$ levels. The rest, on the other hand, may contain $k \leq n$ levels. This way, the entire tree indeed has

$n + 1$ levels, as required. Still, it is not necessarily uniform (symmetric): different subtrees may have a different number of levels, smaller than or equal to $n$. (Later on, we'll also design more symmetric trees.)

From the general tree, we can now deduce a few examples. The small two-level tree in Figure 7.1 is easy enough to construct. For this purpose, set $n = 2$. From the head, issue just two branches. At the end of each branch, stick a trivial one-level tree: just a leaf.

The three-level tree in Figure 7.6, on the other hand, is a little more complicated. To construct it, set $n = 3$. From the head, issue three branches. To the middle branch, attach a leaf. To the left and right branches, on the other hand, attach a new two-level subtree.

### 7.6.2 Arithmetic Expression

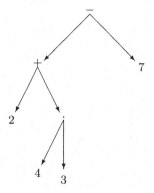

**Fig. 7.7.** Storing the arithmetic expression $2 + 4 \cdot 3 - 7$ in a four-level tree, and calculating its value, bottom to top. The top-priority arithmetic operation, $4 \cdot 3$, is carried out at the bottom. The next operation, $2 + (4 \cdot 3)$, is carried out at the second level. Finally, the least prior operation, $(2 + 4 \cdot 3) - 7$, is carried out at the top.

What is the tree good for? Well, it could be used to store an arithmetic expression (Figure 7.7). In the leaves, store the operands: some numbers. In the rest of the nodes, on the other hand, store arithmetic symbols like '+', '−', '·', or '/'.

Where to place the arithmetic symbols? Well, this depends on their priority. In the head, place the symbol of *least* priority: either '+' or '−'. If your expression contains many such symbols, then pick the latter instance, and place it in the head. After all, it is least prior. This splits the original expression into two subexpressions: one on the left (stored recursively in the left subtree), and the other on the right (stored recursively in the right subtree). This is indeed mathematical induction. After all, the induction hypothesis tells us how to store a shorter subexpression in its own subtree.

Where to store the subtrees? For this purpose, issue two branches from the head. Attach the left subtree to the left branch, and the right subtree to the right branch.

If, however, there is no '+' or '−' in the entire expression, then do the same with higher-priority symbols: '·' or '/', and so on. This completes the entire tree.

### 7.6.3 How to Calculate?

How to calculate the numerical value of the arithmetic expression? First, calculate the values of each subexpression recursively. This way, you obtain two values. Apply to them the arithmetic operation at the head of the tree. That's it: you're done!

### 7.6.4 How to Check Correctness?

How do we know that this is correct? By mathematical induction, of course! After all, each subtree contains less levels than the entire tree. Thus, the induction hypothesis tells us that the subexpressions have been calculated properly. Finally, check that the symbol in the head was applied properly.

This highlights an important principle: mathematical induction mirrors recursion, and is actually the same thing! We'll come back to this later.

For example, the arithmetic expression

$$2 + 4 \cdot 3 - 7$$

is stored in a four-level tree (Figure 7.7). In this expression, what operation is least prior? Clearly, this is the subtraction at the end. Place the '−' symbol at the head, to be applied last.

What operation has higher priority? This is the addition at the beginning. Place the '+' symbol at the next lower level.

Finally, what operation has the highest priority? This is the multiplication at the middle. Place the '·' symbol at the next lower level, to be applied to the numbers in the leaves. This way, the calculation will be carried out bottom to top, as required.

### 7.6.5 Virtual Tree

How does the computer calculate such an arithmetic expression? Fortunately, it uses a tree only virtually and implicitly.

Indeed, in practice, no concrete tree is needed. Instead, the computer uses a recursive algorithm (list of instructions):

1. Scan the arithmetic expression backwards: from right to left, as in Hebrew.
2. Look for the arithmetic symbol of least priority: either '+' (addition) or '−' (subtraction), in between two subexpressions. For example, in $2 + 4 \cdot 3 - 7$, the minus symbol is found first.
3. Once such an arithmetic symbol has been found,
    a) use it to split the original expression into two disjoint subexpressions.
    b) Calculate each of them recursively, using the same algorithm itself. This gives two subresults.
    c) Apply to them the relevant arithmetic operation. (In the above example, subtract them from each other.)
    d) This is your result: you are done.
4. If, however, there is no addition or subtraction in the entire expression, then start from the beginning, and do the above scanning once again, only this time look for arithmetic symbols of higher priority: either '·' (multiplication) or '/' (division).

5. If, however, no such arithmetic symbol can be found either, then look at the first symbol in the expression (on the left).
6. If this is '−', then this is no subtraction, but just the negative of the number that follows.
7. Finally, if there are no arithmetic symbols at all, then this must be a number. So, we're at the innermost recursion, and this number is our result. Later on, we'll see that a number is often written as a polynomial: sum of powers of 10. We'll also see a few ways to calculate it.

Fortunately, this algorithm mirrors the underlying tree, without constructing it explicitly. Indeed, each recursive call opens a new virtual subtree.

What operation is carried out first? Well, first of all, the algorithm carries out the innermost recursive call: just evaluate an individual number, as required.

Let's go ahead and mirror this in a Boolean expression as well.

### 7.6.6 Boolean Expression

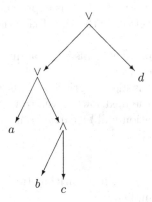

**Fig. 7.8.** Storing the Boolean expression $a \vee b \wedge c \vee d$ in a four-level tree. The calculation is made bottom to top. The top-priority Boolean operation, $b \wedge c$, is made at the bottom. The next operation, $a \vee (b \wedge c)$, is made at the second level. Finally, the least prior operation, $(a \vee b \wedge c) \vee d$, is made at the top.

Boolean expressions mirror arithmetic expressions: they have the same form and structure, with just a different interpretation. Here, the "and" operation (denoted by '$\wedge$') is prior to the "or" operation (denoted by '$\vee$'). The "not" operation, on the other hand, is prior to both. This operation is different: it is a unary operation, applied to one operand only. In this respect, it mirrors the minus sign in arithmetic.

Instead of numbers, we now have Boolean variables, which may take just two possible values: either 1 for true, or 0 for false. As a result, the value of the entire expression must also be either 1 (true) or 0 (false). As before, it is calculated recursively, bottom to top. Indeed, the example in Figure 7.8 mirrors the example in Figure 7.7.

# 7.7 Binary Tree

### 7.7.1 Full Binary Tree

So far, we looked at a general tree: from each node, issue as many branches as you like. Now, let's look at a (full) binary tree: from each node, issue either zero or two branches.

What does "full" mean here? It means that all nodes must have the same number of branches. In our case, this number is two. Indeed, all nodes have two branches. Only leaves have a different number: zero. As a matter of fact, we've already seen two such examples in Figures 7.7–7.8. In what follows, we deal with full trees only, so there is no need to keep writing "full."

How could a (full) binary tree look like? Well, there are two options. It could be really small: no branches at all — just one dangling leaf, and that's it. On the other hand, it could look like a real tree. For this, it must have two branches, issuing from the head. To the left branch, attach a binary subtree, defined recursively. To the right branch, on the other hand, attach yet another binary subtree, which may have a different number of levels. This way, the entire binary tree may be uneven and nonsymmetric: deeper on one side than on the other.

What have we done above? Just mathematical induction! Indeed, we've defined a one-level binary tree: just a single node, and that's it. Now, for $n = 1, 2, 3, 4, \ldots$, look at a fixed $n$. Assume that we already know how to define a $k$-level binary tree $(1 \le k \le n)$. This is indeed the induction hypothesis.

How to define a slightly bigger binary tree, with $n + 1$ levels? First, place a new node at the head. From it, issue two new branches. At the end of one branch, stick a new $n$-level binary subtree. At the end of the other branch, stick a new $k$-level binary subtree, for some $1 \le k \le n$. Because $k$ might be smaller than $n$, the entire binary tree may be uneven and nonuniform.

We've already seen two examples of a (full) binary tree (Figures 7.7–7.8). Why didn't we mention this at the time? Because arithmetic and Boolean expressions should better be stored in a general tree, not a full tree. After all, in many cases, they may need to use a unary operation: the minus sign, or the "not" operation.

As discussed above, the binary tree is not necessarily even or uniform: it may contain leaves not only at the bottom level but also at an intermediate level. In Figures 7.7–7.8, for example, there are leaves not only in the fourth level but also in the third level (the left node), and even the second level (the right node). A perfect tree, on the other hand, is much more even and symmetric.

### 7.7.2 Perfect Binary Tree

In the trees discussed so far, all levels may contain leaves. This is why they are uneven and nonuniform: deeper on one side than on the other. A perfect tree, on the other hand, is much more uniform: only one level may contain leaves: the bottom level. In Figure 7.9, for example, there are leaves in the fourth level only.

How to define a perfect binary tree? As before, use mathematical induction. A one-level tree is very small: just one dangling node, and that's it. Next, let's define more interesting trees. For $n = 1, 2, 3, 4, \ldots$, assume that we already know how to define an $n$-level perfect binary tree. With this knowledge on hand, we can now go

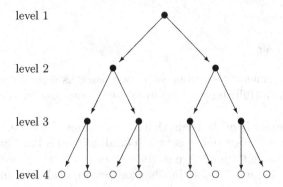

**Fig. 7.9.** A four-level perfect binary tree. The arrows stand for branches. The bullets stand for nodes at intermediate levels. The circles, on the other hand, stand for leaves at the bottom.

ahead and define a new $(n+1)$-level perfect binary tree as well: place a new node at the head, and issue two branches from it. At the end of the left branch, stick an $n$-level perfect binary subtree. At the end of the right branch, on the other hand, stick yet another $n$-level perfect binary subtree. This way, the new $(n+1)$-level tree is indeed even and symmetric, as required. To prove this, use mathematical induction once again.

### 7.7.3 Mathematical Induction and Recursion

What is the difference between mathematical induction and recursion? Well, they mirror each other, and can be viewed as two sides of the same coin. Mathematical induction is the theoretical principle. It goes forward, and engineers the process from scratch onwards. Recursion, on the other hand, puts it into practice. It works the other way around: it goes backwards, and reverse-engineers the process.

To construct a perfect binary tree, for example, the induction step uses two $n$-level subtrees. But how is a subtree constructed in the first place? Recursively, of course! For this purpose, one must use two $(n-1)$-level subtrees, and so on. This is indeed recursion.

Although recursion does the hard work, mathematical induction gets all the glory. After all, it is useful to prove theorems. Our perfect binary tree is a good example for this. Thanks to its multilevel structure, it has some nice properties:

- As in a general tree, the bottom level contains leaves only.
- The rest of the levels, on the other hand, contain no leaf any more. On the contrary: every node in them has exactly two branches.
- This is indeed perfect: even and uniform, as in Figure 7.9.
- The number of nodes doubles from level to level: in the first level, there are $2^0 = 1$ nodes. In the next lower level, there are $2^1 = 2$ nodes. In the $i$th level, there are $2^{i-1}$ nodes $(1 \le i \le n)$.
- The total number of nodes is, thus,

$$\sum_{i=1}^{n} 2^{i-1} = \sum_{i=0}^{n-1} 2^i = \frac{2^n - 1}{2 - 1} = 2^n - 1$$

(Section 7.5.2). To prove these properties, use mathematical induction. In the exercises below, we'll also calculate the total number of branches.

## 7.8 Application in Chaos Theory

### 7.8.1 Exponential Growth

Thus, the number of nodes increases exponentially: it doubles from level to level. This is a very rapid growth: at the 11th level, for example, there are as many as

$$2^{11-1} = 2^{10} = 1024 > 1000 = 10^3$$

nodes. The index $i$, on the other hand, grows much more slowly: linearly. Thus, while the index grows moderately from $i$ to $i+1$, the number of nodes doubles from $2^{i-1}$ to $2^i$.

Such a sharp jump is not always benign. For example, an algorithm (or method) that requires exponential time or storage is often considered as impractical. This is indeed a "bad" exponential growth: it leaves us with no control. This is indeed in the heart of chaos theory.

### 7.8.2 Relative vs. Absolute Error

To introduce chaos, let's consider a simple algebraic process. It starts from a given parameter: $v_0 \neq 0$. How does it continue? Step by step. At each step, it multiplies by yet another given parameter: $\alpha \neq 0$. In other words, for $n = 0, 1, 2, \ldots$, define

$$v_{n+1} \equiv \alpha v_n.$$

As a matter of fact, this is just mathematical induction.

This is called a linear process: in each step, $v_n$ is multiplied by a constant number: $\alpha$. Later on, we'll also see a nonlinear process, in which $v_n$ is multiplied by itself as well.

Thanks to mathematical induction, $v_n$ also has a more explicit form:

$$v_n = \alpha v_{n-1} = \alpha^2 v_{n-2} = \cdots = \alpha^n v_0.$$

Assume now that $v_n$ is some physical quantity, measured to some finite precision only. So, the $v_n$'s can never be known precisely, but only within a small error.

As a matter of fact, the error may appear already in the initial data. To account for this, assume that $v_0$ is available only approximately, as

$$v_0 + \delta_0,$$

where the error $\delta_0$ is small in magnitude relative to $v_0$:

$$|\delta_0| \ll |v_0|.$$

Under these circumstances, we can no longer compute $v_n$ exactly, but only approximately: with an error $\delta_n$. Indeed, mathematical induction can still be used to define the approximations

$$
\begin{aligned}
v_n + \delta_n &\equiv \alpha(v_{n-1} + \delta_{n-1}) \\
&= \alpha^2(v_{n-2} + \delta_{n-2}) \\
&= \cdots = \alpha^n(v_0 + \delta_0) \\
&= \alpha^n v_0 + \alpha^n \delta_0 \\
&= v_n + \alpha^n \delta_0.
\end{aligned}
$$

Thus, the error $\delta_n$ undergoes the same linear process as well:

$$
\delta_n = \alpha^n \delta_0.
$$

If $|\alpha| > 1$, then the absolute error $|\delta_n|$ may grow exponentially with $n$. Fortunately, what matters is not the absolute but the relative error, which doesn't grow at all:

$$
\left| \frac{\delta_n}{v_n} \right| = \left| \frac{\alpha^n \delta_0}{\alpha^n v_0} \right| = \left| \frac{\delta_0}{v_0} \right|.
$$

This is indeed a stable process: the relative error remains small, so $v_n + \delta_n$ approximate $v_n$ well. For this reason, $v_n + \delta_n$ is good enough in practice. In a nonlinear process, on the other hand, this is not necessarily the case.

## 7.9 Stability vs. Instability

### 7.9.1 Unstable Quadratic Process

To describe a nonlinear process, we need yet another parameter: $\beta \neq 0$. With it, we can now define a quadratic process:

$$
v_{n+1} \equiv \alpha v_n + \beta v_n^2.
$$

Let's write this as

$$
v_{n+1} - v_n = y(v_n),
$$

where

$$
y(x) \equiv (\alpha - 1)x + \beta x^2
$$

is the quadratic function that tells us how the process progresses.

Let us distinguish between two major cases. Assume first that

$$
\alpha \leq 1.
$$

For simplicity, assume also that

$$
\beta > 0.
$$

(Otherwise, just work with $-v_n$ and $-\beta$.) This way, the graph of $y(x)$ is a straight parabola, as in Figure 7.10.

Do the $v_n$'s converge to any limit? Well, let $v_\infty$ be such a hypothetical limit. It must satisfy

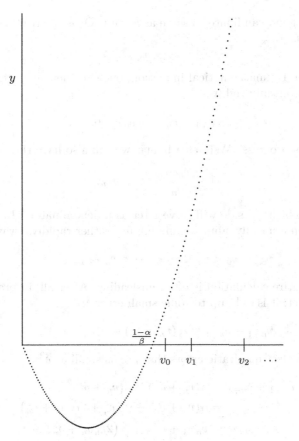

**Fig. 7.10.** An unstable quadratic process. The $v_n$'s increase monotonically towards infinity, away from $(1 - \alpha)/\beta$. The calculation is unstable and meaningless.

$$
\begin{aligned}
0 &= v_\infty - v_\infty \\
&= \lim_{n \to \infty} v_n - \lim_{n \to \infty} v_n \\
&= \lim_{n \to \infty} v_{n+1} - \lim_{n \to \infty} v_n \\
&= \lim_{n \to \infty} (v_{n+1} - v_n) \\
&= \lim_{n \to \infty} y(v_n) \\
&= y(v_\infty).
\end{aligned}
$$

Thus, $v_\infty$ must solve the quadratic equation

$$
y(x) = 0.
$$

We already have one solution: $x = 0$. Still, there is a more interesting solution:

$$
v_\infty \equiv \frac{1 - \alpha}{\beta} \geq 0.
$$

This is indeed a good candidate to serve as a limit. Or is it? Well, assume that we start to the right of $v_\infty$:

$$v_0 > v_\infty,$$

as in Figure 7.10. By mathematical induction, one can then show that the process progresses monotonically rightwards:

$$v_{n+1} - v_n = y(v_n) > 0.$$

How rapid is this progress? Well, as a bonus, we can also have the ratio

$$1 \le \frac{v_{n+1}}{v_n} = \alpha + \beta v_n.$$

(This positive right-hand side will serve later as a denominator.) In summary, the $v_n$'s increase monotonically, approaching infinity rather rapidly, away from $v_\infty$:

$$v_\infty < v_0 < v_1 < v_2 < \cdots < v_n \rightarrow_{n \to \infty} \infty.$$

Yet worse: the entire calculation is often misleading. After all, in practice, the calculation is inexact: it is only up to some small error $\delta$:

$$v_{n+1} + \delta_{n+1} \equiv \alpha(v_n + \delta_n) + \beta(v_n + \delta_n)^2.$$

For simplicity, neglect quadratic errors that are as small as $\delta^2$:

$$\begin{aligned}
v_{n+1} + \delta_{n+1} &= \alpha(v_n + \delta_n) + \beta(v_n + \delta_n)^2 \\
&= \alpha(v_n + \delta_n) + \beta\left(v_n^2 + 2v_n\delta_n + \delta_n^2\right) \\
&= v_{n+1} + \alpha\delta_n + \beta\left(2v_n\delta_n + \delta_n^2\right) \\
&\doteq v_{n+1} + \alpha\delta_n + 2\beta v_n\delta_n \\
&= v_{n+1} + (\alpha + 2\beta v_n)\,\delta_n.
\end{aligned}$$

Thus, the new error is approximately

$$\delta_{n+1} \doteq (\alpha + 2\beta v_n)\,\delta_n.$$

Usually, $\delta_n \ne 0$, so we can also divide:

$$\frac{\delta_{n+1}}{\delta_n} \doteq \alpha + 2\beta v_n.$$

In calculus, this is actually the derivative of $v_{n+1}$ at $v_n$, denoted by $v'_{n+1}(v_n)$. It tells us how the error grows with $n$.

Thanks to the above, we can also go ahead and approximate the relative error as well:

$$\frac{\delta_{n+1}}{v_{n+1}} \doteq \frac{(\alpha + 2\beta v_n)\,\delta_n}{v_{n+1}} = \frac{(\alpha + 2\beta v_n)\,\delta_n}{\alpha v_n + \beta v_n^2} = \frac{\alpha + 2\beta v_n}{\alpha + \beta v_n} \cdot \frac{\delta_n}{v_n}.$$

Thus, upon advancing from $n$ to $n+1$, the relative error multiplies by the approximate factor

$$\frac{\alpha + 2\beta v_n}{\alpha + \beta v_n} = 2 - \frac{\alpha}{\alpha + \beta v_n} \ge \begin{cases} 2 & \text{if } \alpha \le 0 \\ 2 - \frac{\alpha}{\alpha + \beta v_0} & \text{if } \alpha > 0. \end{cases}$$

This is bad news. Indeed, in either case, the relative error grows exponentially fast:

$$\left|\frac{\delta_n}{v_n}\right| \geq \left(\frac{|\alpha| + 2\beta v_0}{|\alpha| + \beta v_0}\right)^n \left|\frac{\delta_0}{v_0}\right|.$$

Thus, the computed value $v_n + \delta_n$ is no good: it is no approximation to the desirable value $v_n$ at all. Let's move on to a more meaningful case.

### 7.9.2 The Logistic Equation

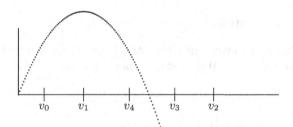

**Fig. 7.11.** A stable quadratic process. The $v_n$'s approach $v_\infty$. The calculation is now stable and meaningful.

So far, we've assumed that $\alpha \leq 1$. Unfortunately, the result was an unstable process, which can never be calculated in practice.

Let us now move on to a fundamentally different case [32]. For this purpose, assume now that

$$\alpha > 1.$$

For simplicity, assume also that

$$\beta < 0.$$

(Otherwise, just work with $-v_n$ and $-\beta$.) This way, the graph of $y(x)$ is now an upside-down parabola, as in Figure 7.11. As before, it still meets the $x$-axis at

$$v_\infty \equiv \frac{1 - \alpha}{\beta} \geq 0.$$

This solves the so-called logistic equation:

$$y(v_\infty) == (\alpha - 1)v_\infty + \beta v_\infty^2 = 0.$$

In terms of stability and convergence, the situation is now much better than before. Indeed, if we start from some $0 < v_0 < v_\infty$, then we have

$$v_1 - v_0 = y(v_0) > 0.$$

If $v_1$ is still smaller than $v_\infty$, then we still have

$$v_2 - v_1 = y(v_1) > 0.$$

But $v_2$ could be too large: it could skip $v_\infty$, as in Figure 7.11. In this case, we have

$$v_3 - v_2 = y(v_2) < 0,$$

so $v_3$ may again approach $v_\infty$, from the right.

This is not the end of the story: $v_4$ may skip $v_\infty$ once again, and be on its left again. Fortunately, $v_5$ may still approach $v_\infty$ from the left, and so on. In summary, the process may approach $v_\infty$ from alternating sides.

### 7.9.3 Control Parameter

By now, it seems that our $v_n$'s do approach their limit: $v_\infty$. Or do they? To check on this, assume that they are already pretty close to $v_\infty$, within a new error $\delta$. This new $\delta$ is not the same as before: it stands now for the distance from the limit. We can now write

$$
\begin{aligned}
v_\infty + \delta_{n+1} &= v_{n+1} \\
&= \alpha v_n + \beta v_n^2 \\
&= \alpha(v_\infty + \delta_n) + \beta(v_\infty + \delta_n)^2 \\
&= \alpha(v_\infty + \delta_n) + \beta\left(v_\infty^2 + 2v_\infty\delta_n + \delta_n^2\right) \\
&= v_\infty + \alpha\delta_n + \beta\left(2v_\infty\delta_n + \delta_n^2\right) \\
&\doteq v_\infty + \alpha\delta_n + 2\beta v_\infty\delta_n \\
&= v_\infty + \alpha\delta_n + 2(1 - \alpha)\delta_n \\
&= v_\infty + (2 - \alpha)\delta_n.
\end{aligned}
$$

In summary,

$$\delta_{n+1} \doteq (2 - \alpha)\delta_n.$$

Assuming that $\delta_n \neq 0$, we can also divide:

$$\frac{\delta_{n+1}}{\delta_n} \doteq 2 - \alpha.$$

In the language of calculus, this is the derivative of $v_{n+1}$ at $v_\infty$, denoted by $v'_{n+1}(v_\infty)$.

Is the process stable? Well, it depends on the original parameter $\alpha$: if it is not too large,

$$1 < \alpha \leq 3,$$

then the error (the distance from the limit) remains small:

$$|\delta_{n+1}| \doteq |(2 - \alpha)\delta_n| = |2 - \alpha| \cdot |\delta_n| \leq |\delta_n|.$$

In this case, the relative error remains small as well:

$$\left|\frac{\delta_{n+1}}{v_\infty}\right| \leq \left|\frac{\delta_n}{v_\infty}\right|,$$

so the calculation is indeed meaningful. This is independent of $\beta$: $\alpha$ alone tells us whether the process is stable or not. This is why $\alpha$ is called the control parameter. Still, $\beta$ is also important: its sign tells us how the parabola looks like.

## 7.10 Chaos: Find the Hidden Order!

### 7.10.1 Bifurcation Points

We can now discuss different kinds of $\alpha$'s. In particular, the special point

$$\alpha_0 \equiv 3$$

distinguishes "good" $\alpha$'s from "bad" $\alpha$'s. This is why $\alpha_0$ is called a bifurcation point. For $\alpha$'s larger than $\alpha_0$, there is no convergence any more.

Still, even for these "bad" $\alpha$'s, there is still some hope. In fact, there is yet another bifurcation point:

$$\alpha_1 \doteq 3.45.$$

For those $\alpha$'s in between

$$\alpha_0 < \alpha < \alpha_1,$$

although the original process no longer converges, it still splits into two subprocesses that do converge: the even-numbered items converge to one limit,

$$v_0, v_2, v_4, v_6, \ldots, v_{2n} \to_{n \to \infty} v_{2\infty},$$

whereas the odd-numbered items converge to another limit:

$$v_1, v_3, v_5, v_7, \ldots, v_{2n+1} \to_{n \to \infty} v_{2\infty+1}.$$

Here, $v_{2\infty}$ and $v_{2\infty+1}$ are just convenient notations for two different numbers. And what happens beyond $\alpha_1$? Well, the next bifurcation point is

$$\alpha_2 \doteq 3.55.$$

For those $\alpha$'s in between

$$\alpha_1 < \alpha < \alpha_2,$$

there are four disjoint subprocesses, converging to four different limits:

$$v_{4n} \rightarrow_{n\to\infty} v_{4\infty}$$
$$v_{4n+1} \rightarrow_{n\to\infty} v_{4\infty+1}$$
$$v_{4n+2} \rightarrow_{n\to\infty} v_{4\infty+2}$$
$$v_{4n+3} \rightarrow_{n\to\infty} v_{4\infty+3},$$

and so on.

### 7.10.2 Bifurcation Tree

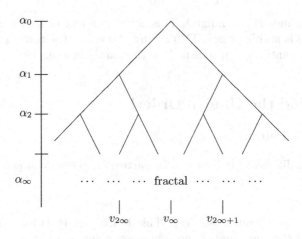

**Fig. 7.12.** The bifurcation tree.

This goes on: one could go on defining infinitely many bifurcation points:

$$\alpha_0 < \alpha_1 < \alpha_2 < \alpha_3 < \alpha_4 < \cdots < \alpha_m \rightarrow_{m\to\infty} \alpha_\infty \doteq 3.57.$$

What happens at $\alpha = \alpha_\infty$? Chaos! No subsequence converges any more. In other words, the original process has no accumulation point any more (Chapter 4, Section 4.4.4).

Fortunately, even at chaos, there is still some hope. Indeed, from deep chaos, a beautiful geometrical pattern emerges: a new bifurcation tree (Figure 7.12). At its "bottom," it approaches a new fractal, of dimension less than 1.

### 7.10.3 Binary Tree and Its Fractal

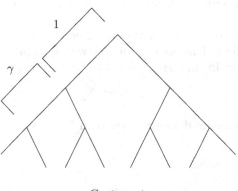

$\cdots \quad \cdots \quad \cdots$ Cantor set $\cdots \quad \cdots \quad \cdots$

**Fig. 7.13.** At its bottom, the binary tree approaches a Cantor set, of dimension less than 1.

How does the bifurcation tree look like from below? To visualize this, let's turn to a simpler model: a binary tree (Figure 7.13). Assume that it has infinitely many levels: at the bottom, the levels "shrink," and get more and more crowded. In other words, upon advancing from one level to the next lower level, the branches get shorter by the constant factor

$$0 < \gamma < \frac{1}{2}.$$

How does this look like? Look at the left edge. The top branch, issuing from the head, is of length 1. The next lower branch, on the other hand, is of length $\gamma$. The next lower branch, on the other hand, is of length $\gamma^2$, and so on (Figure 7.13).

These are the leftmost branches. Together, they make the left edge of the entire tree: from the head at the top, down to the lower-left corner. What is the total length of the entire edge? This is an infinite power series (Section 7.5.3):

$$1 + \gamma + \gamma^2 + \gamma^3 + \cdots = \sum_{i=0}^{\infty} \gamma^i = \frac{1}{1-\gamma}.$$

By now, we've measured the entire left side of the tree: from the head to the lower-left corner. Now, let's drop the first (top-left) branch, and start from the second one. Without the first branch, the length is now

$$\gamma + \gamma^2 + \gamma^3 + \cdots = \sum_{i=1}^{\infty} \gamma^i = \gamma \sum_{i=0}^{\infty} \gamma^i = \frac{\gamma}{1-\gamma}.$$

This is $\gamma$ times as short. This splits the entire left side into two parts: a long part (one branch only), and a shorter part (the rest). The bottom could now split in the same proportion (as in Figure 5.1). Thus, the bottom is indeed as sparse and transparent as the original Cantor set.

To see this, cover the entire bottom with small intervals (as in Section 7.4.1). There is just a slight change: start with intervals of length $\gamma$ (times the distance from the lower-left corner to the lower-right corner). How many are needed? Clearly, two.

Now, turn to the next finer scale, and use intervals $\gamma$ times as short. How many are needed? Clearly, four. This goes on and on: with intervals $\gamma^n$ times as short, we need $2^n$ times as many. In summary, since $2\gamma < 1$,

$$2^n < \left(\gamma^{-1}\right)^n,$$

so this is indeed a fractal, of dimension less than 1.

### 7.10.4 Feigenbaum's Constant

In the above binary tree, the levels "shrink" by a constant factor: $\gamma$. Indeed, upon advancing from one level to the next lower level, the branch gets $\gamma$ times as short. The original bifurcation tree in Figure 7.12 is nearly the same: its levels "shrink" by a nearly constant factor. Indeed,

$$\frac{\alpha_m - \alpha_{m-1}}{\alpha_{m+1} - \alpha_m} \to_{m \to \infty} \delta \doteq 4.669.$$

Here, $\delta$ is a new constant: Feigenbaum's constant. It governs many phenomena in physics and other sciences.

In summary, for $\alpha \geq \alpha_\infty$, the quadratic process gets chaotic: it makes a new fractal. This is indeed the aim of chaos theory: to find some pattern even in a complete disorder.

This is quite practical. Consider, for example, weather forecast. For this purpose, a new field was developed: computational fluid dynamics. It offers a few complex models to help make a reliable and stable forecast. This is quite tricky: there are highly nonlinear interactions between unknowns like velocity, temperature, and pressure.

Due to its chaotic nature, weather can be predicted for a very short time: five days ahead, or so. For more than that, no prediction can be made: there is too much sensitivity to initial data. The difficulty is mathematical, not computational: even a supercomputer would fail. This is why Edward Lorenz said: "the flap of a butterfly wing in Brazil may set off a tornado in Texas!" This is what happens when one attempts to predict the future...

## 7.11 Some Philosophical Remarks

### 7.11.1 Can We Predict the Future?

In his science-fiction trilogy "Original Foundation," Isaac Asimov describes the state of humanity a few centuries ahead in time. Mankind is no longer limited to the Earth: there are many settlements on other planets as well, where people are better off. People who stay here on the Earth, on the other hand, are not so well off. This drives a social gap, and eventually a severe social crisis.

Fortunately, the great sociologist is about to give a lecture about this. Everyone awaits the rare opportunity to learn some new hints or perhaps even a solution to the acute problem.

Of course, the great sociologist is long gone. Fortunately, before his death, he left a sealed video cassette. It is time to open it, and listen to what he has to say.

Upon opening the cassette, everyone is shocked. It turns out that the great sociologist failed to foresee the future. He talks about a completely different state, in which the entire humanity is still confined to the Earth, and never left. How could his thoughts be relevant?

Fortunately, some people in the audience still manage to benefit from the talk. Although the great sociologist didn't predict the present state, his speech still offers some deep insight and comprehension in broad terms. After all, the detailed circumstances are not so important. What is important is the fundamental principles that characterize and motivate sociological processes.

In the ancient (and middle) ages, people used to look for answers in the Bible or Plato's or Aristotle's scripts. Fortunately, in the Bible, the prophets rarely predict the concrete future. More often, they just set moral standards and guidelines, relevant at any time and age. Thanks to this advice, one could have more peace of mind, and make better decisions.

Likewise, Asimov's sociologist doesn't know all. On the contrary: he only knows what's before his eyes. From this, he could never predict a concrete disaster, say a global pandemic. Still, he may well have foreseen a more general event that may force us to change our way of life. To cope with this, he uses his own wisdom and intuition, and offers us a general theory. This is indeed induction. It is even better than a specific solution: it leaves us some work to do. Indeed, it is up to us to read between the lines, examine the advice we hear, and act upon it here and now. This is indeed deduction. (Sections 7.1.1–7.1.2.)

### 7.11.2 The Phase Plane

Our society is a highly complex system. This is too complicated. To simplify things, let's focus on one simple phenomenon in physics, economics, sociology, or any other field. How to model it geometrically? Well, there may still be a lot of unknown variables that control or affect or govern it. For simplicity, assume that there are just two unknowns: $p$ and $q$.

So far, we've only considered a scalar process: $v$. Here, on the other hand, $v$ is a two-dimensional vector, with two components, or degrees of freedom:

$$v \equiv (p, q)$$

(Figure 7.14). During the process, the scalar unknowns $p$ and $q$ may interact nonlinearly with each other, and change. This produces a nonlinear curve (trajectory) of states: start from the initial state ($v_0 \equiv (p_0, q_0)$), and advance gradually to slightly different states that may take place in the future.

How could one say anything about the future? Well, for this purpose, a prophet (or an academic) could only use the information available at the present: $p_0$ and $q_0$.

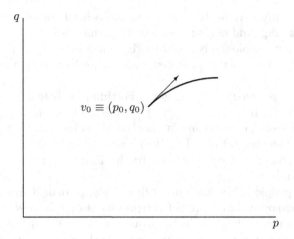

**Fig. 7.14.** A stable nonlinear process. From the initial state $v_0 \equiv (p_0, q_0)$, issue a smooth curve of new states. The tangent vector points at the direction of the best prediction available.

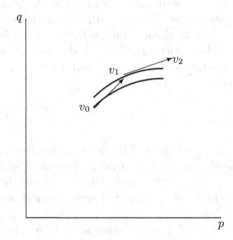

**Fig. 7.15.** The discrete process $v_0, v_1, v_2, \ldots$ approximates the smooth nonlinear curve. At $v_0$, a tangent vector is used to advance to $v_1$. At $v_1$, on the other hand, a new tangent vector is used to advance to $v_2$, and so on.

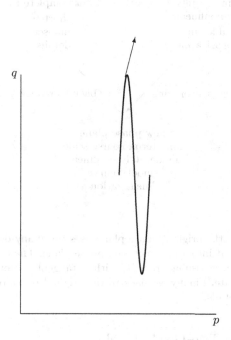

**Fig. 7.16.** A highly nonlinear (unstable) process. The curve changes too sharply, so linear prediction is useless.

To predict the future, one must linearize: look straight ahead, along the tangent vector in Figure 7.14.

Still, the tangent vector is inaccurate: it only approximates the true nonlinear curve. Even the best prediction must contain an error, due to uncertainty. After all, only a vague idea is available: the first derivative, or the first-order approximation.

How strong is the nonlinear interaction between $p$ and $q$? Well, if it is moderate, then the curvature is rather small, so the relative error is kept small as well. In this case, one may step along a discrete path: $v_0, v_1, v_2, \ldots$, approximating the original curve rather well (Figure 7.15).

In the highly nonlinear case, on the other hand, the curve may change too sharply: no discrete path is useful any more (Figure 7.16). This means that the original phenomenon is too uncertain, chaotic, and unstable: no linearization could work any more.

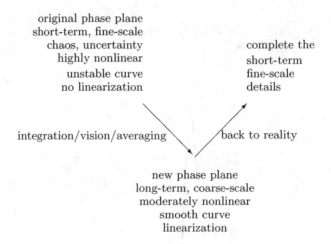

**Fig. 7.17.** The V-cycle: the original phase plane uses too many degrees of freedom: the scale is too fine. Average it into a new (coarser) phase plane. There, pick a suitable initial state, with a smooth curve issuing from it, with a tangent vector that points towards the desired long-term state. Finally, go back to the original phase plane, to work out the remaining short-term details.

### 7.11.3 Leadership: a Two-Level Model

Could the leader impose his/her own will upon the future, and shape it in his/her own vision? Could he/she focus on the long term, and work out the short-terms details later, without spoiling the long-term goal?

To do this, better work with two models: fine and coarse (Figure 7.17). The fine model is just our original phase plane. Now, average on it. This gives a new model: a new (coarser) phase plane, with a new (smoother) curve, suitable for linearization. Only after working in it should one go back to the original model, to work out the short-term details, without interacting with the coarse scale too much, and without spoiling the work done so far.

So, vision is not enough. To realize his/her view, the leader could also nominate a hierarchy of reliable managers to carry out his/her original plan, and make sure that the fine details are kept well under control, and don't undermine the entire project.

In a way, the leader uses a longer time scale: a special kind of "glasses," to look ahead into the deep future, disregarding any detail that may obscure the big picture. In this new (approximate) phase plane, the leader can pick a suitable initial state, with a sufficiently smooth curve issuing from it. This could indeed be linearized: the tangent vector may then point in the right direction — towards the desired state. All that is left to do is to go back to the fine model, to sort out the short-term details, without introducing any instability that could harm the entire goal. This could be viewed as an induction–deduction tree. (See Figure 7.17, and compare also to Figure 7.2.)

### 7.11.4 Feedback and Instability

In the above, we have just two unknowns: $p$ and $q$. Now, one could also introduce many more unknowns, and work not only in a plane but also in a (multidimensional) hyperspace:

$$v \equiv (p, q, \ldots).$$

Like the sociologist and the prophet, the leader can only set guidelines: long-term ideas or principles. Geometrically, this means averaging on reality: producing a coarse (inexact) phase plane, where curves may get smoother, and easier to linearize, with less instability. Once this has been done, we may be well on track towards our ultimate goal, at least in broad terms. Still, the details must be worked out carefully, to fit in the same direction.

So, can the leader shape the future? This is discussed in a series of talks by Gershon Hacohen. He says that, by setting a clear vision, the leader may generate just enough faith, motivation, and power of will in the hearts of the people, to get them right on track to overcome any difficulty, and complete the original task successfully. After all, faith is the key to making the initial step in the right direction.

Still, there must be some symmetry: a constant interaction and communication between the leader and his/her people. After the initial step, the people may indeed give important feedback to the leader, to update the original plan, if necessary. After all, the fine short-term details, however tiny, may still interact with the long-term process, and affect it quite substantially.

Geometrically, this is illustrated in the curvature. If it is moderate, then it could be smoothed out in the coarse scale. This is indeed stability. In this case, the feedback could be positive, and even build up: generate more and more motivation and will, to follow the original plan, or even improve on it.

## 7.12 Exercises: Complexity

### 7.12.1 Binary Tree and Its Complexity

1. Show that, in a general tree, every node but the head has exactly one incoming branch (or edge) pointing at it.
2. Conclude that the total number of edges is smaller by 1 than the total number of nodes. Hint: map each node (but the head) to the incoming branch leading to it.
3. Show that, in a perfect binary tree, the $i$th level contains $2^{i-1}$ nodes ($i \geq 1$). Hint: use mathematical induction on $i = 1, 2, 3, \ldots$.
4. Conclude that, in an $n$-level perfect binary tree, there are

$$\sum_{i=1}^{n} 2^{i-1} = \sum_{i=0}^{n-1} 2^i = \frac{2^n - 1}{2 - 1} = 2^n - 1$$

   nodes.
5. Prove this in yet another way. Hint: use mathematical induction on $n = 1, 2, 3, \ldots$.

6. Conclude that, in an $n$-level perfect binary tree, there are $2^n - 2$ branches (edges). Hint: we've already established that the total number of edges is smaller by 1 than the total number of nodes, even in a general tree.

7. Prove this in yet another way. Hint: use mathematical induction on $n = 1, 2, 3, \ldots$.

8. Write an algorithm that embeds a given arithmetic expression in a tree, and calculates its value. Hint: use recursion. The solution can be found in Chapter 1 in [45].

9. Write an algorithm that embeds a given Boolean expression in a tree, and calculates its true value (1 or 0).

10. Write an algorithm that transforms a given natural number from its decimal representation (in base 10) to its binary representation (in base 2).

11. How many arithmetic operations are needed?

12. This is the cost. It grows with the total number of digits in the original number. How fast is this growth? Exponential or linear?

13. Write an algorithm that transforms a given natural number from its binary representation back to its decimal representation.

14. How many arithmetic operations are needed?

15. This is the cost. It grows with the total number of digits in the original number. How fast is this growth? Exponential or linear?

16. Improve both algorithms, and make sure that their cost grows only linearly, not exponentially. Hint: use recursion. The solution can be found in Chapter 1 in [45].

17. In Figure 7.14, where is the time variable? Why is it missing? Hint: this is a phase plane. In it, time is implicit: it is just a parameter that "pushes" along the curve. To see this more vividly, make a movie that shows the advancement from state to state along the curve.

# 8

# Entropy and Information

Consider a closed system: say, a closed box, full of warm water. What is entropy? This tells us how uniform the system is: to what extent all molecules have the same (kinetic) energy.

As time goes by, the system gets more and more "boring:" the molecules get more and more similar, with nearly the same energy. Heat tends to spread out, equally in the entire box. It is quite uncommon to see one spot getting suddenly hot, while its surroundings remaining colder.

This is the second law of thermodynamics: in a closed system, entropy can only increase, but never decrease. Nature favors maximum entropy and balance. No particular spot should be special, or different from the rest of the box.

Why? Because this would require more information: to specify this spot, and distinguish it from the rest of the box. We don't have the resources required to support this kind of extra information. In fact, the system can only *lose* information, but never gain. This is how the second law of thermodynamics is relevant in computer science too.

## 8.1 Random Variable and Its Probability

### 8.1.1 Throw a Coin

Let's play a game: throw a coin. At probability $\alpha$, it would show a head. At probability $\beta$, on the other hand, it would show a tail. Clearly, the probabilities must sum up to 1:

$$\alpha + \beta = 1.$$

After all, the coin must show something. An event that must happen has probability 1. In our game, at probability 1, the coin must show head or tail. This is deterministic.

If this is a good coin, then it is symmetric: equally likely to show head or tail:

$$\alpha = \beta = \frac{1}{2}.$$

This gives us a discrete probability function (or distribution), telling us how likely we are to get head or tail (Figure 8.1).

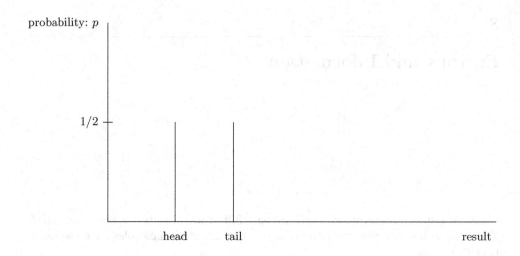

**Fig. 8.1.** Throwing a symmetric coin: at probability 1/2, it would show a head. At probability 1/2, it would show a tail.

### 8.1.2 Throw a Dice

Now, let's play another game: throw a dice. What would it show? Well, we can predict: at probability 1/6, it would show 1. At probability 1/6, it would show 2. At probability 1/6, it would show 3, and so on. Again, the probability is discrete and uniform (Figure 8.2).

### 8.1.3 Random Variable

This is a random variable: we don't know what value it would take. It could take either 1 or 2 or 3 or 4 or 5 or 6.

If this is a good coin, then these values are equally likely. The probability is uniform: either value could be obtained at the same probability: 1/6.

## 8.2 Continuous Probability Function

### 8.2.1 Probability and Its Distribution

The probability function in Figure 8.2 is discrete: it is defined on six points only. This means that only six results are possible, and have a positive probability. All the rest, on the other hand, are impossible: have zero probability.

More complicated random variables, on the other hand, could take infinitely many results, and even a continuum of potential results: say, every result from 0 to 1.

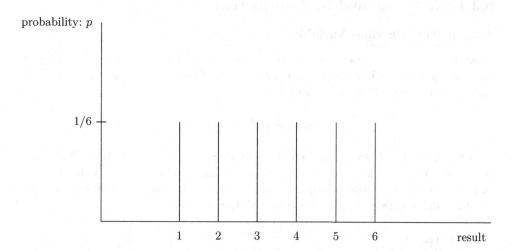

**Fig. 8.2.** Throwing a dice: at probability 1/6, it would show 1. At probability 1/6, it would show 2. At probability 1/6, it would show 3, and so on.

The probability function (distribution) should then be a (nonnegative) integrable function, defined on the entire interval $[0, 1]$. Its integral there must be 1:

$$p([0,1]) = \int_0^1 p(x)dx = 1.$$

After all, at probability 1, the random variable must take some result between 0 and 1. This is deterministic.

How to use the probability function? Given two potential results $0 \le a < b \le 1$, we can now calculate the probability to obtain a result between $a$ and $b$:

$$p([a,b]) = \int_a^b p(x)dx.$$

### 8.2.2 Null Set: Zero Probability

In particular, how likely are we to get $a$ exactly? This could never happen! In other words, this is unrealistic:

$$p([a,a]) = \int_a^a p(x)dx = 0.$$

After all, the degenerate interval $[a, a]$ is a null set of zero measure (Chapter 1, Section 1.14.1).

## 8.3 Distribution and Its Information

### 8.3.1 A New Random Variable

Now, let's work the other way around: assume that we are given a probability function: a positive integrable distribution $p(x)$. Could we define an interesting random variable from scratch? Let's try

$$\log_2 \left( \frac{1}{p} \right) = - \log_2 p.$$

This is a new (nonphysical) random variable: it exists in our mind only. What is its meaning? Well, it tells us how rare a certain result is. For example, look at some potential result $x$ (of the original random variable). It has probability $p(x) > 0$. Assume that $x$ is rare, so $p(x)$ is small, and $1/p(x)$ is large.

### 8.3.2 Surprise!

So, if we obtain $x$, how surprised should we be? As surprised as if we threw a coin many times, and always obtained a head, and never a tail! How many? As many as $- \log_2 p(x)$. Indeed, the probability to obtain a head so many times is as small as

$$\left( \frac{1}{2} \right)^{- \log_2 p(x)} = 2^{\log_2 p(x)} = p(x),$$

which is the same as the probability to obtain $x$.

### 8.3.3 Information

What is the meaning of this new random variable? Well, it is completely abstract and fictitious: it came from the distribution only, with no physical meaning whatsoever. Still, it has an important job: if we obtain $x$ (in our original random variable), then it tells us how surprised we should be.

In particular, if we obtain a rare $x$, then we should be really surprised. How surprised? As surprised as $- \log_2 p(x)$. This is also the amount of information encapsulated in the news that $x$ was obtained. The rarer $x$ is, the smaller $p(x)$ is, and the more information we get from this news.

## 8.4 Von-Neumann's Entropy

### 8.4.1 Information and Its Expectation

Still, this is rare: it doesn't happen too often. Thus, to be fair, this should be multiplied by $p(x)$. This gives us some balance: for more common $x$'s, their surprise (or information) is lower, but is also multiplied by their (higher) probability. This is fair: it gives us the weighted average of our new random variable: each potential surprise (or amount of information) is multiplied by a suitable weight: its probability. Upon integrating over all possible $x$'s, we obtain the expectation of our new random variable:

$$- \int p(x) \log_2 p(x) dx.$$

### 8.4.2 Von-Neumann's Entropy

This is Von-Neumann's entropy. It tells us not only the amount of information encapsulated in a particular $x$, but also the information we can expect in advance, before designing an experiment to measure $x$. After all, before the experiment, we don't know what result we'd get: we could in theory obtain just any $x$, at probability $p(x)$. Thus, to have a fair estimate of the expected information, we should multiply $x$'s surprise by $x$'s probability, and integrate over all possible $x$'s, as in the above integral.

### 8.4.3 Zero Probability

Here, one may ask: what if $x$ is never obtained:

$$p(x) = 0?$$

In this case, is the surprise infinite:

$$-\log_2 p(x) = -\log_2 0 = \infty?$$

Fortunately, in the entropy, we're only interested in the product

$$-p(x)\log_2 p(x) = -\frac{\log_2 p(x)}{1/p(x)}.$$

As $p(x) \to 0$, this is denoted by $\infty/\infty$. To calculate the limit, use L'Hopital's rule: differentiate both numerator and denominator with respect to $p(x)$:

$$\lim_{p(x)\to 0} \frac{\log_2 p(x)}{1/p(x)} = \lim_{p(x)\to 0} \frac{1/p(x)}{-1/p(x)^2} = -\lim_{p(x)\to 0} p(x) = 0.$$

This could be denoted by

$$0 \cdot \log_2 0 = 0.$$

Thus, such $x$'s contribute nothing to the above integral.

## 8.5 Application in Computer Science

### 8.5.1 Boolean Expression

This has an interesting application in computer science. Consider a symmetric Boolean expression of the form

$$(a \vee b) \vee (c \vee d)$$

(Figure 8.3). In what sense is it symmetric? Well, it forms a perfect binary tree, with full levels. The Boolean variables $a$, $b$, $c$, and $d$ are not yet specified: they could be either 0 or 1. In this sense, they could also be viewed as random variables, which could take two possible values, at equal probability. This way, the entire expression is a random variable as well: it could be either 0 or 1.

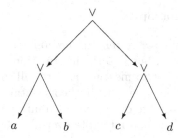

**Fig. 8.3.** A symmetric Boolean expression: $(a \vee b) \vee (c \vee d)$. Here, $a$, $b$, $c$, or $d$ are reached at probability $1/4$ each. Therefore, they have surprise $\log_2 4 = 2$ each. (This is why they are at level 2.) Thus, the entropy is $2/4 + 2/4 + 2/4 + 2/4 = 2$.

What is its Von-Neumann's entropy? Well, thanks to symmetry, $a$, $b$, $c$, and $d$ have the same status: each could be reached at probability $1/4$. Therefore, each has the same surprise:

$$-\log_2\left(\frac{1}{4}\right) = \log_2 4 = 2.$$

This can also be seen geometrically in Figure 8.3: $a$, $b$, $c$, and $d$ are all placed at level 2. We are now ready to calculate the entropy:

$$\text{entropy} = \sum_{a,b,c,d} \text{surprise} \cdot \text{probability} = \sum_{a,b,c,d} 2\frac{1}{4} = \frac{2}{4} + \frac{2}{4} + \frac{2}{4} + \frac{2}{4} = 2.$$

Is this good? Well, let's compare this to a more nonsymmetric tree.

### 8.5.2 Nonsymmetric Boolean Expression

Compare this to a more nonsymmetric expression:

$$a \vee (b \vee c) \vee d.$$

In what sense is this nonsymmetric? Well, it occupies an uneven tree (Figure 8.4). This is why it has a different entropy.

Indeed, at the bottom, $b$ and $c$ could be reached at probability $1/8$ each. This is why they have a big surprise:

$$-\log_2\left(\frac{1}{8}\right) = \log_2 8 = 3.$$

This can also be seen geometrically in Figure 8.4: both $b$ and $c$ are at level 3.

$a$, on the other hand, could be reached at a higher probability: $1/4$. This is why it has a lower surprise: 2. Finally, $d$ could be reached at probability $1/2$, and has surprise 1 only. We are now ready to calculate the entropy:

$$\text{entropy} = \frac{3}{8} + \frac{3}{8} + \frac{2}{4} + \frac{1}{2} = 1.75.$$

This is less than before. Why?

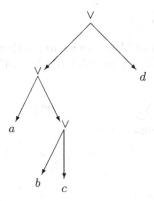

**Fig. 8.4.** An uneven Boolean expression: $a \lor (b \lor c) \lor d$. At the bottom, $b$ and $c$ could be reached at probability $1/8$ each. Therefore, they have a big surprise: $\log_2 8 = 3$. (This is why they are at level 3.) Furthermore, $a$ has probability $1/4$, and surprise 2. Finally, $d$ has probability $1/2$, and surprise 1 only. Thus, the entropy is $6/8 + 2/4 + 1/2 = 1.75$.

### 8.5.3 Computer Code and Its Entropy

Each Boolean expression could also be viewed as a computer code. In fact, each '$\lor$' could stand for an if-else question. On average, the former code (Figure 8.3) is expected to tell us more than the latter (Figure 8.4). This is why it has a higher entropy.

### 8.5.4 Before Running the Code

Before running the code, however, things were the other way around. Before running the former code, we probably knew less than before running the latter. This is why we needed a more informative code to tell us the missing news. Thus, the higher the entropy, the less we know before running the code (or, in physics, before running an experiment to measure the actual value of our random variable).

## 8.6 System and Its Entropy

### 8.6.1 System of $n$ Coins

Consider now a system of $n$ independent coins. Each coin is equally likely to show 0 (head) or 1 (tail). As a whole, the entire system shows a complete configuration of 0's and 1's, say

$$0110011101010100000111100000011110101010101011101001.$$

### 8.6.2 Configuration of Bits

This is a list of $n$ bits: 0 or 1. How likely are we to see this particular configuration? Well, the probability for this is $2^{-n}$. To check on this, let's sum up. Since there are $2^n$ different configurations,

$$\sum_{\text{all configurations}} 2^{-n} = 2^n 2^{-n} = 1,$$

as required.

### 8.6.3 Entropy of the System

We are now ready to calculate the Von-Neumann entropy of the entire system. Still, our system is discrete, not continuous. Therefore, we need to sum, not to integrate. For this purpose, scan the configurations, one by one. Each configuration contributes a product: probability times surprise. Fortunately, this is uniform — all contributions are the same:

$$\text{entropy} = -\sum_{\text{all configurations}} 2^{-n} \log_2 \left(2^{-n}\right)$$
$$= -2^n 2^{-n} \log_2 \left(2^{-n}\right)$$
$$= -\log_2 \left(2^{-n}\right)$$
$$= \log_2 \left(2^n\right)$$
$$= n.$$

How is this relevant to physics?

## 8.7 Application in Thermodynamics

### 8.7.1 Closed System: Box of Gas

How to use this in thermodynamics? Consider a box of gas of volume $V$. Assume that it is closed: no force or energy enters, and no force or energy leaks. Is this similar to a system of coins? It sure is. It is only a little bit more complicated: the molecules are not static but dynamic. At each individual time, each molecule can move at some velocity

$$v = \|v\| \cdot \frac{v}{\|v\|}.$$

This factors the velocity as a product: speed $\|v\|$ times direction $v/\|v\|$.

### 8.7.2 Random Variable: Kinetic Energy

What is the kinetic energy of one molecule? This is

$$\text{kinetic energy} = \text{mass} \cdot \frac{\|v\|^2}{2}.$$

This depends on speed only, not direction. Furthermore, all molecules have the same mass. Therefore, the probability to have a certain kinetic energy can be written as $p(\|v\|)$ (uniformly for all molecules alike). To serve as a legitimate probability function, it must have integral 1:

$$\int_0^\infty p(\|v\|)d\|v\| = 1.$$

This is an integral over all possible speeds: $0 \leq \|v\| < \infty$. Later on, we'll also write $p$ more directly, as a function of energy rather than speed.

### 8.7.3 Classical vs. Quantum Energy

In quantum mechanics, a particle may have a few possible energies at the same time, each at a certain probability. This is called superposition: a sum of a few legitimate possibilities. Later on, we'll see that this makes a few parallel universes, each uses one of these possible energies, and could be entered at some probability.

Here, on the other hand, things are much more calm and prosaic. At each individual time, each molecule has its own concrete energy. Still, since there are so many molecules, we don't care to measure their concrete energy, but leave this as an (unknown) random variable. This will give us more insight about energy and its behavior, in stochastic terms. This is not because we can't be more accurate, but because we don't want to: it won't serve any purpose, and won't tell us any more about the system and its physics.

In quantum mechanics, on the other hand, even nature remains puzzled: it can't make up its mind how much energy a particle has, and must leave it vague and uncertain: all energies are still possible, and are still included in the superposition. Only upon observation does nature decide what energy to assign, and what universe to enter at the next moment. We'll come back to this later. Now, let's go back to our (classical) box of gas.

### 8.7.4 The Edge

In nature, molecules are discrete. Still, mathematically, it makes sense to model them as a continuum. This way, we don't sum up but integrate over the entire box (and also over all possible speeds).

Here, one may ask: what about the edge of the box? After all, molecules that lie next to the edge are not free to move in all directions: they are confined to move inward only, to keep the system closed. Shouldn't their probability depend on $v/\|v\|$ too? Yes, it should, but this is negligible: this has no effect on the integral, since the edge has no volume at all in 3-D.

### 8.7.5 Bit vs. Molecule

Recall again our original bit. Could it model a real system in physics? Well, a bit is very simple: it takes only two possible values: 0 or 1. Thus, it has a constant probability:

$$p(0) = p(1) = \frac{1}{2}.$$

Although simple, it can still model a molecule of gas. Likewise, a system of $n$ bits, whose entropy is simply $n$ (Section 8.6.3), can also model a complete box of gas.

### 8.7.6 Entropy: Integration

Thus, we can expect the box of gas to have entropy proportional to the total number of molecules, or indeed to the volume. Indeed,

$$
\begin{aligned}
\text{entropy} &= -\int_0^\infty \int \int \int_{\text{box}} p(\|v\|) \log_2 \left( p(\|v\|) \right) d\|v\| dx dy dz \\
&= -\int_0^\infty p(\|v\|) \log_2 \left( p(\|v\|) \right) d\|v\| \int \int \int_{\text{box}} dx dy dz \\
&= -V \int_0^\infty p(\|v\|) \log_2 \left( p(\|v\|) \right) d\|v\|.
\end{aligned}
$$

Is this familiar? This is just a product: the volume times the entropy of one molecule. In particular, this is proportional to the volume, as expected. This shows once again how suitable our coins are to model molecules of gas in thermodynamics.

## 8.8 Additivity

### 8.8.1 Two Systems

Consider now two independent systems. On the first system, define one random variable, with value $x$. The probability to obtain this value is, say, $p_1(x)$. On the second system, define an independent random variable, with value $y$. The probability to obtain this value is, say, $p_2(y)$. How to design a joint system, with a joint random variable? Easy: consider a pair of elements, one from the first system, and another from the second system. On it, define a joint random variable, whose value is a pair of the form $(x, y)$. The probability to obtain this value is

$$
p(x, y) = p_1(x) p_2(y)
$$

(because the original random variables are independent). Is this a legitimate probability function? Let's check: it is indeed nonnegative, and has integral 1:

$$
\int \int p(x, y) dx dy = \int \int p_1(x) p_2(y) dx dy = \int p_1(x) dx \int p_2(y) dy = 1 \cdot 1 = 1.
$$

Now, what is the entropy of the joint system?

### 8.8.2 Entropy is Additive

Let's calculate the entropy of the joint system:

$$
\begin{aligned}
\text{entropy} &= -\int \int p(x, y) \log_2 \left( p(x, y) \right) dx dy \\
&= -\int \int p_1(x) p_2(y) \log_2 \left( p_1(x) p_2(y) \right) dx dy \\
&= -\int \int p_1(x) p_2(y) \left( \log_2 \left( p_1(x) \right) + \log_2 \left( p_2(y) \right) \right) dx dy
\end{aligned}
$$

$$= -\int\int p_1(x)p_2(y)\log_2\left(p_1(x)\right)dxdy - \int\int p_1(x)p_2(y)\log_2\left(p_2(y)\right)dxdy$$

$$= -\int p_2(y)dy\int p_1(x)\log_2\left(p_1(x)\right)dx - \int p_1(x)dx\int p_2(y)\log_2\left(p_2(y)\right)dy$$

$$= -\int p_1(x)\log_2\left(p_1(x)\right)dx - \int p_2(y)\log_2\left(p_2(y)\right)dy.$$

What is this? This is the sum: the entropy of the first system plus the entropy of the second one. Thus, entropy is additive. How to use this in practice?

### 8.8.3 Back to System of Coins

How to use additivity? Well, we've already used this implicitly in a system of coins.

Our original coin (or bit) is symmetric: it has only two possible values, each of probability $1/2$. This is why its entropy is

$$-\frac{1}{2}\log_2\left(\frac{1}{2}\right) - \frac{1}{2}\log_2\left(\frac{1}{2}\right) = \frac{1}{2}\log_2 2 + \frac{1}{2}\log_2 2 = \frac{1}{2} + \frac{1}{2} = 1.$$

Now, add another coin. This makes a system of two independent coins. Thanks to additivity, it has entropy 2. Next, add yet another coin, to make a system of three independent coins, with entropy 3, and so on. By mathematical induction, a system of $n$ coins indeed has entropy $n$, in agreement with Section 8.6.3.

### 8.8.4 Back to Box of Gas

Likewise, in a box of gas, entropy is proportional to the number of molecules. In fact, it is the number of molecules times the entropy of one molecule.

In nature, gas is made of discrete molecules. Mathematically, however, we often model it as a continuum in 3-D. In this model, each individual point stands for one molecule. Instead of a sum, we now have an integral in 3-D. This is why, in Section 8.7.6, we indeed integrated over the box. This gave us the entropy of the entire box as a product: volume times the entropy of one molecule. This is additivity again, in a continuum.

## 8.9 Laws of Thermodynamics

### 8.9.1 Energy is Conserved Even in an Open System

The laws of thermodynamics talk about a closed system (Chapter 6, Sections 6.3.4–6.3.5). On such a system, no external force acts from the outside. Moreover, the first law of thermodynamics may work even in an open system. Indeed, this law tells us that energy is conserved. Now, this is true not only in a closed system but also in an open system, provided that potential energy is also included. After all, in this case, the force is not really external: it enters through potential difference. (See exercises below.)

### 8.9.2 Closed System: Entropy Increases

The second law, on the other hand, works in a closed system only. It tells us that the entropy in the system can only increase, but never decrease. Why?

Well, we already know why. After all, more entropy means that our random variable is expected to give us more surprise and news. Thus, *before* running an experiment to measure the random variable, there was a lot of room for surprise, because the state probably contained only little information. In summary, more entropy means more uncertainty: less information in the original state. This makes sense: the system can't gain information from nothing, but only lose: get more and more disordered and vague. Nature seeks maximum entropy.

## 8.10 Maxwell's Demon

### 8.10.1 Closed System: Cup of Coffee

Or does it? Consider a cup of coffee. Assume that it is completely isolated, so no heat leaks from it to the air. This makes a closed system.

### 8.10.2 Entropy Increases

Initially, the coffee is hot, and the cup itself is colder. This means low entropy: the system must contain a lot of data to distinguish between the coffee and the cup, and tell us that only the coffee is hot, but the cup is not.

As time goes by, the coffee gets colder: heat flows from it to the cup, which gets warmer. This means more entropy: things get more and more balanced and gradual. In the middle, the coffee is still hot. Closer to the cup, on the other hand, the coffee is now colder than before. Information is now much more vague than before: no longer strictly hot or cold, but also warm. Reality is not black or white any more, but mostly gray. Entropy got higher.

### 8.10.3 Equilibrium: Maximum Entropy

Nature seeks higher and higher entropy. This is why the coffee gets colder and colder. The coffee in the middle never gets hot again. The system gets more and more balanced and uniform. Nature seeks equilibrium: maximum entropy. This is as uniform as a full perfect tree, with all leaves at the same level (Figure 8.3).

### 8.10.4 Maxwell's Demon

Or is it? Let's do a thought experiment. Imagine a demon who sits at the middle of our cup of coffee, and wants to violate the second law of thermodynamics. For this purpose, the demon looks at the molecules, and measures their speed. (Assume that this costs no energy.) Those of high speed are sent leftwards, and those of low speed are sent rightwards. In the end, our coffee gets half hot half cold: hot on the left, and cold on the right. We have now more information than before: a strict dichotomy between hot and cold coffee. Entropy has decreased! How come?

This is Maxwell's demon. Isn't this a paradox? How could entropy decrease? Isn't this in violation of the second law of thermodynamics?

## 8.11 Data Unit: Shannon's Bit

### 8.11.1 Maxwell's Demon and Data

So far, we assumed that Maxwell's demon needs no extra energy to look at the individual molecules, measure their speed, and decide whether to send them leftwards or rightwards. Still, he needs memory to record and store these data. After all, this is what entropy is all about: information.

This memory is part of our system. At some point, the demon will run out of memory, and will have to remove some old data, to make room for new data about new molecules and their speed. This is irreversible: the demon will "forget" this forever. At this time, some information will get lost forever: entropy will increase, as in the second law of thermodynamics.

### 8.11.2 Shannon's Bit

Shannon introduced a new unit of data: a bit: 0 or 1. When your computer runs a code, it may fetch data from the memory, calculate in the processor, and show the answer on the screen. So long as this process is reversible, entropy remains the same. When does entropy increase? When some old data is removed forever. Only then does the system lose information, or gain entropy.

## 8.12 Maxwell-Boltzmann Distribution

### 8.12.1 Random Variable: Energy

How is this used in nature? In Section 8.7.2, we've already seen a closed system: a box of gas. In such a system, the relevant random variable is (kinetic) energy. After all, molecules of the same energy are essentially the same. Now, the molecules move all the time. Still, since they have the same mass, their energy depends on speed only (not direction):

$$\text{energy} = \text{mass} \cdot \frac{\|v\|^2}{2}.$$

This is why we can write the probability to find a molecule of a certain energy as a (composite) function of speed: $p(\|v\|)$. Here, on the other hand, we take a more direct approach, and write the probability as a simple function of energy: $p(E)$.

### 8.12.2 Maxwell-Boltzmann Statistics

How should $p$ look like? This is the Maxwell-Boltzmann distribution (or statistics):

$$p(E) = \frac{\exp\left(-\frac{E}{K_B T}\right)}{K_B T}.$$

This is the probability to find a molecule of energy $E$, when the overall temperature is $T$ (in Kelvin's scale), and the average energy is therefore $K_B T$ (where $K_B$ is Boltzmann's constant). Thanks to this definition, $p$ is indeed a legitimate probability function: it is positive, and has integral 1, as required.

### 8.12.3 Surprise!

What is the surprise? For simplicity, let's use the natural logarithm:

$$-\log{(p(E))} = \frac{E}{K_B T} + \log K_B + \log T.$$

Look at the former term: $E/(K_B T)$. It tells us how surprised we should be to discover a molecule of energy $E$. The higher the energy, the lower the probability, and the more surprised we should be. After all, it is quite a pleasant surprise to find a fast molecule, particularly at a low temperature. Thus, it was indeed a good idea to define the distribution in this way in the first place.

## 8.13 Exercises: Open vs. Closed System

### 8.13.1 System: Open or Closed?

1. Consider an aircraft, dropping a bomb. Look at the bomb on its own. Does it make a closed system? Hint: no — it is not isolated, but attracted to the Earth.
2. Why? Hint: there is an outer force: gravity.
3. So, the bomb makes an open system. Why then is its total energy conserved? Hint: although gravity makes an outer force, its effect is still included in the system: the potential.
4. Now, look at the bomb and the Earth together. Do they make a closed system? Hint: yes, they attract each other.
5. In this system, what is the total momentum? Hint: zero. Indeed, while the bomb accelerates downwards, the entire Earth also accelerates upwards. Although this is tiny, the Earth has a very big mass. Thus, its momentum cancels the momentum of the bomb, leaving the total momentum at zero all the time.
6. Why? Hint: at each given time, the bomb is attracted down, and its momentum increases downwards. Thanks to Newton's third law, there is a reaction: the entire Earth is attracted up, and its momentum increases upwards.
7. Write this in a more general (differential) form. Hint: let the momentum $p$ and the force $F$ be vector functions, defined in a closed system $V \subset \mathbb{R}^3$. Since the total force vanishes, the total momentum is conserved:

$$\frac{d}{dt} \int \int \int_V p\, dx dy dz = \int \int \int_V \frac{dp}{dt} dx dy dz = \int \int \int_V F\, dx dy dz = \mathbf{0} \in \mathbb{R}^3.$$

The Binomial Formula
and Quantum Statistical Mechanics

# The Binomial Formula and Quantum Statistical Mechanics

As discussed above, a binary tree grows exponentially: the next level contains twice as many nodes. This is quite a rapid growth. This is why it is useful to model instability.

There is, however, a more moderate growth. Pascal's triangle, for example, grows only linearly. Indeed, the next level contains just one more cell. Still, even in this moderate growth, there is a hidden exponential growth. Indeed, in Pascal's triangle, one could draw all sorts of geometrical paths. Together, these paths make a familiar binary tree. This will lead to the (extended) binomial formula, useful in quantum statistical mechanics.

In this part, we introduce the binomial formula from two points of view: algebraic and geometrical. Indeed, thanks to mathematical induction, we get to learn a lot about the binomial coefficients and their algebraic properties. Still, they also have a geometrical face: they are mirrored by geometrical paths in Pascal's triangle. This leads to an interesting application in statistical physics: Brownian motion.

Furthermore, in its extended form, the binomial formula is useful in quantum mechanics as well. Indeed, thanks to conservation of mass and energy, we can go ahead and estimate the number of particles in each energy level. This leads to three types of distributions: molecules are distributed in Maxwell-Boltzmann statistics, bosons in Bose-Einstein statistics, and electrons in Fermi-Dirac statistics.

# 9

# Newton's
# Binomial and Trinomial
# Formulas

So far, we've used binary trees to introduce chaos theory from a geometrical point of view. We've seen that the (infinite) binary tree is transparent from below. Indeed, at its bottom, it approaches a fractal: a Cantor set, of dimension less than 1. This helps model a bifurcation tree in chaos theory.

Here, on the other hand, we use binary trees for yet another purpose: count geometrical paths in Pascal's triangle. This leads to two important formulas: the binomial and trinomial formulas. Later on, we'll extend them even more, and use them in physics.

What is a binary tree? It is a good example of a multilevel hierarchy, with an exponential growth. Indeed, the next level contains twice as many nodes (and branches).

In this chapter, on the other hand, we study a more modest growth: linear. Indeed, in Pascal's triangle, the next lower level contains just one more cell. Still, we soon get our good old binary tree back again: in Pascal's triangle, one could draw all sorts of geometrical paths. Together, they make a new binary tree, with its familiar exponential growth. This way, we can sort these paths, and count them. This leads to an important combinatorial result: Newton's binomial formula. To show how useful it is, we use it in a stochastic process in statistical physics.

This approach gives a lot of geometrical insight. Still, there is also a more algebraic proof, using mathematical induction only. This approach is more general: it helps extend the original binomial coefficient, and develop yet more advanced formulas. Later on, we'll use them in quantum mechanics and relativity.

## 9.1 Geometrical Point of View

### 9.1.1 Pascal's Triangle

The binomial formula is most useful in quantum mechanics and relativity. To introduce it properly, we need some geometrical-combinatorial background. For this purpose, we start from Pascal's triangle.

Pascal's triangle is made of oblique cells, listed row by row (Figure 9.1). Later on, each cell will contain a number (entry).

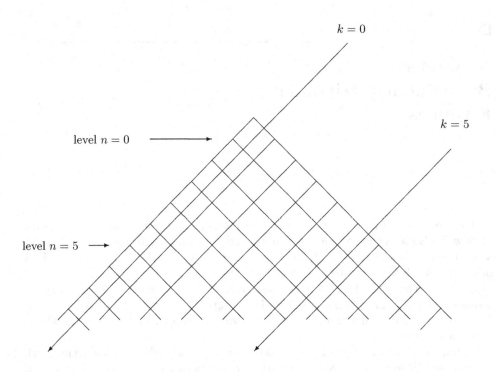

**Fig. 9.1.** Pascal's triangle: to index the levels (or rows), use $n = 0, 1, 2, \ldots$. To index the individual cells in the $n$th level, use $k = 0, 1, 2, \ldots, n$.

How are the cells indexed? By two indices: $n$ and $k$. The rows (or levels) are indexed by $n = 0, 1, 2, 3, \ldots$, top to bottom. For example, the zeroth level (at the top) contains just one cell (and one entry in it). The next lower level, on the other hand, contains two cells, indexed by $k = 0, 1$. The next lower level, on the other hand, contains three cells, indexed by $k = 0, 1, 2$, and so on.

How to define the entries in the cells? By mathematical induction, of course! In the top cell, place the entry 1 (Figure 9.2). This fills the zeroth level, as required. Now, for $n = 0, 1, 2, 3, \ldots$, assume that the $n$th level has already been filled with entries, as required. (This is the induction hypothesis.) How to fill the $(n + 1)$th level just below it?

This is done as follows. In each cell in the $(n + 1)$th level, place the sum of two entries: one from the upper-left cell, and the other from the upper-right cell.

Clearly, there are two exceptions. The leftmost cell has no upper-left neighbor. Therefore, the entry in it must be the same as the upper-right entry: 1. Likewise, the rightmost cell has no upper-right neighbor. Thus, the entry in it must be the same as the upper-left entry: 1.

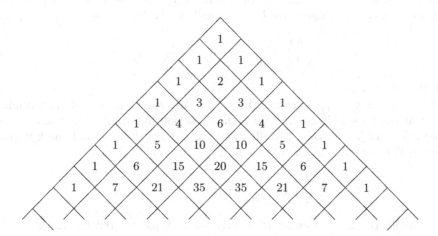

**Fig. 9.2.** The entries in Pascal's triangle: each entry is the sum of the upper-left and upper-right entries.

So, at all levels, the leftmost and rightmost entries are just 1. The intermediate entries, on the other hand, are greater than 1. This can be proved easily by mathematical induction.

The above definition is constructive: it actually offers a practical algorithm to fill Pascal's triangle with entries: recursively, level by level. Still, what if we only need one particular entry? Is it possible to calculate it on its own, without filling the entire triangle?

### 9.1.2 The Binomial Coefficient

How to calculate an individual entry, without calculating all the others? For this purpose, recall that the factorial function is defined recursively by

$$n! \equiv \begin{cases} 1 & \text{if } n = 0 \\ (n-1)! \cdot n & \text{if } n > 0. \end{cases}$$

In Pascal's triangle, how are the cells indexed? Recall: the levels are indexed top to bottom: $n = 0, 1, 2, 3, \ldots$ In each particular level, the cells are indexed left to right: $k = 0, 1, 2, 3, \ldots, n$. This way, each cell is indexed by two indices: $n$ and $k$ ($n \geq k \geq 0$). Let's use these indices.

In the $n$th level, let's focus on the $k$th entry (from the left). How does it look like algebraically? Let's prove that it looks like this:

$$\binom{n}{k} \equiv \frac{n!}{k!(n-k)!}.$$

This is the binomial coefficient.

How to prove this? By mathematical induction, of course! Indeed, at the zeroth level, where $n = 0$, we must have $k = 0$ as well, so the formula is indeed correct:

$$\binom{0}{0} = \frac{0!}{0! \cdot (0-0)!} = \frac{1}{1 \cdot 1} = 1,$$

as in Pascal's triangle.

Now, for $n = 1, 2, 3, \ldots$, assume that we already know that the above formula is correct at the $(n-1)$th level. (This is the induction hypothesis.) Is it also correct at the $n$th level just below it? Well, let's look at the $n$th level, and check it entry by entry. For the leftmost entry, for which $k = 0$, we indeed have

$$\binom{n}{0} = \frac{n!}{0!(n-0)!} = \frac{n!}{1 \cdot n!} = 1,$$

as in Pascal's triangle. Likewise, let's check the rightmost entry, for which $k = n$:

$$\binom{n}{n} = \frac{n!}{n!(n-n)!} = \frac{n!}{n! \cdot 1} = 1,$$

as in Pascal's triangle. Finally, let's check those intermediate entries, for which $0 < k < n$. Fortunately, we already know that the formula holds at the previous level just above. After all, this is the induction hypothesis. Now, in the $n$th level, how was the $k$th entry defined in the first place? It was defined as the sum of the upper-left and upper-right entries:

$$\binom{n-1}{k-1} + \binom{n-1}{k} = \frac{(n-1)!}{(k-1)!(n-k)!} + \frac{(n-1)!}{k!(n-k-1)!}$$

$$= \frac{k}{n} \cdot \frac{n!}{k!(n-k)!} + \frac{n-k}{n} \cdot \frac{n!}{k!(n-k)!}$$

$$= \frac{k+n-k}{n} \cdot \binom{n}{k}$$

$$= \binom{n}{k},$$

as the formula indeed says. This shows that the formula is correct at the $n$th level as well. This completes the induction step.

Thus, the formula is true in general: each individual entry is the same as the corresponding binomial coefficient.

### 9.1.3 Geometrical Paths in Pascal's Triangle

In Pascal's triangle, consider a path: start from the top, and "jump" from level to level, down the triangle. In each "jump," move to the next lower level: either down-left, or down-right (Figures 9.3–9.4).

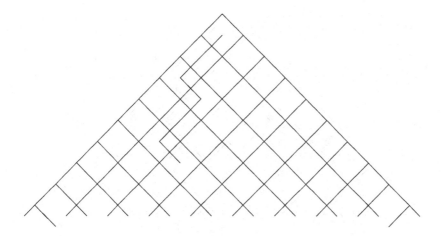

**Fig. 9.3.** A path leading to cell $k = 2$ in level $n = 6$ could be modeled by the binary vector $(0, 0, 1, 0, 0, 1)$: the third and sixth moves are down-right, and the rest are down-left.

Now, in the $n$th level, focus on the $k$th cell from the left ($0 \leq k \leq n$). How many paths lead to it? Well, the answer is already written in the cell! To prove this, use mathematical induction, and the original definition of the entries. Indeed, to reach our cell, the path must approach it, either from the upper-left, or from the upper-right.

How many paths approach from the upper-left, and how many from the upper-right? Fortunately, thanks to the induction hypothesis, the answers are already written in the previous level. So, we just need to sum the upper-left and upper-right entries (if exist), to obtain the correct answer. Fortunately, this is exactly what is already written in the cell.

## 9.2 Combinatorial Point of View

### 9.2.1 Binary Vectors

Not convinced? Let's prove this in yet another way: not only geometrically, but also combinatorially. A path as above contains $n$ moves: either down-right, or down-left. Let's index them from 1 to $n$:

$$v_1, \ v_2, \ v_3, \ \ldots, \ v_n.$$

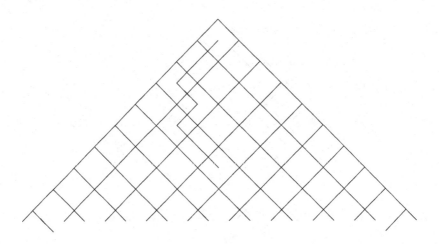

**Fig. 9.4.** A path leading to cell $k = 3$ in level $n = 6$ could be modeled by the binary vector $(0, 0, 1, 0, 1, 1)$: the third, fifth, and sixth moves are down-right, and the rest are down-left.

For $1 \le i \le n$, the $i$th move is modeled by

$$v_i \equiv \begin{cases} 1 & \text{if the } i\text{th move is down-right} \\ 0 & \text{if the } i\text{th move is down-left.} \end{cases}$$

This way, the entire path is modeled by a binary $n$-dimensional vector: a zero component stands for a down-left move, whereas 1 stands for a down-right move (Figures 9.3–9.4).

There is one degenerate case: for $n = 0$, the empty vector $\emptyset$ models the static "path," with no move at all. Let's look at more interesting cases, with $n > 0$.

### 9.2.2 Subsets

Let's get some more insight from set theory. Indeed, what is this binary vector? It could be interpreted as a characteristic function, which maps $n$ natural numbers, either to 0 or to 1:

$$v: \{1, 2, 3, \ldots, n\} \to \{0, 1\}.$$

Which numbers are mapped to 1? In other words, what is the origin of 1? This makes a new subset of the form

$$S \subset \{1, 2, 3, \ldots, n\},$$

containing those $i$'s for which $v_i \neq 0$:

$$i \in S \iff v_i = 1, \quad 1 \leq i \leq n.$$

In Figure 9.3, for example,

$$v = (0, 0, 1, 0, 0, 1),$$

so

$$S = \{3, 6\} \subset \{1, 2, 3, 4, 5, 6\}.$$

In Figure 9.4, on the other hand,

$$v = (0, 0, 1, 0, 1, 1),$$

so

$$S = \{3, 5, 6\} \subset \{1, 2, 3, 4, 5, 6\}.$$

Our original question was geometrical: how many paths lead to the $k$th cell in the $n$th level? Thanks to the above, it is now combinatorial: how many binary vectors have $k$ 1's and $n - k$ 0's? Or, in set-theory language, what is the cardinality (size) of the set

$$\{v \equiv (v_1, v_2, \ldots, v_n) \in \{0, 1\}^n \mid v_1 + v_2 + \cdots + v_n = k\}?$$

Or, in terms of subsets, what is the cardinality of the set

$$\{S \subset \{1, 2, 3, \ldots, n\} \mid |S| = k\}?$$

### 9.2.3 Subsets and the Binomial Coefficient

In Section 9.1.3, we've already studied Pascal's triangle, and counted the geometrical paths in it. Thanks to the language of subsets, we can now do this in yet another way, and obtain the same result once again: the binomial coefficient.

In fact, our original question takes now a pure combinatorial form: how many different subsets of the form

$$S \equiv \{s_1, s_2, s_3, \ldots, s_k\}$$

could be extracted from $\{1, 2, 3, \ldots, n\}$? Well, to design such a subset, we need to pick $k$ different natural numbers, not larger than $n$.

How to do this? Well, let's pick $s_1$. It could be either 1, or 2, or 3, or ..., or $n$. So, there are $n$ options to pick $s_1$.

For each such an option, there are now $n - 1$ options to pick $s_2$. After all, $s_2$ could be either smaller or larger than $s_1$, but not the same. So, in total, there are $n(n-1)$ options to pick $s_1$ and $s_2$.

Or are there? After all, in Figure 9.3,

$$\{3, 6\} = \{6, 3\}$$

are the same subset. There is no need to count it twice.

So, the above count contains some duplication. To avoid this, we must divide by 2. This means that we adopt a new convention: write $\{s_1, s_2\}$ in a unique writing style, in which the elements are written from small to large:

$$\{3,6\} \quad \text{rather than} \quad \{6,3\}.$$

Thanks to this convention, there is now no duplication any more, so there are only

$$\frac{n(n-1)}{2} = \binom{n}{2}$$

different options to pick

$$s_1 < s_2,$$

to help design each subset of the form $\{s_1, s_2\}$ only once, not twice.

For each such option, we now have $n - 2$ options to pick $s_3$. After all, $s_3$ could be either smaller than $s_1$, or larger than $s_2$, or in between $s_1$ and $s_2$, but mustn't be equal to $s_1$ or $s_2$.

Unfortunately, there is still some duplication. For example, to design the path in Figure 9.4, one should pick

$$\{s_1, s_2, s_3\} = \{3,6,5\} = \{3,5,6\} = \{5,6,3\}.$$

In other words, one could pick

$$s_1 = 3 \quad \text{and} \quad s_2 = 6$$

(as in Figure 9.3), and then

$$s_3 = 5.$$

But this is not the only way: one could equally well pick

$$s_1 = 3, \quad s_2 = 5, \quad \text{and} \quad s_3 = 6,$$

or

$$s_1 = 5, \quad s_2 = 6, \quad \text{and} \quad s_3 = 3.$$

To avoid this kind of duplication, stick to our convention: use the unique option in which

$$s_1 < s_2 < s_3,$$

and drop the other two. With no duplication, there are just

$$\frac{n(n-1)}{2} \cdot \frac{n-2}{3} = \frac{n!}{(n-3)!} \cdot \frac{1}{3!} = \binom{n}{3}$$

different options to design $\{s_1, s_2, s_3\}$.

By repeating this process (or by mathematical induction on $k = 1, 2, 3, \ldots, n$), we have

$$\binom{n}{k-1} \cdot \frac{n-k+1}{k} = \frac{n!}{(k-1)!(n-k+1)!} \cdot \frac{n-k+1}{k} = \frac{n!}{k!(n-k)!} = \binom{n}{k}$$

different options to pick

$$s_1 < s_2 < s_3 < \cdots < s_k,$$

to help design each subset of the form $\{s_1, s_2, \ldots, s_k\}$ exactly once: uniquely, as required.

### 9.2.4 Monotonically Increasing Sequences

By now, we've counted how many sequences of the form

$$s_1 < s_2 < s_3 < \cdots < s_k$$

could be extracted from $\{1, 2, 3, \ldots, n\}$. Still, these strong inequalities are a little inconvenient: the $s_i$'s must lie in different intervals:

$$1 \le s_1 \le n - k + 1$$
$$2 \le s_2 \le n - k + 2$$
$$3 \le s_3 \le n - k + 3$$
$$\vdots$$
$$k \le s_k \le n.$$

How to transform them to new numbers that do lie in the same interval? For this purpose, let's define the new numbers

$$t_1 \equiv s_1 - 1$$
$$t_2 \equiv s_2 - 2$$
$$t_3 \equiv s_3 - 3$$
$$\vdots$$
$$t_k \equiv s_k - k.$$

Fortunately, the $t_i$'s do lie in the same interval:

$$0 \le t_1 \le n - k$$
$$0 \le t_2 \le n - k$$
$$0 \le t_3 \le n - k$$
$$\vdots$$
$$0 \le t_k \le n - k.$$

Furthermore, they make a weakly-increasing sequence:

$$0 \le t_1 \le t_2 \le t_3 \le \cdots \le t_k \le n - k.$$

Fortunately, this transformation is invertible. Indeed, given a weakly-increasing sequence like this, we can always produce a strongly-increasing sequence from it, by defining

$$s_1 \equiv t_1 + 1$$
$$s_2 \equiv t_2 + 2$$
$$s_3 \equiv t_3 + 3$$
$$\vdots$$
$$s_k \equiv t_k + k.$$

To simplify, let's introduce a new number:

$$m \equiv n - k.$$

(What is the geometrical nature of $m$? To see this, see the exercises below.) This way, both $m$ and $k$ have the same status. Indeed, since $n \geq k \geq 0$, both $m$ and $k$ must be nonnegative:

$$m, k \geq 0.$$

Thanks to this new definition, our weakly-increasing sequence lies in the new interval $[0, m]$:

$$0 \leq t_1 \leq t_2 \leq t_3 \leq \cdots \leq t_k \leq m.$$

How many such sequences could be extracted from $\{0, 1, 2, 3, \ldots, m\}$? We already know the answer:

$$\binom{m + k}{k}.$$

Does this formula work even in degenerate cases? Yes, it does! For $k = 0$, there is just one possible sequence: the empty sequence $\emptyset$. This goes well with the formula. For $m = 0$, there is just one possible sequence:

$$0 \leq 0 \leq 0 \leq \cdots \leq 0.$$

This agrees hand-in-hand with the formula as well.

## 9.3 Geometrical Problem

### 9.3.1 Discrete Simplex

By now, we've managed to avoid strong (or strict) inequalities, and replace them by weak inequalities. This way, our new $t_i$'s lie in a more uniform interval: $[0, m]$. Still, by now, the discussion got too combinatorial. How to make it geometrical again, easy to illustrate and visualize?

To answer this question, let's split it into two subquestions. First, how to get rid of the inequalities altogether? Second, our $t_i$'s are still extracted from an existing set: $\{0, 1, 2, 3, \ldots, m\}$. How to get rid of this requirement altogether?

For this purpose, let's define new differences:

$$d_1 \equiv t_1$$
$$d_2 \equiv t_2 - t_1$$
$$d_3 \equiv t_3 - t_2$$
$$\vdots$$
$$d_k \equiv t_k - t_{k-1}.$$

What algebraic conditions must the $d_i$'s satisfy? First, they must be nonnegative:

$$d_i \geq 0, \quad 1 \leq i \leq k.$$

Next, their sum mustn't exceed $m$. After all, it is equal to $t_k$:

$$\sum_{i=1}^{k} d_i = d_1 + d_2 + d_3 + \cdots + d_k = t_k \leq m.$$

Thus, the $t_i$'s have been transformed to the new $d_i$'s. This is an invertible transformation. After all, given $d_i$'s that satisfy the above conditions, one could always calculate their partial sums, and obtain a new weakly-increasing sequence:

$$t_1 \equiv d_1$$
$$t_2 \equiv d_1 + d_2$$
$$t_3 \equiv d_1 + d_2 + d_3$$
$$\vdots$$
$$t_k \equiv d_1 + d_2 + d_3 + \cdots + d_k.$$

In summary, we got what we wanted: instead of our original $t_i$'s, we can now work with the new $d_i$'s, which are no longer extracted from any existing set, but just satisfy two geometrical conditions:

$$d_i \geq 0, \quad 1 \leq i \leq k,$$

and

$$\sum_{i=1}^{k} d_i \leq m.$$

In what sense are these conditions geometrical? To see this, just place the $d_i$'s in a new $k$-dimensional vector:

$$(d_1, d_2, d_3, \ldots, d_k) \in \left(\mathbb{Z}^+\right)^k,$$

where

$$\mathbb{Z}^+ \equiv \mathbb{N} \cup \{0\}$$

contains the nonnegative integer numbers.

Where is this new vector? Well, it is in the $k$-dimensional infinite grid. Furthermore, its components are nonnegative, and their sum is at most $m$. In summary, our vector is in the discrete simplex of size $m$:

$$\left\{ (d_1, d_2, \ldots, d_k) \in \left(\mathbb{Z}^+\right)^k \ \Big| \ \sum_{i=1}^{k} d_i \leq m \right\}.$$

### 9.3.2 Discrete Triangle

How does the discrete simplex look like? For $k > 3$, this is not easy to see. For $k = 2$, on the other hand, this is just a discrete triangle. In Figure 9.5, we illustrate a discrete triangle of size $m = 3$: each edge contains just four grid points.

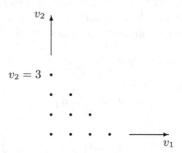

**Fig. 9.5.** The case $k = 2$: a discrete triangle of size $m = 3$. It is placed in the two-dimensional $v_1$-$v_2$ grid. Inside the triangle, we have $v_1 \geq 0$, $v_2 \geq 0$, and $v_1 + v_2 \leq m$.

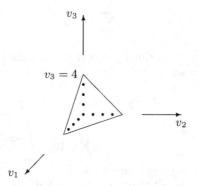

**Fig. 9.6.** The case $k = 3$: a discrete tetrahedron of size $m = 4$. It is placed in the three-dimensional $v_1$-$v_2$-$v_3$ grid. Inside the tetrahedron, we have $v_1 \geq 0$, $v_2 \geq 0$, $v_3 \geq 0$, and $v_1 + v_2 + v_3 \leq m$.

### 9.3.3 Discrete Tetrahedron

For $k = 3$, on the other hand, this is a discrete tetrahedron. In Figure 9.6, we illustrate a discrete tetrahedron of size $m = 4$: each edge contains five grid points.

Let's go ahead and count the points in a discrete triangle or tetrahedron. Better yet, let's consider a discrete simplex of just any dimension, and count the points in it.

### 9.3.4 How Many Points?

How many points are there in the discrete $k$-dimensional simplex of size $m$? We already know the answer:

$$\left| \left\{ (d_1, d_2, \ldots, d_k) \in (\mathbb{Z}^+)^k \mid \sum_{i=1}^{k} d_i \leq m \right\} \right| = \binom{m+k}{k}.$$

Does this formula work even in degenerate cases? Let's check: for $k = 0$, there is just one possible vector: the empty vector $\emptyset$, with no components at all. This goes well with the formula. For $m = 0$, there is just one possible vector — the origin:

$$(0, 0, 0, \ldots, 0).$$

This agrees with the formula as well.

### 9.3.5 Subsimplex of Lower Dimension

Thanks to the discrete simplex, we have a geometrical insight back again. Why is this so important? Because it gives us not only a better intuition but also a practical bonus. This is relevant for every nontrivial simplex, in one dimension or more: $k \geq 1$.

In our discrete simplex, the $d_i$'s must sum up to $m$ or less. Here, on the other hand, they sum up to $m$ exactly:

$$d_1 + d_2 + d_3 + \cdots + d_k = m.$$

What does this mean geometrically? Well, instead of looking at the entire simplex, look at its upper face only.

Is this a new simplex in its own right? Yes, it is! Indeed, what are we doing here? We just assume that the last coordinate, $d_k$, can no longer be picked freely. On the contrary: it must be defined as

$$d_k \equiv m - \sum_{i=1}^{k-1} d_i.$$

Fortunately, we are still free to pick $d_1, d_2, \ldots, d_{k-1}$. These should satisfy two geometrical conditions:

$$d_i \geq 0, \quad 1 \leq i \leq k - 1,$$

and

$$\sum_{i=1}^{k-1} d_i \leq m.$$

In what sense are these conditions geometrical? Well, they make a familiar geometrical structure: a discrete $(k-1)$-dimensional simplex of size $m$. In summary, all we did here amounts to reducing the dimension from $k$ to $k - 1$.

How many points are there in this new subsimplex? Well, we already know the answer:

$$\left| \left\{ (d_1, d_2, \ldots, d_k) \in \left( \mathbb{Z}^+ \right)^k \mid \sum_{i=1}^{k} d_i = m \right\} \right|$$

$$= \left| \left\{ (d_1, d_2, \ldots, d_{k-1}) \in \left( \mathbb{Z}^+ \right)^{k-1} \mid \sum_{i=1}^{k-1} d_i \le m \right\} \right|$$

$$= \binom{m + k - 1}{k - 1}.$$

### 9.3.6 Splitting the Discrete Simplex

Thanks to the above bonus, we can now split (or slice) the original discrete simplex into $m + 1$ disjoint subsimplices (or oblique slices), and obtain a new combinatorial formula:

$$\binom{m + k}{k}$$

$$= \left| \left\{ (d_1, d_2, \ldots, d_k) \in \left( \mathbb{Z}^+ \right)^k \mid \sum_{i=1}^{k} d_i \le m \right\} \right|$$

$$= \left| \cup_{j=0}^{m} \left\{ (d_1, d_2, \ldots, d_k) \in \left( \mathbb{Z}^+ \right)^k \mid \sum_{i=1}^{k} d_i = j \right\} \right|$$

$$= \sum_{j=0}^{m} \left| \left\{ (d_1, d_2, \ldots, d_k) \in \left( \mathbb{Z}^+ \right)^k \mid \sum_{i=1}^{k} d_i = j \right\} \right|$$

$$= \sum_{j=0}^{m} \binom{j + k - 1}{k - 1}.$$

This is indeed how our new geometrical insight is used to develop a new interesting combinatorial formula. In the exercises below, we'll prove it in yet another way. Now, let's see a few applications in physics.

## 9.4 Application in Statistical Physics

### 9.4.1 Brownian Motion

The binomial coefficients, and the geometrical paths associated with them, can be used in yet another interesting application: Brownian motion. Consider a particle confined to the real axis, moving step by step. In each step, the particle "jumps" along the real axis: either rightwards or leftwards.

In the beginning, the particle starts from the origin: 0. In each step, it moves exactly one unit: either rightwards (from some $l$ to $l + 1$) or leftwards (from $l$ to $l - 1$).

The process is random (or stochastic, or nondeterministic): in each individual step, we don't know in what direction the particle will jump, or where it will land. Fortunately, we still do know how *likely* it is to move in each direction. To move leftwards, the probability is $a$ (for some fixed $0 \le a \le 1$). To move rightwards, on the other hand, the probability is $b$ (for some fixed $0 \le b \le 1$). Since there is no other option, the probabilities must sum up to 1:

$$a + b = 1.$$

After all, the particle must move, either rightwards, or leftwards. We'll come back to this later, in our introduction to quantum mechanics.

The interesting question is: where would the particle be after $n$ steps? Of course, we can never tell this for sure. Fortunately, we can still tell how likely it is to be at any specific point on the real axis.

In $n$ steps, the particle must make $k$ leftward moves and $n - k$ rightward moves, for some $0 \le k \le n$. Thus, after $n$ steps, the particle could reach the point

$$-k + (n - k) = n - 2k$$

on the real axis. How likely is the particle to get there? We already know the answer: there are $\binom{n}{k}$ distinct ways to pick $k$ leftward moves and $n - k$ rightward moves (Section 9.1.3). Furthermore, since the steps are independent of each other, all these ways have the same probability: $a^k b^{n-k}$. Thus, after $n$ steps, the particle would reach $n - 2k$ at probability

$$\binom{n}{k} a^k b^{n-k}.$$

These probabilities must sum up to 1. After all, after $n$ steps, the particle must lie somewhere. Later on, we'll also prove this algebraically.

### 9.4.2 Symmetric Diffusion

Consider, for example, the symmetric case, in which the particle is equally likely to move rightwards or leftwards:

$$a = b = \frac{1}{2}.$$

In this case, the above process models a symmetric one-dimensional diffusion.

After $n$ steps, how likely is the particle to reach $n - 2k$? The above formula already tells us the answer:

$$\binom{n}{k} \left(\frac{1}{2}\right)^k \left(\frac{1}{2}\right)^{n-k} = \frac{1}{2^n} \binom{n}{k}.$$

This distribution is illustrated in three instances: $n = 5$, 6, and 7 (Figures 9.7–9.9).

The nonsymmetric case $a < b$, on the other hand, models convection-diffusion, with a slight wind, blowing rightwards. Likewise, the case $a > b$ models a wind blowing leftwards.

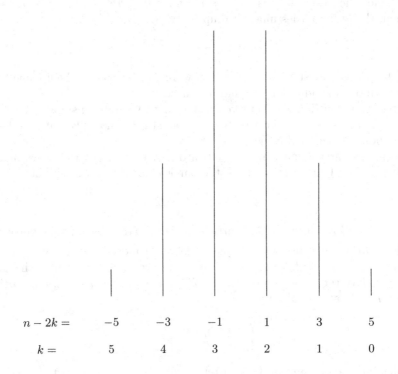

| $n - 2k =$ | $-5$ | $-3$ | $-1$ | $1$ | $3$ | $5$ |
|---|---|---|---|---|---|---|
| $k =$ | $5$ | $4$ | $3$ | $2$ | $1$ | $0$ |

**Fig. 9.7.** Symmetric Brownian motion ($a = b = 1/2$): distribution after $n = 5$ steps. The columns in the diagram stand for the probability of the particle to reach the point $n - 2k$ ($0 \le k \le n$) after $n = 5$ steps. (This requires $k$ leftward and $n - k$ rightward moves.)

### 9.4.3 Closed System: Entropy Must Increase

In a closed physical system, entropy measures the amount of uniformity: to what extent everything is the same? For example, in the above model, assume that all particles are initially concentrated at the origin: 0. This means minimum entropy. Indeed, there is no uniformity at all: the origin, on one hand, is full of matter, whereas the rest of the real axis is completely empty.

Afterwards, thanks to diffusion, things get better: the particles start to move, rightwards or leftwards, independent of each other. The distribution of the particles gets gradually more and more spread out, as in Figures 9.7–9.9. This illustrates how entropy indeed increases monotonically: the diagrams get more and more spread out, uniform, and flat.

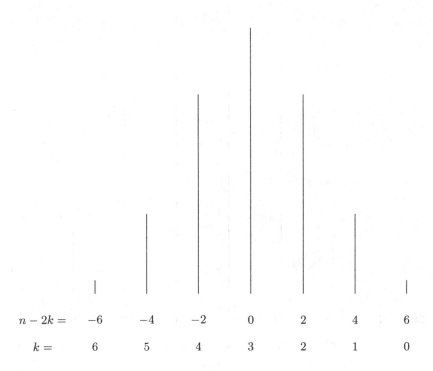

**Fig. 9.8.** Symmetric Brownian motion: distribution after $n = 6$ steps. The columns stand for the probability to reach the point $n - 2k$ after $n = 6$ steps.

### 9.4.4 Big Bang and Inflation

This is indeed the second law of thermodynamics: on average, entropy must always increase. At the very beginning of the universe, at the big bang, entropy was at its minimum: all matter and energy had been concentrated at one singular point. Since then, on average, entropy has only increased.

In one theory, in the inflation that followed, the universe had expanded exponentially fast. Afterwards, matter and energy continued to spread out, increasing entropy even more.

Still, not everything expanded. Thanks to gravity, when the universe cooled down, stars and galaxies formed. Still, this didn't decrease entropy. On the contrary: this generated more and more heat, spreading energy around, and increasing entropy even more on average.

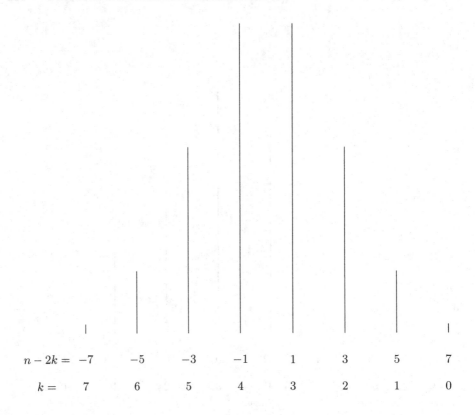

| $n - 2k =$ | $-7$ | $-5$ | $-3$ | $-1$ | $1$ | $3$ | $5$ | $7$ |
|---|---|---|---|---|---|---|---|---|
| $k =$ | $7$ | $6$ | $5$ | $4$ | $3$ | $2$ | $1$ | $0$ |

**Fig. 9.9.** Symmetric Brownian motion: distribution after $n = 7$ steps. The columns stand for the probability to reach the point $n - 2k$ after $n = 7$ steps.

## 9.5 The Binomial Formula: Mathematical Induction

### 9.5.1 Newton's Binomial Formula

By now, we've used the binomial coefficients for a combinatorial purpose: counting increasing sequences (or subsets). This task also has a geometrical version: counting geometrical paths in Pascal's triangle, or points in a discrete simplex. This leads to our main objective: the binomial formula.

Let $n$ be a nonnegative integer number. Let $a$ and $b$ be some numbers. Consider the $n$th power of $a + b$:

$$(a + b)^n \equiv (a + b)(a + b)(a + b) \cdots (a + b) \quad (n \text{ times}).$$

In this product, there are $n$ factors of the same form: $a + b$. In other words, the same factor $a + b$ multiplies $n$ times.

How to open parentheses? From each factor of the form $a + b$, pick either $a$ or $b$. This way, you must pick $k$ $a$'s and $n - k$ $b$'s (for some $0 \leq k \leq n$). Upon opening parentheses, this produces a new product: $a^k b^{n-k}$.

How many times is this product produced? In other words, how many different ways are there to pick $k$ $a$'s from $k$ factors of the form $a + b$? Well, each such way can be modeled by a distinct $n$-dimensional binary vector: 1 stands for picking $a$, and 0 for picking $b$. As we've seen above, there are

$$\left| \left\{ v \in \{0,1\}^n \ \Big| \ \sum_{i=1}^{n} v_i = k \right\} \right| = \binom{n}{k}$$

such vectors.

Thus, once the parentheses open, each product of the form $a^k b^{n-k}$ appears $\binom{n}{k}$ times in the resulting sum:

$$(a + b)^n = \sum_{k=0}^{n} \binom{n}{k} a^k b^{n-k}.$$

This is Newton's binomial formula. Not convinced? In a moment, we'll prove this more directly.

How to calculate these binomial coefficients in practice? Easy: just fill Pascal's triangle, level by level, until the $n$th level. For this purpose, you only need to add numbers, and never multiply. This is rather cheap and efficient.

### 9.5.2 Mathematical Induction in the Binomial Formula

To prove Newton's binomial formula, we've studied the binomial coefficients, gaining a lot of insight about their geometrical and combinatorial nature. Still, there is yet another approach. After all, we are now much more experienced in mathematical induction. Why not use it directly?

Indeed, for $n = 0$, the binomial formula is true:

$$(a + b)^0 = 1 = \binom{0}{0} a^0 b^0 = \sum_{k=0}^{0} \binom{0}{k} a^k b^{0-k}.$$

Now, for $n = 1, 2, 3, \ldots$, assume that the induction hypothesis holds:

$$(a + b)^{n-1} = \sum_{k=0}^{n-1} \binom{n-1}{k} a^k b^{n-1-k}.$$

Using this hypothesis and the formula in Section 9.1.2, we can now make the induction step:

$$(a + b)^n = (a + b)(a + b)^{n-1}$$

$$= (a + b) \sum_{k=0}^{n-1} \binom{n-1}{k} a^k b^{n-1-k}$$

$$= a \sum_{k=0}^{n-1} \binom{n-1}{k} a^k b^{n-1-k} + b \sum_{k=0}^{n-1} \binom{n-1}{k} a^k b^{n-1-k}$$

$$= \sum_{k=0}^{n-1} \binom{n-1}{k} a^{k+1} b^{n-(k+1)} + \sum_{k=0}^{n-1} \binom{n-1}{k} a^k b^{n-k}$$

$$= \sum_{k=1}^{n} \binom{n-1}{k-1} a^k b^{n-k} + \sum_{k=0}^{n-1} \binom{n-1}{k} a^k b^{n-k}$$

$$= a^n + \sum_{k=1}^{n-1} \binom{n-1}{k-1} a^k b^{n-k} + \sum_{k=1}^{n-1} \binom{n-1}{k} a^k b^{n-k} + b^n$$

$$= a^n + \sum_{k=1}^{n-1} \left( \binom{n-1}{k-1} + \binom{n-1}{k} \right) a^k b^{n-k} + b^n$$

$$= a^n + \sum_{k=1}^{n-1} \binom{n}{k} a^k b^{n-k} + b^n$$

$$= \sum_{k=0}^{n} \binom{n}{k} a^k b^{n-k},$$

as required. This completes the induction step, and indeed the entire proof.

## 9.6 Extended Formulas

### 9.6.1 Extended Formula: Mathematical Induction

Let's extend the binomial formula yet further. For this purpose, let $a$ and $n$ be two nonnegative integer numbers:

$$a, n \geq 0.$$

(This new $a$ has nothing to do with the old $a$ used in $a + b$ above.) Define a ratio of two factorials:

$$C_{a,n} \equiv \begin{cases} \frac{a!}{(a-n)!} & \text{if } a \geq n \\ 0 & \text{if } a < n. \end{cases}$$

How to calculate this? You could follow the original definition: first, calculate the numerator. Then, calculate the denominator. Finally, divide.

But isn't this a bit silly? After all, why calculate the entire numerator explicitly? After all, it contains a lot of factors that appear in the denominator as well, and are soon going to cancel out anyway!

How to avoid this? The original definition should be used for theoretical purposes only. In practice, on the other hand, better use a new (recursive) definition:

$$C_{a,n} \equiv a(a-1)(a-2) \cdots (a - (n-1)) = \begin{cases} 1 & \text{if } n = 0 \\ aC_{a-1,n-1} & \text{if } n \geq 1. \end{cases}$$

This way, we also get a bonus: $a$ could now be just any number, integer or not, positive or not. Furthermore, for every two numbers $a$ and $b$, we can now mirror Newton's binomial formula:

$$C_{a+b,n} = \sum_{k=0}^{n} \binom{n}{k} C_{a,k} C_{b,n-k}.$$

To prove this, use mathematical induction. Indeed, the formula is clearly true for $n = 0$. Now, for $n \geq 1$, assume that the induction hypothesis holds: for every $a$ and $b$,

$$C_{a+b,n-1} = \sum_{k=0}^{n-1} \binom{n-1}{k} C_{a,k} C_{b,n-1-k}.$$

Using this hypothesis and the formula in Section 9.1.2, we can now make the induction step:

$$
\begin{aligned}
C_{a+b,n} &= (a+b)C_{a+b-1,n-1} \\
&= aC_{(a-1)+b,n-1} + bC_{a+(b-1),n-1} \\
&= a\sum_{k=0}^{n-1} \binom{n-1}{k} C_{a-1,k} C_{b,n-1-k} + b\sum_{k=0}^{n-1} \binom{n-1}{k} C_{a,k} C_{b-1,n-1-k} \\
&= \sum_{k=0}^{n-1} \binom{n-1}{k} C_{a,k+1} C_{b,n-(k+1)} + \sum_{k=0}^{n-1} \binom{n-1}{k} C_{a,k} C_{b,n-k} \\
&= \sum_{k=1}^{n} \binom{n-1}{k-1} C_{a,k} C_{b,n-k} + \sum_{k=0}^{n-1} \binom{n-1}{k} C_{a,k} C_{b,n-k} \\
&= C_{a,n} + \sum_{k=1}^{n-1} \binom{n-1}{k-1} C_{a,k} C_{b,n-k} + \sum_{k=1}^{n-1} \binom{n-1}{k} C_{a,k} C_{b,n-k} + C_{b,n} \\
&= C_{a,n} + \sum_{k=1}^{n-1} \left( \binom{n-1}{k-1} + \binom{n-1}{k} \right) C_{a,k} C_{b,n-k} + C_{b,n} \\
&= C_{a,n} + \sum_{k=1}^{n-1} \binom{n}{k} C_{a,k} C_{b,n-k} + C_{b,n} \\
&= \sum_{k=0}^{n} \binom{n}{k} C_{a,k} C_{b,n-k},
\end{aligned}
$$

as required. This completes the proof.

## 9.6.2 The Extended Binomial Coefficient

How to use this formula fully? Well, in Section 9.1.2, the binomial coefficient has been defined with integer numbers only. Let's extend it to every number $a$, integer or not, positive or not:

$$\binom{a}{k} \equiv \frac{C_{a,k}}{k!}.$$

As in the original definition, $k$ still has to be a nonnegative integer number. $a$, on the other hand, can now be just any number: even complex!

Now, look at this new definition. Is it consistent with the old one? Let's check: what happens in the special case in which $a$ happens to be integer? Well, there are two options: if $a \geq k$, then the new definition indeed agrees with the old one. Indeed, both give the same familiar binomial coefficient. If, on the other hand, $0 \leq a < k$, then there is no old definition at all. Fortunately, the new definition tells us that, in this case, the (extended) binomial coefficient vanishes.

Now, look again at the formula proved in Section 9.6.1. To make it look better, divide it by $n!$. This gives

$$\binom{a+b}{n} = \frac{1}{n!} C_{a+b,n}$$

$$= \frac{1}{n!} \sum_{k=0}^{n} \binom{n}{k} C_{a,k} C_{b,n-k}$$

$$= \sum_{k=0}^{n} \frac{1}{k!(n-k)!} C_{a,k} C_{b,n-k}$$

$$= \sum_{k=0}^{n} \frac{C_{a,k}}{k!} \cdot \frac{C_{b,n-k}}{(n-k)!}$$

$$= \sum_{k=0}^{n} \binom{a}{k} \binom{b}{n-k}.$$

This formula mirrors the binomial formula: it gives $\binom{a+b}{n}$ as a sum rather than a product. Moreover, in the special case in which $a = b = n$, we have

$$\binom{2n}{n} = \sum_{k=0}^{n} \binom{n}{k} \binom{n}{n-k} = \sum_{k=0}^{n} \binom{n}{k}^2.$$

This way, the middle binomial coefficient (the highest column in Figure 9.8) is written as a sum rather than a product.

### 9.6.3 The Trinomial Formula

In the binomial formula, there are two numbers: $a$ and $b$. In the trinomial formula, on the other hand, there is yet another number: $c$.

To obtain the trinomial formula, let's apply the binomial formula twice:

$$(a+b+c)^n = ((a+b)+c)^n$$

$$= \sum_{k=0}^{n} \binom{n}{k} (a+b)^k c^{n-k}$$

$$= \sum_{k=0}^{n} \binom{n}{k} \left( \sum_{l=0}^{k} \binom{k}{l} a^l b^{k-l} \right) c^{n-k}$$

$$= \sum_{k=0}^{n} \sum_{l=0}^{k} \binom{n}{k} \binom{k}{l} a^l b^{k-l} c^{n-k}$$

$$= \sum_{k=0}^{n} \sum_{l=0}^{k} \frac{n!}{k!(n-k)!} \cdot \frac{k!}{l!(k-l)!} a^l b^{k-l} c^{n-k}$$

$$= \sum_{k=0}^{n} \sum_{l=0}^{k} \frac{n!}{l!(k-l)!(n-k)!} a^l b^{k-l} c^{n-k}$$

$$= \sum_{0 \leq l,j,m \leq n, \; l+j+m=n} \frac{n!}{l!j!m!} a^l b^j c^m.$$

In the latter sum, we just introduce two new indices: $j \equiv k - l$, and $m \equiv n - k$. This way, the new indices $l$, $j$, and $m$ have the same status. This is as in the transformation in Section 9.3.1.

This is the trinomial formula. Let's go ahead and mirror it, to produce a yet more advanced formula. For this purpose, let's apply the formula in Section 9.6.1 twice:

$$C_{a+b+c,n} = C_{(a+b)+c,n}$$

$$= \sum_{k=0}^{n} \binom{n}{k} C_{a+b,k} C_{c,n-k}$$

$$= \sum_{k=0}^{n} \binom{n}{k} \left( \sum_{l=0}^{k} \binom{k}{l} C_{a,l} C_{b,k-l} \right) C_{c,n-k}$$

$$= \sum_{k=0}^{n} \sum_{l=0}^{k} \binom{n}{k} \binom{k}{l} C_{a,l} C_{b,k-l} C_{c,n-k}$$

$$= \sum_{k=0}^{n} \sum_{l=0}^{k} \frac{n!}{k!(n-k)!} \cdot \frac{k!}{l!(k-l)!} C_{a,l} C_{b,k-l} C_{c,n-k}$$

$$= \sum_{k=0}^{n} \sum_{l=0}^{k} \frac{n!}{l!(k-l)!(n-k)!} C_{a,l} C_{b,k-l} C_{c,n-k}$$

$$= \sum_{0 \leq l,j,m \leq n, \; l+j+m=n} \frac{n!}{l!j!m!} C_{a,l} C_{b,j} C_{c,m}.$$

To make this formula look better, divide it by $n!$. This gives

$$\binom{a+b+c}{n} = \sum_{0 \leq l,j,m \leq n, \; l+j+m=n} \binom{a}{l} \binom{b}{j} \binom{c}{m}.$$

In the latter sum, how many terms are there? Thanks to Section 9.3.5, we already know the answer:

$$\binom{n+3-1}{3-1} = \binom{n+2}{2} = \frac{(n+1)(n+2)}{2}.$$

## 9.7 Exercises: Pascal's Triangle

### 9.7.1 Oblique Lines in Pascal's Triangle

1. Show that, for every nonnegative integer number $n \geq 0$,

$$\binom{n}{0} = \binom{n}{n} = 1.$$

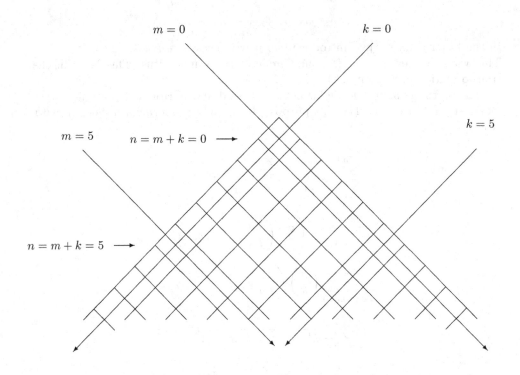

**Fig. 9.10.** Pascal's triangle: an oblique $m$-$k$ grid. To index the oblique down-left lines, use $k = 0, 1, 2, \ldots$, as before. To index the oblique down-right lines, on the other hand, use the new index $m = 0, 1, 2, \ldots$. Finally, to index the horizontal levels, use $n = m+k = 0, 1, 2, \ldots$, as before.

2. Show that, for every two integer numbers $n \geq k \geq 0$, the binomial coefficient is symmetric in the sense that

$$\binom{n}{k} = \binom{n}{n-k}.$$

3. What does this mean geometrically? Hint: Figures 9.7–9.9 are indeed symmetric.
4. Show that, for every two natural numbers $n > k > 0$,

$$\binom{n}{k} = \binom{n-1}{k} + \binom{n-1}{k-1}.$$

Hint: see Section 9.1.2.
5. In Pascal's triangle, draw the lines $n =$ const. How do they look like geometrically? Hint: these are horizontal rows (Figure 9.1).
6. Next, draw the lines $k =$ const. How do they look like geometrically? Hint: these are oblique lines, pointing down-left (Figures 9.1 and 9.10).

7. Define the new index $m \equiv n - k$. What is its geometrical meaning? Hint: it indexes the oblique down-right lines.
8. Use this new index to draw the lines $m =$const. How do they look like geometrically? Hint: these are oblique lines, pointing down-right (Figure 9.10).

9. View Pascal's triangle as an oblique (infinite) $m$-$k$ grid.
10. In this oblique grid, draw the lines $m + k =$const. How do they look like geometrically? Hint: these are horizontal rows (Figure 9.10).
11. Use the result in Section 9.1.2 recursively, time and again:

$$\binom{n}{k} = \binom{n-1}{k-1} + \binom{n-1}{k} = \binom{n-1}{k-1} + \binom{n-2}{k-1} + \binom{n-2}{k} = \cdots.$$

12. How does this process look like geometrically? Hint: in Pascal's triangle, look at the oblique line $k - 1 =$const. March along it, cell by cell, and sum the entries up. Stop just before hitting the $n$th level. On your left, you'd then see the same sum.
13. To help simplify this, avoid the old index $n$. Instead, use the new index $m = n - k$.
14. Use the above to prove once again the combinatorial formula in Section 9.3.6. Hint: use mathematical induction on $m = n - k = 0, 1, 2, \ldots$.
15. Write this mathematical induction as a recursive algorithm.
16. What is the geometrical meaning of this process? Is this familiar? Hint: we've already seen it before. In Pascal's triangle, look at the oblique line $k - 1 =$const. March along it, cell by cell, and sum the entries up. Stop just before hitting the $n$th level. On your left, you'd then see the same sum.
17. In terms of a discrete simplex, what is the geometrical meaning of this formula? Hint: look at the discrete $k$-dimensional simplex of size $m$. Count the points in it, face by face.
18. Conclude that it was indeed a good idea to view Pascal's triangle as an infinite $m$-$k$ grid.

### 9.7.2 Binary Vectors and Characteristic Functions

1. Define the set of $n$-dimensional binary vectors:

$$V \equiv V(n) \equiv \{(v_1, v_2, \ldots, v_n) \mid v_i = 0 \text{ or } v_i = 1, \ 1 \le i \le n\}.$$

2. Note that each binary vector could also be viewed as a characteristic (or binary) function on the set $\{1, 2, 3, \ldots, n\}$. Hint: define the associated characteristic function simply as $v(i) \equiv v_i$.
3. Write $V$ as the set of all characteristic (binary) functions on the set $\{1, 2, 3, \ldots, n\}$. Hint: in the notation of set theory,

$$V \equiv V(n) \equiv \{0, 1\}^{\{1, 2, \ldots, n\}}.$$

4. Note that $V$ depends on $n$. Still, we often drop the '$(n)$.'
5. What happens in the degenerate case $n \equiv 0$? How to define $V(0)$? Hint: $V(0)$ contains just one "vector:" the empty "vector" $\emptyset$, which contains no component at all:

$$V(0) \equiv \{\emptyset\}.$$

6. What is the cardinality of $V(n)$? In other words, how many vectors are there in $V(n)$?

7. Use mathematical induction on $n \geq 0$ to show that $V \equiv V(n)$ contains

$$|V| \equiv |V(n)| = 2^n$$

distinct vectors.

8. Let $n \geq 1$ be fixed. Let $V_k \subset V$ contain those vectors with exactly $k$ nonzero components:

$$V_k \equiv \left\{ v \in V \mid \sum_{i=1}^{n} v_i = k \right\}.$$

9. Use mathematical induction on $k = 0, 1, 2, \ldots, n$ to show that $V_k$ contains

$$|V_k| = \binom{n}{k}$$

distinct vectors. Hint: see Section 9.2.3.

10. Show that, for $k \neq l$, $V_k$ and $V_l$ are disjoint:

$$V_k \cap V_l = \emptyset.$$

11. Conclude that $V$ could be written as the union of $n$ disjoint subsets:

$$V = \cup_{k=0}^{n} V_k.$$

12. Conclude that

$$2^n = |V| = \sum_{k=0}^{n} |V_k| = \sum_{k=0}^{n} \binom{n}{k}.$$

13. Does this agree with the binomial formula?

14. In Pascal's triangle, how many distinct paths lead from the top cell (the zeroth level) to the $k$th cell in the $n$th level ($0 \leq k \leq n$)? Hint: see Section 9.1.3.

15. Why must this number be the same as $|V_k|$? Hint: model each path by a unique vector in $V_k$.

16. Prove more directly (by mathematical induction on $n$ rather than $k$) that $V_k$ indeed contains

$$|V_k| = \binom{n}{k}$$

distinct vectors. Hint: split $V_k$ into two disjoint subsets: the first containing those vectors ending with 0, and the second containing those vectors ending with 1. Then, use the induction hypothesis, and the formula in Section 9.1.2.

17. In the algebraic expression $(a + b)^n$, open the parentheses. The result is a sum. In this sum, look at one particular term: $a^k b^{n-k}$ (where $k$ is a fixed number between 0 and $n$). How many duplicate copies of this term appear in the above sum? Hint: upon opening parentheses, to obtain this term, pick exactly $k$ $a$'s from exactly $k$ factors of the form $a + b$. To model this pick, use a unique vector in $V_k$.

18. Conclude that there are

$$|V_k| = \binom{n}{k}$$

duplicate copies of that term.

19. Conclude that, in Newton's binomial formula, the term $a^k b^{n-k}$ must indeed have the coefficient

$$\binom{n}{k}.$$

# 10

## Applications
## in Quantum
## Statistical Mechanics

The binomial formula is particularly useful in statistical mechanics. In Chapter 9, Sections 9.4.1–9.4.3, we've already seen one example: Brownian motion. Here, we see more. For this purpose, we need to extend the binomial and trinomial formulas, to help model a system of molecules. This leads to the Maxwell-Boltzmann statistics.

Molecules are distinguishable from each other. Bosons, on the other hand, are not. Fortunately, thanks to our original study of discrete simplex, we can now model a system of bosons as well. This leads to the Bose-Einstein statistics.

Electrons are indistinguishable as well. Still, they are different from bosons: they obey Pauli's exclusion principle. Still, thanks to our earlier study in discrete math, we can now model them as well. This leads to the Fermi-Dirac statistics [8, 21, 27, 43]. In summary, discrete math helps model three different kinds of physical systems.

## 10.1 The Quadrinomial Formula

### 10.1.1 The Trinomial Coefficient

To obtain the present applications in quantum statistical mechanics, recall a few results from discrete math. The trinomial formula tells us that

$$(a+b+c)^n = \sum_{0 \le l,j,m \le n,\ l+j+m=n} \frac{n!}{l!j!m!} a^l b^j c^m$$

(Chapter 9, Section 9.6.3). What is the combinatorial meaning of this? Well, upon opening parentheses on the left-hand side, we obtain a sum of many products. Each product may contain $l$ $a$'s, $j$ $b$'s, and $m$ $c$'s, where $l$, $j$, and $m$ are some nonnegative integers that sum up to $n$:

$$l + j + m = n.$$

This produces a product of the form

$$a^l b^j c^m,$$

as on the right-hand side.

How many duplicate copies of this product are there? In other words, how many different ways are there to pick $l$ $a$'s, $j$ $b$'s, and $m$ $c$'s? The answer is the trinomial coefficient:

$$\frac{n!}{l!j!m!}.$$

Let's go ahead and extend this.

### 10.1.2 Extending the Trinomial Formula

Let's extend the above yet further:

$$(a+b+c+d)^n = \sum_{0 \leq l,j,m,k \leq n,\ l+j+m+k=n} \frac{n!}{l!j!m!k!} a^l b^j c^m d^k.$$

This is the quadrinomial formula. What is its combinatorial meaning? Well, upon opening parentheses on the left-hand side, we obtain a sum of products of the form

$$a^l b^j c^m d^k,$$

where $l$, $j$, $m$, and $k$ are some nonnegative integers that sum up to $n$:

$$l + j + m + k = n.$$

To obtain such a product, how many different ways are there to pick $l$ $a$'s, $j$ $b$'s, $m$ $c$'s, and $k$ $d$'s? The answer is the quadrinomial coefficient:

$$\frac{n!}{l!j!m!k!}.$$

Let's go ahead and use this in physics.

## 10.2 Mass and Energy

### 10.2.1 Closed System: Constant Energy

To use the above in physics, we need four new indices:

$$n_1 \equiv l$$
$$n_2 \equiv j$$
$$n_3 \equiv m$$
$$n_4 \equiv k.$$

Consider now a system of $n$ molecules. Assume that the system is closed: isolated and insulated. This way, the system has constant energy.

### 10.2.2 Energy Levels

In quantum mechanics, energy is not continuous but discrete: it comes in discrete energy levels:

$$0 < E_1 < E_2 < E_3 < E_4 < E_5 < \cdots.$$

Only these $E_i$'s are allowed: the system can never have any other energy in between. In this sense, energy is different from position and time, which remain continuous: at each individual time, the molecule can still lie in any position in space.

### 10.2.3 Conservation of Mass

Now, what are the $a$'s? Let's interpret them as molecules of energy $E_1$. Likewise, let's interpret the $b$'s, the $c$'s, and the $d$'s as molecules of energy $E_2$, $E_3$, and $E_4$, respectively. This way, in our original system of $n$ molecules, $n_i$ molecules have energy $E_i$ $(1 \le i \le 4)$. This splits the original system into four disjoint subsets:

$$\sum_{i=1}^{4} n_i = n_1 + n_2 + n_3 + n_4 = n.$$

This is actually conservation of mass: uniting the subsets produces the original system back again.

### 10.2.4 Configuration — Distribution

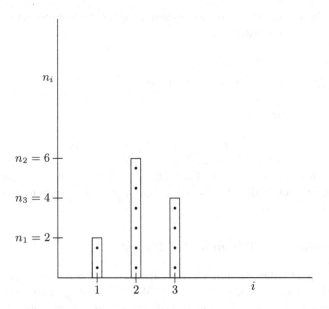

**Fig. 10.1.** Distribution of molecules: there are $n_1$ molecules of energy $E_1$, $n_2$ molecules of energy $E_2$, and so on.

As a matter of fact, why stop at 4? After all, there could be many more subsets of yet higher energy levels: $n_5$ molecules of energy $E_5$, $n_6$ molecules of energy $E_6$, and so on. This way, the above sum takes the new form

$$\sum_{i=1}^{\infty} n_i = n.$$

This is not really an infinite sum. After all, only a few $n_i$'s are nonzero, and all the rest vanish. Still, because we don't know which $i$'s are redundant, we leave them all in.

The $n_i$'s make a new configuration or distribution. This is illustrated in a column diagram (Figure 10.1): the $i$th column contains $n_i$ molecules of energy $E_i$. This is how the original $n$ molecules split into separate energy levels.

### 10.2.5 Conservation of Energy

So far, we've used one constraint only: conservation of mass. This law is more mathematical than physical. After all, the total number of molecules must be $n$. Still, we also have a more fundamental physical law — conservation of energy:

$$\sum_{i=1}^{\infty} n_i E_i = E_{\text{total}}$$

(the total energy in the entire system). This is indeed the first law of thermodynamics.

Let's use these constraints to design a new optimization problem. For this purpose, we need one more formula.

## 10.3 The Multinomial Coefficient

### 10.3.1 Random Variables

In our original system, $n$ and $E_{\text{total}}$ are given in advance. Likewise, the individual $E_i$ are available in advance as well. The $n_i$'s, on the other hand, are not: they are random variables, not yet known. Furthermore, they depend on each other, as in the above constraints. Still, let's focus on some particular configuration, with some specific $n_i$'s.

### 10.3.2 Extending the Trinomial Coefficient

To design this particular configuration, how many different ways are there to pick the molecules and group them in subsets? Well, there are many ways to pick the first $n_1$ molecules of energy $E_1$. On top of that, there are also many ways to pick $n_2$ molecules of energy $E_2$, and so on. In total, there are

$$n! \frac{1}{n_1!} \cdot \frac{1}{n_2!} \cdot \frac{1}{n_3!} \cdots$$

different ways. This is just a finite product. After all, only a few $n_i$'s are nonzero. Still, this product could be quite long. This is indeed the multinomial coefficient.

### 10.3.3 Degeneracy

In our configuration, there are $n_i$ molecules of energy $E_i$ ($i \geq 1$). Still, they are not necessarily in the same state. On the contrary: they split into $d_i$ different states (or degeneracies), where $d_i$ is a constant natural number. This is illustrated in Figure 10.2: the $i$th column splits into $d_i$ subsquares.

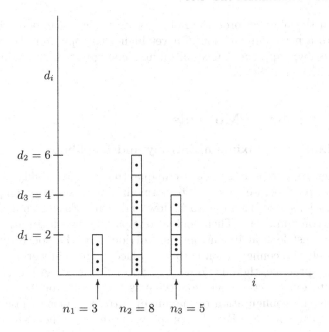

**Fig. 10.2.** At energy level $E_i$, the original $n_i$ molecules split now into $d_i$ different states (degeneracies), each containing a few molecules.

How many different ways are there to split the original $n_i$ molecules into $d_i$ disjoint subsets? Clearly, the first molecule could be placed in either of the $d_i$ subsets. On top of that, the second molecule could be placed in either of the $d_i$ subsets as well, and so on. Thus, the answer is

$$\left|\{1,2,3,\dots,d_i\}^{\{1,2,3,\dots,n_i\}}\right| = d_i^{n_i}.$$

Thus, the product in Section 10.3.2 should be modified to read

$$n!\frac{d_1^{n_1}}{n_1!}\cdot\frac{d_2^{n_2}}{n_2!}\cdot\frac{d_3^{n_3}}{n_3!}\cdots.$$

This is the total number of different ways to design our configuration.

### 10.3.4 Configuration and Its Entropy

This product is called the entropy of the configuration. It tells us how likely the configuration is to take place: the higher it is, the more options the configuration has to form, and the more likely it is to appear in reality. As time goes by, the system may change, but only in one direction: our system could only "jump" from one particular configuration to yet "better" configuration, with more entropy, and a larger probability to occur. Indeed, thanks to the second law of thermodynamics, entropy can only increase, but never decrease. (This is discussed throughout our study in discrete math: in Chapter 6, Section 6.3.5, and Chapter 9, Section 9.4.3.)

As a result, in the physical process in our closed system, a configuration can only transform into a new configuration of a yet higher entropy and likelihood. In the end, the process will approach an equilibrium: a configuration of maximum entropy (among those that are allowed).

## 10.4 Distinguishable Molecules

### 10.4.1 Equilibrium: Maximum Entropy and Likelihood

Let's use the entropy to obtain a new useful quantity: the probability of the configuration. For this purpose, we only need to normalize: make sure that all probabilities of all configurations indeed sum up to 1. How to do this? Easy: sum up all entropies of all possible configurations. Then, look at one particular configuration. To obtain its probability, just look at its entropy, and divide it by the above sum. This will tell us how likely the configuration is to take place in the real world.

Actually, in practice, there is no need to normalize at all. After all, this makes no difference to our practical question: what is the most probable configuration? Clearly, this is the configuration of maximum entropy (among those that are allowed). How to uncover it? For this purpose, better work not with the entropy itself, but with its natural logarithm. After all, the logarithm is a monotonically-increasing function. Thus, maximizing it is the same as maximizing the original entropy.

We must also work under two constraints: conservation of mass and energy. This way, we'll obtain our solution: the configuration that maximizes entropy among those configurations that obey conservation of mass and energy. This will indeed be equilibrium, approached by the physical process in our closed system. This is called the Maxwell-Boltzmann distribution (Chapter 8, Section 8.12.2).

### 10.4.2 Maxwell-Boltzmann Statistics

To obtain the probability of a specific configuration, we looked at its entropy, and divided it by the sum of all entropies of all possible configurations. This was only implicit: in practice, it is unnecessary.

As time goes by, the system will eventually arrive at its equilibrium: the optimal configuration. How does it look like? Well, this is just a column diagram: at energy level $i$, there are $n_i$ molecules of energy $E_i$ (Figure 10.1). In other words, this is a discrete distribution. To make it continuous, we must first normalize. This, however, is a different kind of normalization. (After all, by now, we already have only one configuration, not many.) This new normalization has one purpose only: to make the columns sum up to 1, to make a proper probability function.

Thus, in the Maxwell-Boltzmann distribution, look at the $i$th column, and normalize it. In other words, look at the number of molecules of energy $E_i$, and divide by the total number of molecules. This defines a new probability function:

$$p(E_i) \equiv \frac{n_i(E_i)}{n}.$$

Now, if you pick a molecule from the system at random, how likely is it to have energy $E$? We now have the answer: the probability for this is $p(E)$. Fortunately,

$p(E)$ doesn't depend on the discrete index $i$ any more, but on the (continuous) energy $E$ only.

Later on, in quantum mechanics, we'll see that energy is not really continuous, but only discrete: it comes in discrete lumps only. Still, these lumps are so tiny that they could make a continuous graph, to illustrate the behavior of $p$ as a function of $E$. For this purpose, however, $p$ must renormalize, to make sure that the area under the graph is 1, as required in a continuous distribution.

## 10.5 Indistinguishable Bosons

### 10.5.1 Indistinguishable Bosons

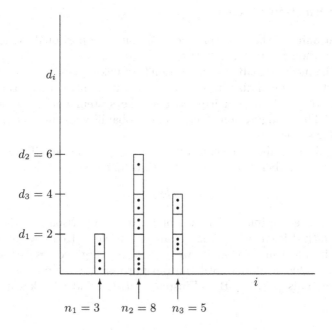

**Fig. 10.3.** Bosons, on the other hand, are indistinguishable: it doesn't matter which $n_1$ bosons to place in the first energy level, which $n_2$ bosons to place in the next energy level, and so on.

Instead of molecules, consider now $n$ bosons, in one particular configuration. These are indistinguishable particles: there is no way to distinguish one boson from another. Therefore, it doesn't matter which $n_1$ bosons are placed in the first energy level. Likewise, it doesn't matter which $n_2$ bosons are placed in the second energy level, and so on.

Still, there are many ways to split the first $n_1$ bosons among the $d_1$ possible states in the first energy level (Figure 10.3). How many? Well, to split $n_1$ bosons into $d_1$ disjoint states, we need to decide how many are in state 1, how many in state

2, and so on. (After all, since the bosons are indistinguishable, it doesn't matter which ones, but only how many.) For this purpose, we need to pick $d_1$ nonnegative integers that sum up to $n_1$. How many ways are there to do this? We already know the answer:

$$\binom{n_1 + d_1 - 1}{d_1 - 1}$$

(Chapter 9, Sections 9.3.5–9.3.6). The same could be done in the other energy levels as well. In summary, to have our original configuration, there are

$$\binom{n_1 + d_1 - 1}{d_1 - 1} \binom{n_2 + d_2 - 1}{d_2 - 1} \binom{n_3 + d_3 - 1}{d_3 - 1} \cdots$$

possible ways.

### 10.5.2 Bose-Einstein Statistics

What is the meaning of the above product? Well, it is proportional to the probability of our configuration to appear in nature.

What is the most probable configuration? As before, take the logarithm of the above product, and maximize it, subject to a constraint: conservation of energy. (Conservation of mass is often irrelevant — the system may contain arbitrarily many bosons.) The configuration of maximum probability is called the Bose-Einstein distribution (or statistics).

The Bose-Einstein distribution is a discrete column diagram. To make it continuous, look at the number of bosons in the $i$th column:

$$n_i (E_i) .$$

This is written as a continuous function of $E_i$. It is proportional to the probability that a boson picked from the system at random would indeed have energy $E_i$.

Later on, in quantum mechanics, we'll see that energy comes in discrete lumps only: not every value is allowed. Still, for our purpose, we assume that $E_i$ may take just any value. This way, the Bose-Einstein statistics indeed takes its continuous face.

## 10.6 Indistinguishable Electrons

### 10.6.1 Pauli Exclusion Principle

Instead of bosons, let us now consider $n$ electrons, arranged in our original configuration. Like bosons, electrons are indistinguishable. Furthermore, they must also obey Pauli's exclusion principle: in each energy level, each state may contain one electron at most. For this reason, the number of electrons in this level can never exceed the number of states:

$$n_i \le d_i, \quad i \ge 1.$$

Under these circumstances, how to split $n_i$ electrons into $d_i$ disjoint states? For this purpose, look at the $i$th column in Figure 10.4. It contains $d_i$ subsquares. Out of

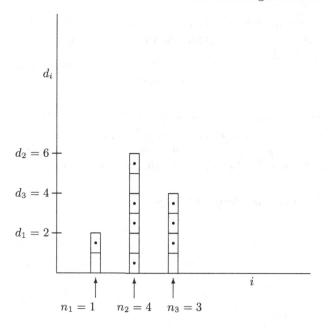

**Fig. 10.4.** Electrons, on the other hand, must obey Pauli's exclusion principle: each state (degeneracy) may contain one electron at most, so $n_i \leq d_i$.

these, pick just $n_i$ subsquares to contain an electron, leaving $d_i - n_i$ subsquares empty. (After all, because electrons are indistinguishable, it doesn't matter which electrons are picked, but only which states are picked.) How many ways are there to do this? We already know the answer:

$$\binom{d_i}{n_i}.$$

Thus, to produce our original configuration, there are

$$\binom{d_1}{n_1}\binom{d_2}{n_2}\binom{d_3}{n_3}\cdots$$

different ways.

## 10.6.2 Fermi-Dirac Statistics

What is the meaning of the above product? Well, it is proportional to the probability of our configuration to take place in reality.

What is the most probable configuration? As before, take the logarithm of the above product, and maximize it, subject to two constraints: conservation of mass and energy. The configuration of maximum probability is called the Fermi-Dirac distribution (or statistics).

The Fermi-Dirac distribution is a discrete column diagram. To make it continuous, look at the number of electrons in the $i$th column, and normalize it. This defines a new probability function:

$$p\left(E_i\right) \equiv \frac{n_i\left(E_i\right)}{n}.$$

This could also be written as a continuous function of $E_i$. (For this purpose, however, there is a need to renormalize, as discussed below.) How to use it? Well, if you pick an electron from the system at random, how likely is it to have energy $E$? The probability for this is $p(E)$. Let's see how this function looks like.

### 10.6.3 Fermi Level and Fermi Energy

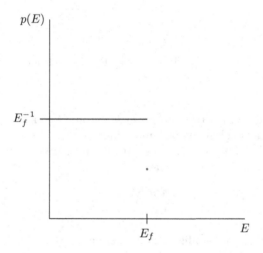

**Fig. 10.5.** As $T \to 0^+$, $p$ may take two constant values only: below the Fermi energy, $p$ is a positive constant (full levels). Above the Fermi energy, on the other hand, $p = 0$ (empty levels). For this reason, the Fermi energy is the maximal energy in the entire system: no electron may have more energy.

It turns out that $p$ has a closed analytic form [21], in terms of the exponent function:

$$p(E) = \frac{E_f^{-1}}{\exp\left(\frac{E - E_f}{K_B T}\right) + 1},$$

where

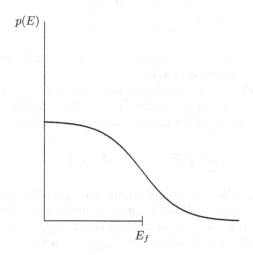

**Fig. 10.6.** As $T$ increases, $E_{\text{total}}$ increases as well, so some electrons may get enough energy to "jump" from a level just below the Fermi level to a higher level just above the Fermi level. This way, $p$ remains symmetric around $E_f$.

- $f$ is a fixed natural number: the Fermi level,
- $E_f$ is the Fermi energy: the energy at the Fermi level,
- $T$ is the temperature (in Kelvin's scale),
- and $K_B$ is Boltzmann's constant.

This probability function should be used only weakly, not strongly. In other words, it should never be interpreted pointwise, but only in some interval. For instance, let $b > a \geq 0$ be two nonnegative numbers. Now, pick an electron from the system at random. How likely is the electron to have its energy in between $a$ and $b$? The probability for this is

$$\int_a^b p(E)dE.$$

This is no longer a sum, but an integral. Therefore, a new kind of normalization is needed, to make sure that the total area under the integral is 1:

$$\int_{-\infty}^{\infty} p(E)dE = 1,$$

as required in a continuous distribution.

How does $p$ look like? Let's start from the easy case, in which $p$ is uniform. This is the limit case, in which $E_{\text{total}}$ is minimal, and $T$ approaches the absolute zero:

$$T \to 0^+ = (-273°C).$$

At this limit,

$$p(E) \to \begin{cases} E_f^{-1} & \text{if} \quad E < E_f \\ 0 & \text{if} \quad E > E_f \\ E_f^{-1}/2 & \text{if} \quad E = E_f \end{cases}$$

(Figure 10.5). This tells us the true physical meaning of the Fermi level: the first nonfull level, or the last nonempty level.

In fact, all levels below $f$ are full of electrons, and all levels above $f$ are empty. The energy at $f$ is called the Fermi energy: $E_f$. This is the maximal possible energy in the entire system. Indeed, no electron can have more energy:

$$\int_0^{E_f} p(E)dE = E_f^{-1}E_f = 1.$$

Now, let's begin heat up the system, and pump more and more energy into it. This way, both $E_{total}$ and $T$ increase. What happens quantum-mechanically? Well, some electron must "jump" from some nonempty level, say level $f$, to the next higher (nonfull) level, say level $f + 1$. This increases $E_{total}$ by a tiny bit of

$$E_{f+1} - E_f > 0.$$

As $T$ increases further, more electrons may "jump" as well, say from level $f - 1$ to level $f + 2$. This way, $p(E)$ will always remain symmetric around $E_f$, as in Figure 10.6.

## 10.7 Exercises: Spin

### 10.7.1 Spin-Up and Spin-Down

1. Extend the trinomial formula, and obtain the quadrinomial formula:

$$(a + b + c + d)^n = \sum_{0 \le l,j,m,k \le n, \ l+j+m+k=n} \frac{n!}{l!j!m!k!} a^l b^j c^m d^k.$$

Hint: extend the proof in Chapter 9, Section 9.6.3.

2. How many terms are there in this sum? Hint:

$$\binom{n+4-1}{4-1} = \binom{n+3}{3} = \frac{(n+1)(n+2)(n+3)}{6}$$

(Chapter 9, Sections 9.3.5–9.3.6).

3. In the above sum, look at the quadrinomial coefficient:

$$\frac{n!}{l!j!m!k!}.$$

What is its combinatorial meaning? Hint: this is the total number of different ways to pick $l$ $a$'s, $j$ $b$'s, $m$ $c$'s, and $k$ $d$'s.

4. What is its physical meaning? Hint: this is the total number of different ways to pick $l$ molecules of energy $E_1$, $j$ molecules of energy $E_2$, $m$ molecules of energy $E_3$, and $k$ molecules of energy $E_4$ from a system of $n = l + j + m + k$ molecules.

5. What would happen if, instead of molecules, we had bosons? Hint: unlike molecules, bosons are indistinguishable. Therefore, all of the above ways would coincide, and become one and the same.

6. What would happen if, instead of bosons, we had electrons? Hint: electrons are indistinguishable as well. Thus, there would be no change: all of the above ways still coincide.

7. Consider a system of $n$ electrons. In the first energy level, there are just two possible states (degeneracies): either spin-up, or spin-down. In other words,

$$d_1 = 2.$$

How many ways are there to have $n_1 = 2$ electrons in this energy level? Hint: there is only one way: one electron must have spin-up, and the other must have spin-down.

8. Is this as expected from our formula? Hint: yes. Indeed, in this case, there is only

$$\binom{d_1}{n_1} = \binom{2}{2} = 1$$

way.

9. One student gave a different answer: "there are two different ways:
   • either the first electron has spin-up and the second has spin-down,
   • or the first electron has spin-down and the second has spin-up."

What's wrong with this answer? Hint: these ways coincide, because there is no way to distinguish between the "first" and "second" electrons.

10. How many ways are there to have $n_1 = 1$ (just one electron in the first energy level)? Hint: two: either spin-up, or spin-down.

11. Is this as expected from our formula? Hint: yes. Indeed, in this case, there are

$$\binom{d_1}{n_1} = \binom{2}{1} = 2$$

ways.

12. Conclude that $n_1 = 1$ is more likely than $n_1 = 2$.

13. Does this mean that, in the Fermi-Dirac distribution, the first energy level must always contain one electron only? Hint: no! The Fermi-Dirac distribution must obey two constraints: conservation of mass and energy:

$$E_{\text{total}} = n_1 E_1 + n_2 E_2 + n_3 E_3 \cdots$$

(for some given parameter $E_{\text{total}}$). Under these constraints, if the most probable configuration has $n_1 = 2$, then the first level is full:

$$n_1 (E_1) = 2.$$

In practice, this is often the case: the first energy level is usually full (Figures 10.5–10.6).

14. Consider now a system of bosons rather than electrons. In the first energy level, there are now three possible states (degeneracies): either spin-up, or spin-down, or spin-zero. In other words,

$$d_1 = 3.$$

How many ways are there to have $n_1$ bosons in this energy level? Hint:

$$\binom{n_1 + 3 - 1}{3 - 1} = \binom{n_1 + 2}{2} = \frac{(n_1 + 1)(n_1 + 2)}{2}.$$

15. Later on, in the context of Lie algebras, we'll learn more about bosons and electrons and their spin.

# Part V

Towards
General Relativity
and Quantum Mechanics

# Towards General Relativity and Quantum Mechanics

A physical phenomenon should better be independent of the mathematical coordinates used to describe it. After all, the laws of nature must be consistent. You must always observe the same laws, no matter what coordinates you use to carry out your experiment.

But what if you are on a spaceship, travelling away from the Earth at a constant speed? In your spaceship, you may have a lamp. Look at its light, and measure its speed. The result must be $300,000$ kilometer per second.

I on the Earth, on the other hand, also want to calculate the speed of light. But I'm a bit naive: I take the speed of light that you've just measured, and add to it yet another speed: the speed of the spaceship with respect to the Earth. So, I think that light is faster than $300,000$ kilometer per second!

Fortunately, we know better: light travels at the same speed in all systems, moving or not. Indeed, this was already proved in many experiments. How to solve the above paradox? Well, speeds should not be added linearly, but nonlinearly.

This is how special relativity improves on Newton's original laws. Fortunately, in practical engineering problems, this effect is usually negligible, so Newtonian mechanics is good enough.

Still, in modern technology, this can no longer be ignored. To account for this, relativity theory designs a new geometry: it takes the time axis, and sticks it to space, to form a new four-dimensional hyperspace: spacetime. In it, we get to see two new effects most vividly: time dilation, and length contraction. Later on, these effects will be extended to general relativity as well.

In general relativity, spacetime takes a new geometrical form: a four-dimensional differentiable manifold. How to design it? Around each event in spacetime, draw a little chart: a local $t$-$x$-$y$-$z$ coordinate system, to help carry out physical experiments locally.

Thanks to Zorn's lemma, there exists a maximal atlas of charts. Furthermore, thanks to the axiom of choice, at each event in spacetime, we can pick one coordinate system, to help describe our local "lab." This is also a good preparation work for Lie groups, used often in quantum mechanics.

Thanks to this, we can envelope spacetime with tangent spaces all over. These are Riemann normal coordinates: local, not global. This is indeed Einstein's equivalence principle, in its geometrical face. This was indeed the starting point of general relativity.

# 11

## Spacetime
## and Its Local Coordinates

What is spacetime? So far, spacetime was a four-dimensional hyperspace, spanned by the $x$, $y$, $z$, and $t$ axes. Here, on the other hand, spacetime is more general: a four-dimensional manifold. Furthermore, it is also differentiable: at each event in spacetime, we have a local "lab:" a little chart, made of three-dimensional time levels. This makes a new $t$-$x$-$y$-$z$ coordinate system to help carry out physical experiments locally.

Thanks to Zorn's lemma, there exists a maximal atlas of charts. Moreover, thanks to the axiom of choice, at each event in spacetime, we can pick one individual chart: some coordinate system, to help describe our experiment in our local "lab." This is also a good preparation work for Lie groups, used often in quantum mechanics.

We can now go ahead and envelope spacetime with tangent spaces all over. These are Riemann normal coordinates: local, not global. This is indeed Einstein's equivalence principle in its geometrical face. This was indeed the starting point of general relativity.

In Chapter 7, we've already used a binary tree to model a bifurcation tree in chaos theory. From below, this tree is completely transparent: it looks like Cantor's null set.

In general relativity, on the other hand, a binary tree is no longer good enough. From each node in the tree, we need to issue not only two but four branches. This makes a quaternary tree, useful to model a tensor.

Thanks to their recursive nature, quaternary trees are suitable to model big tensors with arbitrarily many indices, as required in numerical relativity. Indeed, it is now easy to implement all sorts of algebraic operations between tensors: addition, contraction, etc., no matter how many indices are used. This is indeed a good example of using a multilevel hierarchy efficiently.

Tensors are particularly useful in general relativity. Thanks to them, one can model the curvature in spacetime. The key to the curvature is the metric that tells us how spacetime stretches, and in what direction [4, 5, 11, 42].

How does the metric do this? At each event in spacetime, the metric specifies four directions, orthogonal to each other. Three of these directions are space-like: to each of them, the metric assigns a positive weight. The fourth direction, on the other hand, is time-like: to it, the metric assigns a negative weight. Together, these weights tell us the (local) possible speed: how much one could advance in space, compared to the advance in time.

Unfortunately, the metric is not yet known. Fortunately, Einstein equations tell us an important thing about the metric: it can never come from nothing. On the contrary: it must come from a physical source: the stress tensor. This produces the Einstein equations, whose solution is the metric itself. They will be introduced elsewhere (Chapter 15 in [46]). Here, on the other hand, we are more interested in their geometrical meaning. For this, we'll use tensors for yet another important purpose: to discuss Riemann normal coordinates in spacetime. Thanks to these (theoretical, implicit, local) coordinates, we'll be able to see Einstein's equivalence principle geometrically.

## 11.1 Some Background

### 11.1.1 Special Relativity — No Gravity

In special relativity, one assumes no gravity at all. This way, an object can fly undisturbed at a constant speed, and never accelerate or change direction.

In real systems such as the solar system, on the other hand, gravity can no longer be ignored. On the contrary: it must be explained by a new mathematical model, independent of the coordinates that happen to be used. After all, the coordinates are just mathematical artifacts to help model nature. As such, they should never affect nature or the laws that govern it.

To put general relativity in its proper context, let's start with some historical background.

### 11.1.2 Flat Geometry

The ancient Greeks introduced Euclidean geometry for one main purpose: to model static shapes in the two-dimensional plane — triangles, circles, and so on. Later on, this theory was also extended to the three-dimensional space. This was quite useful to calculate volume, surface area, and more.

This was still rather static. Dynamics was still missing.

## 11.2 Physics and Philosophy

### 11.2.1 Newton and Plato

Newton, on the other hand, introduced a new time axis on top, to help model not only static but also dynamic shapes. This was indeed a breakthrough: a new force can now be applied to the original shape from the outside, to accelerate its original velocity, and even change its direction.

This fits well in Plato's philosophy. To refer to a geometrical shape (or just any general object), we must introduce a new word in our language, to represent not only one concrete instance but also the "godly" spirit behind all possible instances. This way, the word stands behind the deep concept fully. Likewise, in physics, force stands behind the motion, and affects it from the outside.

### 11.2.2 Einstein and Aristotle

Newton viewed time as an auxiliary parameter, which makes a new (nonphysical) axis, perpendicular to the (physical) phase-space, where the original motion takes place. Einstein, on the other hand, threw the time dimension back into the very heart of geometry. This way, time is not different from any other spatial dimension. Once the time axis is united with the original three-dimensional space, we have a new four-dimensional manifold: spacetime.

This is more in the spirit of Aristotle's philosophy. A word in our language takes its meaning not from the outside but from the very inside: the deep nature of the general concept it stands for.

To describe spacetime and gravity, Einstein used differential geometry. Here is some background.

## 11.3 Differentiable Manifold

### 11.3.1 Coordinates on the Sphere: Longitude and Latitude

For a start, consider a two-dimensional manifold: a sphere. How to refer to a particular point on it? For this purpose, the point must have a "name" or label: two coordinates.

Consider, for example, the face of the Earth. On it, consider a particular location. How to refer to it? By two coordinates: latitude and longitude. These are just two numbers that label the location, and give it a name. The longitude is, say, between $0°$ and $180°$, and the latitude between $-90°$ and $90°$. (Later on, we'll exclude the Greenwich longitude, and give it a special treatment.)

### 11.3.2 The Equator: a Level Set

Around each location on the face of the Earth, we can draw a chart: say, horizontal and vertical lines. The equator, for example, makes a horizontal line. On it, the latitude is constant: $0°$. This is the zero level set of the latitude. The Greenwich longitude, on the other hand, makes a vertical line. On it, the longitude is constant: $0°$. This is the zero level set of the longitude.

### 11.3.3 Singularity at the Pole

Unfortunately, this chart is not entirely global. Indeed, it fails at two points: the north and south poles. After all, at the pole, longitude can never be defined uniquely any more. With respect to this chart, the pole is singular. Around it, we must use an alternative chart.

### 11.3.4 Regular Chart in the Arctic

How to design it? This could be done in many different ways. The present approach has an advantage: it sticks to the standard coordinate for as long as possible.

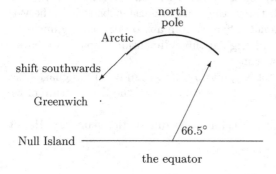

**Fig. 11.1.** The northern hemisphere of the Earth: a view from the east. From the Arctic Circle and south, use the standard coordinates: latitude and longitude. In the Arctic itself, on the other hand, use new coordinates: shift the entire Arctic southwards, until the north pole hits the Null Island. There, use longitude and latitude as a chart for the shifted Arctic. Finally, shift it back north, with its new chart.

For a start, look at the Arctic Circle, at latitude $66.5°$ (Figure 11.1). To the north of it, we have the Arctic, with the north pole at the middle. How to draw a new (regular) chart there?

For this purpose, shift the entire Arctic southwards. In this process, the north pole will shift southwards as well, through Greenwich, until it hits the equator at the origin $(0,0)$ (latitude $0°$, and longitude $0°$). This is the Null Island, in the gulf of Guinea, in west Africa. Around it, we already have a good chart: the familiar longitude and latitude. Use them as a chart for the shifted Arctic. Finally, shift it back north, together with its new chart. This way, the Arctic is now covered with a new regular chart, as required.

### 11.3.5 Local Coordinates

In the new chart in the Arctic, what are the coordinates of the north pole? Well, they are inherited from the Null Island: $(0,0)$. These are local coordinates: they label the north pole uniquely in terms of the new chart in the Arctic. Likewise, each point in the Arctic inherits its new coordinates from the surroundings of the Null Island. Again, these are just local coordinates: they "live" in the Arctic only, not outside it.

Thus, the new coordinates (inherited from the Null Island and its surroundings) map the entire Arctic continuously into $\mathbb{R}^2$, labeling each individual point by two local coordinates. Could this be done in the entire sphere as well, in a consistent way?

### 11.3.6 Compatible Charts on the Arctic Circle

subtropics                    subtropics

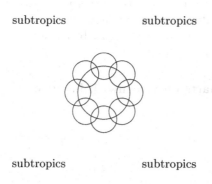

subtropics                    subtropics

**Fig. 11.2.** The northern hemisphere: a view from above. The big circle is the Arctic Circle. It is covered by a list of small overlapping circles, round and round.

subtropics                    subtropics

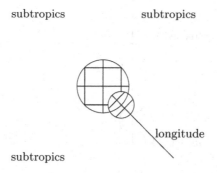

longitude

subtropics

**Fig. 11.3.** The northern hemisphere: a view from above. The Arctic has now a nonstandard chart, borrowed from the Null Island and its surroundings. The small circle on the bottom-right, on the other hand, still sticks to the standard chart: oblique longitude lines (directed to the north pole), and perpendicular latitude lines. In its northern part, where it overlaps with the Arctic, both charts are compatible: smoothly transferable from one to another.

Unfortunately, there is still a problem: what to do on the Arctic Circle itself? To its south, we still use the standard chart: longitude and latitude. To its north, on the other hand, we use our new chart. How to make them match?

For this purpose, let's cover the Arctic Circle with a list of small overlapping circles, round and round (Figure 11.2). (Each circle is open: only its interior, without its boundary.) In each small circle, let's keep the original coordinates: longitude and latitude.

Now, look at one small circle (Figure 11.3). In its southern part, there is no problem: it matches. Likewise, it also matches with its neighbors: the small circles on its east and west. But what about its northern part, where it overlaps with the Arctic? There, the charts don't match any more. Fortunately, they are still compatible: they could be transformed to one another by a smooth mapping, differentiable as many times as you like.

### 11.3.7  Compatible Charts on Greenwich Longitude

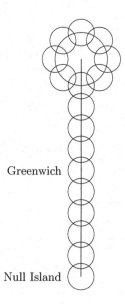

Greenwich

Null Island

**Fig. 11.4.** The Greenwich longitude is also covered by a list of small overlapping circles. This accounts for the jump in longitude: from $1°$, $2°$, $3°$, ... on the east, to $179°$, $178°$, $177°$, ... on the west.

We are not done yet. There is still one more problem: what to do on the Greenwich longitude? After all, there is a discontinuity there: on the east, the longitude is numbered $1°$, $2°$, $3°$, .... On the west, on the other hand, it is numbered $179°$, $178°$, $177°$, .... How to fix this?

We already know how. After all, we already had the same problem in the small circles that cover the Arctic Circle. Indeed, look at the bottom circle in Figure 11.2. In it, we want to use longitude and latitude. Still, to guarantee continuity, we must also allow negative longitude: ..., $-3°$, $-2°$, $-1°$, $0°$, $1°$, $2°$, $3°$, .... This matches on the east, but not on the west. Fortunately, it is still compatible. After all, adding 180° is a very smooth mapping.

The same trick could work on the entire Greenwich longitude as well: cover it with a list of small overlapping circles (Figure 11.4). On each of them, allow negative longitude as well. This way, they match on the east, and also on the west (up to an additive constant). Thus, we still have compatibility on the west as well, as required.

### 11.3.8 Atlas of Compatible Charts

The above design could be mirrored in the southern hemisphere as well. This covers the entire sphere with compatible charts all over. This is called an atlas. Thanks to it, the sphere is about to become a differentiable manifold.

The above is just one example: there are many other legitimate atlases. Which one to pick?

### 11.3.9 Maximal Atlas

Our atlas is still rather small: it contains only a few charts. We can now extend it into a bigger atlas: just introduce a new chart around some point on the sphere. In this step, compatibility must be preserved: the new chart must be compatible to all previous ones. This could be repeated time and again, producing a chain of bigger and bigger atlases, which include one another in a row.

Is there a maximal atlas that can't be extended any more? To answer this, let's use a powerful tool from set theory: Zorn's lemma.

### 11.3.10 Using Zorn's Lemma

To use Zorn's lemma, consider a chain of bigger and bigger atlases that include one another in a row. Does it have an upper bound? In other words, is there a yet bigger atlas that includes all atlases in the chain? Easy: just look at their union. Is it a legitimate atlas? To check on this, pick two charts from it. Are they compatible to one another? They must be. After all, they must belong to some atlas in the chain.

Thus, every chain of atlases indeed has an upper bound: their union. Thanks to Zorn's lemma, there exists a maximal atlas. It has two attractive properties: on one hand, all charts in it are compatible to one another. On the other hand, no new chart could be introduced without violating compatibility.

### 11.3.11 Differentiable Manifold

Thanks to the maximal atlas, we can now consider not only our original coordinates, but also any new system of coordinates, obtained by a smooth invertible transformation. This is benign: the new coordinates could be differentiated with respect to the old ones, and vice versa. Together with this maximal atlas, the sphere indeed makes a legitimate differentiable manifold.

### 11.3.12 Using the Axiom of Choice

Our original atlas contained only a few charts. The maximal atlas, on the other hand, may be much bigger, and even contain infinitely many charts. In fact, each point on the sphere may be contained in many overlapping charts (compatible with each other). Which one to pick?

Fortunately, in set theory, we have the axiom of choice to help do this. Still, the axiom of choice is implicit: it offers no explicit method to pick one suitable chart. This job is left to us. Here is one attractive suggestion.

### 11.3.13 Riemann Normal Coordinates

For each individual point on the sphere, pick a little chart around it. Then, shift the chart a little, so that the point is now labeled by $(0,0)$ in its new (private) chart. These are Riemann normal coordinates. They are only local, not global. Still, they are particularly useful to design a vector field.

## 11.4 Vector Field and Its Basis

### 11.4.1 Continuous and Smooth Real Functions

Consider the space of continuous real functions:

$$C^0(\mathbb{R}) \equiv \{f \mid f \text{ is continuous in } \mathbb{R}\}.$$

In it, consider a smaller subspace of functions: those that are not only continuous but also smooth:

$$C^\infty(\mathbb{R}) \equiv \{f \mid f,\ f',\ f'',\ f''',\ \dots \text{ are continuous in } \mathbb{R}\}.$$

What is so good about this function space? Well, each function in it is differentiable as many times as you like. This is quite useful in physics.

Now, consider a very simple function: a constant function that never changes:

$$f_1 \equiv 1.$$

$f_1$ will tell us that we are on a straight line: the real axis. Still, as a constant function, $f_1$ can never distinguish between different points on this line. This job is left to its (nonconstant) coefficient.

### 11.4.2 How to Span the Real Functions?

Consider now a new real function $g$. How to "span" it, or rewrite it in terms of $f_1$? For this purpose, we must do a silly thing: multiply $f_1$ by a "new" (nonconstant) coefficient, which is actually the same as $g$:

$$g(x) = \lambda_1(x) f_1,$$

where

$$\lambda_1(x) \equiv g(x).$$

What's good about this? It splits our jobs: $\lambda_1$ takes now the job of distinguishing between different $x$'s, whereas $f_1$ takes a more geometrical job: to tell us that we have just one function, not two. This sounds silly, but is not. To see how important this is, let's move on to a higher dimension.

### 11.4.3 Vector Field or Function

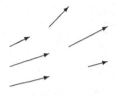

**Fig. 11.5.** A vector field: to each point in the Cartesian plane, attach a small arrow, issuing from it, and pointing in some direction.

Let's move on to a more interesting geometry: the two-dimensional Cartesian plane $\mathbb{R}^2$. In this case, we consider a vector field: a pair of two scalar functions, both defined in the Cartesian plane. Together, they make a new arrow, or direction vector (Figure 11.5).

As a matter of fact, this is not only a vector field but also a vector *function*, or an actual mapping: each point in $\mathbb{R}^2$ is mapped to the arrow issuing from it, which could be a completely different vector in $\mathbb{R}^2$. Still, this is not always the case: later on, we'll see a vector field that is *not* a function.

Consider now a subspace of vector fields (or functions). Denote it by $T$:

$$T \subset \left(\mathbb{R}^2\right)^{\left(\mathbb{R}^2\right)} \equiv \left\{f \mid f : \mathbb{R}^2 \to \mathbb{R}^2\right\}.$$

What is $T$? For now, we only know that it contains some vector fields. How does a vector field look like? Well, to each point in $\mathbb{R}^2$, it attaches a small arrow, issuing from this point, and specifying a new direction vector. Still, $T$ contains only "good" vector fields.

### 11.4.4 Continuous Vector Field

To belong to $T$, the vector field must be continuous: if the original point changes a little, then the arrow issuing from it could change as well, but only a little. For this purpose, the vector field must be made of two continuous (scalar) functions:

$$T \subset C^0\left(\mathbb{R}^2\right) \times C^0\left(\mathbb{R}^2\right) = \left(C^0\left(\mathbb{R}^2\right)\right)^2.$$

### 11.4.5 Smooth Vector Field

Better yet, $T$ should contain only smooth vector fields: as the original point changes, the arrow should change smoothly (Figure 11.6). For this purpose, the vector field must be made of two smooth (real) functions:

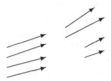

**Fig. 11.6.** A smooth vector field: if the original point changes a little, then the arrow issuing from it may change as well, but this change must be small and smooth in terms of both length and direction.

$$T \equiv C^\infty \left( \mathbb{R}^2 \right) \times C^\infty \left( \mathbb{R}^2 \right) = \left( C^\infty \left( \mathbb{R}^2 \right) \right)^2.$$

This kind of vector field is most useful in physics.

### 11.4.6 Basis of Vector Fields

How to span these vector fields? Again, we better split into two jobs. Our first job is to take care of our new two-dimensional geometry. For this purpose, we need two constant "functions:"

$$f_1 \equiv \begin{pmatrix} 1 \\ 0 \end{pmatrix} \quad \text{and} \quad f_2 \equiv \begin{pmatrix} 0 \\ 1 \end{pmatrix}.$$

This reflects well the two-dimensional nature of the arrows in our vector field. Indeed, $f_1$ spans the first spatial dimension, whereas $f_2$ spans the second. Together, they form a basis for $\mathbb{R}^2$. This is why $f_1$ is often denoted by $\partial/\partial x$, and $f_2$ by $\partial/\partial y$.

### 11.4.7 How to Span a Vector Field?

Still, as constant functions, $f_1$ and $f_2$ can never distinguish between different points in $\mathbb{R}^2$. For this job, they must have new (nonconstant) coefficients. This way, a given (smooth) vector field $g \in T$ can now be represented uniquely as

$$g = \lambda_1 f_1 + \lambda_2 f_2,$$

where $\lambda_1$ and $\lambda_2$ are new (scalar) functions, defined in $\mathbb{R}^2$, and dependent on $g$. This way, they take the job of distinguishing between different points in $\mathbb{R}^2$. This frees the constants $f_1$ and $f_2$ to focus on their geometrical job: to distinguish between the horizontal and vertical dimensions in each arrow.

In the above example, $\mathbb{R}^2$ is spanned by an elegant geometrical basis: $f_1$ and $f_2$. In a more complex geometry, on the other hand, there may be no such basis at all.

# 11.5 Vector Field on a Sphere

### 11.5.1 Riemann Normal Coordinates: Local and Tangent

Consider now a more complicated case: the original points are no longer in the Cartesian plane, but on a sphere. In other words, from each point on the sphere, issue a little arrow, tangent to the sphere. This way, in the tangent plane, we can now draw two axes, to make two local coordinates, fitting with our local chart.

In theory, we could draw them as we like. Still, in the tangent plane, they should better be perpendicular to each other. The $x$-direction is denoted by $(1, 0)$ or $\partial/\partial x$, and the $y$-direction by $(0, 1)$ or $\partial/\partial y$. In terms of these axes, the original point lies at the origin $(0, 0)$, and the arrow leads to $(x, y)$ on the tangent plane (outside of the sphere).

These are Riemann normal coordinates. How to specify them more clearly? This will be our next job.

### 11.5.2 Tangent Plane — Local Chart

Thus, $\partial/\partial x$ is a vector field in its own right. After all, from each individual point on the sphere, it points in some direction, tangent to the sphere. This could be the horizontal direction. Still, this is not a must.

Likewise, $\partial/\partial y$ is a vector field as well. Usually, it is perpendicular to $\partial/\partial x$ in the tangent plane. Still, this is not a must.

Together, $\partial/\partial x$ and $\partial/\partial y$ span the tangent plane: $\partial/\partial x$ points horizontally (aligning with the latitude), and $\partial/\partial y$ is perpendicular to it (aligning with the longitude). There could be many other choices, but this choice makes a lot of sense. This spans the tangent plane: the geometrical realization of our local chart.

So far, we only have vector *fields*: each arrow "loves" in its own tangent plane (or local chart). Now, is it possible to identify all tangent planes with each other, and produce not only a vector field but also a vector *function*?

### 11.5.3 Vector Field: Not a Function!

Let's focus on $\partial/\partial x$. So far, it is only defined locally, in each individual chart, tangent to the sphere. This makes it a legitimate vector field. Still, is it also a legitimate vector *function*?

For this purpose, consider all tangent planes as one and the same. In this case, could $\partial/\partial x$ be defined not only locally but also globally, everywhere on the sphere at the same time? In other words, is

$$\frac{\partial}{\partial x} \in (C^\infty \, (\text{sphere}))^2?$$

No! Indeed, at the poles, this global definition would lead to singularity, and even discontinuity:

$$\frac{\partial}{\partial x} \notin (C^0 \, (\text{sphere}))^2.$$

How to fix this? How to smooth it out at the poles?

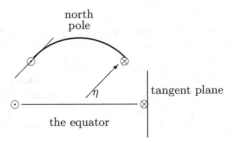

**Fig. 11.7.** The northern hemisphere of the Earth: a view from the side. On the right, there is a vertical tangent plane at the equator. On the upper-left, on the other hand, there is an oblique tangent plane at latitude $0 < \eta < \pi/2$.

### 11.5.4 Tangent Planes at the Equator

Let's focus again on $\partial/\partial x$. At the equator, it could be viewed as a legitimate vector function (Figure 11.7). Indeed, in each tangent plane, it indeed points horizontally, along the equator. This could be viewed as a constant function, just like $f_1$ in Section 11.4.6. As a constant function, it is indeed continuous and smooth, as required. In fact, it spans the first Riemann normal coordinate: the horizontal coordinate. This helps envelope the entire equator with tangent horizontal arrows, round and round. (Together with $\partial/\partial y$, this could even make a rotating coordinate system, spanning the tangent planes that envelope the Earth all over.)

The equator is a level set: the latitude is $\eta = 0$. Consider now a higher latitude of $0 < \eta < \pi/2$. At this latitude, look at a horizontal cross section of the Earth (Figure 11.7). The same approach works here as well: $\partial/\partial x$ is still constant, and helps envelope the entire cross section with horizontal arrows, round and round.

### 11.5.5 Singularity at the North Pole

But what happens as $\eta \to \pi/2$, approaching the north pole? In this case, $\partial/\partial x$ is still valid in the local chart, tangent to the north pole. As such, it still makes a legitimate vector field. However, it can no longer be viewed as a vector function. Indeed, as such, it would becomes discontinuous. Indeed, as $\eta \to \pi/2$, the entire cross section shrinks to just one point: the north pole itself. This singular point could be approached from either side: Upon approaching the north pole from the east (along longitude 90°), $\partial/\partial x$ points in one direction. Upon approaching from the west (along longitude 270°), on the other hand, $\partial/\partial x$ points the other way around. How to smooth this out?

## 11.6 Generating System

### 11.6.1 How to Smooth Out?

To smooth this out, let's define a new vector function $f_1$ that decreases monotonically at higher and higher latitudes, and vanishes smoothly at the north pole:

$$f_1 \equiv \exp\left(-\tan^2(\eta)\right)\frac{\partial}{\partial x}.$$

This new $f_1$ is still horizontal. Still, at higher and higher latitudes, the arrows get shorter and shorter, until vanishing completely at the north pole. The same is mirrored at the south pole as well. Thus, $f_1$ makes not only a vector field but also a legitimate vector function, smooth on the entire sphere.

### 11.6.2 Horizontal Level Sets

For our new $f_1$, the equator is still a level set. Indeed, on the equator, where $\eta = 0$, $f_1$ is the same as $\partial/\partial x$: $(1,0)$.

Likewise, at a fixed altitude $0 < \eta < \pi/2$, the cross section makes a level set for $f_1$ as well. After all, $f_1$ is just a scalar multiple of $\partial/\partial x$ there. As such, it still points horizontally, tangent to the cross section, enveloping it with little horizontal arrows, round and round.

Unfortunately, as $\eta \to \pm\pi/2$, $f_1$ is not very useful any more: it approaches zero. As a matter of fact, the north pole itself makes a degenerate level set: the zero level set. To fix this, we need yet another vector function, which will "live" at the poles as well. Furthermore, it will complete $f_1$: it will point not horizontally but vertically.

### 11.6.3 Oblique Tangent Direction

How to do this? Let $W$ be the rotation that rotates the entire sphere counterclockwise (Figure 11.8). This way, the north pole "slides" southwards along longitude $270°$, until hitting the equator at the Galapagos islands. In this process, each point $p$ on the sphere is mapped to $Wp$, carrying its own arrow $f_1(p)$ with it. In its new place, the arrow gets a new name:

$$f_2(Wp) \equiv f_1(p).$$

This way, $f_2$ points obliquely, not horizontally. Furthermore, its level sets are now vertical loops, not horizontal.

### 11.6.4 Vertical Level Sets

How does a level set of $f_2$ look like? In Figure 11.8, look at a fixed angle $\kappa$. This marks a vertical cross section of the Earth. Around it, $f_2$ makes a vertical loop of arrows, enveloping it with tangent arrows all over, round and round, from south to north and back.

Often, the new level sets of $f_2$ cross those of $f_1$. This way, at each individual point on the globe, they could be used to span the tangent plane, in terms of two new local coordinates. Unfortunately, this could break.

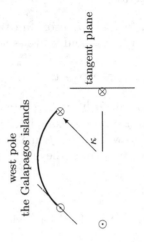

**Fig. 11.8.** The rotation $W$ rotates the entire sphere counterclockwise. This way, the north pole slides along longitude 270°, until it hits the equator at the Galapagos islands. In this process, each point carries with it the horizontal arrow $f_1$. This makes a new arrow $f_2$.

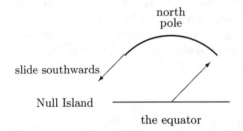

**Fig. 11.9.** The rotation $Q$ rotates the entire sphere: the north pole slides along the Greenwich longitude, until it hits the equator at the Null Island. In this process, each point carries with it the original arrow $f_1$. In its new place, it is called $f_3$.

### 11.6.5 A Third Tangent Direction

Unfortunately, along longitudes $90°$ and $270°$, there is still a problem: $f_2$ aligns with $f_1$: they are both horizontal. So, we still have only one tangent direction. How to design another one?

For this purpose, let $Q$ be yet another rotation of the entire sphere. Here, however, the north pole slides along the Greenwich longitude, until hitting the equator at the Null Island (Figure 11.9). In this process, each point $p$ on the sphere is mapped to $Qp$, carrying its own arrow $f_1(p)$ with it. In its new place, the arrow gets a new name:

$$f_3(Qp) \equiv f_1(p).$$

Like $f_2$, $f_3$ points obliquely, and its level sets are vertical loops. We're particularly interested in longitudes $90°$ and $270°$. This is indeed a vertical level set of $f_3$: it loops the Earth with tangent arrows, round and round, from north to south and back. This aligns neither with $f_1$ nor with $f_2$. On the contrary: it is perpendicular to them there. Thus, this supplies the missing local coordinate in the tangent planes there.

Together, $f_1$, $f_2$, and $f_3$ make a generating system, ready to represent just any vector function. Let's see how.

### 11.6.6 Generating System

By now, at each point on the globe, we already have three level sets. Of these, at least two don't align with one another, and can serve as local coordinates, to help draw the local chart, and span the tangent plane.

Usually, three is too many. After all, in a two-dimensional manifold, two local coordinates should be enough to span the two-dimensional tangent plane. Still, to be on the safe side, we better have three. After all, two of them may align with each other, and point in the same direction, producing one and the same coordinate. Furthermore, at the poles, a level set may even shrink and degenerate completely. In such a case, the third one comes to our aid. Indeed, it is perpendicular to the two others there, and can therefore supply the missing coordinate.

Thus, two local coordinates are good enough locally, to span the local chart and the tangent plane, and indeed represent any given vector field. A vector *function*, on the other hand, is trickier: it is defined not only locally but also globally, in the entire sphere. To represent it, we must therefore have three vector functions.

Indeed, consider a given vector function

$$g \in (C^\infty \, (\text{sphere}))^2 \, .$$

It can now be represented (albeit not uniquely) as

$$g = \lambda_1 f_1 + \lambda_2 f_2 + \lambda_3 f_3,$$

where

$$\lambda_1, \, \lambda_2, \, \lambda_3 \, \in C^\infty \, (\text{sphere})$$

are three scalar functions. These are the coefficients, which depend on $g$. Their job is to distinguish between different points on the globe.

This frees the vector functions $f_1$, $f_2$, and $f_3$ (which don't depend on $g$) to take care of the special geometry of the sphere and its tangent planes. Together, they form a new generating system.

### 11.6.7 Generating System: Not a Basis

Could we drop one of them? Unfortunately not. Indeed, to be smooth on the entire sphere, each of them must vanish somewhere. There, we must also have two others. In summary, the third one is necessary to supply the missing tangent vector wherever the two others align with each other (or degenerate).

The Cartesian plane is simple. This is why it has a basis of two (constant) vector functions (Section 11.4.6). The sphere, on the other hand, is more complicated. This is why it must have a generating system that is no longer a basis: it contains not only two but three vector functions. Together, they take care of the geometry, not only locally but also globally: they can span just any given vector function, once its coefficients are defined properly.

Below, however, we'll focus on vector fields that are *not* functions: they are defined only locally, not globally. To represent these, we don't need three: two are enough.

## 11.7 Time: Can It Ever Stop?

### 11.7.1 Zeno's Paradox

Why did we talk about the sphere so much? Because it could help model a slippery term: time.

To do this, consider again the northern hemisphere (Figure 11.9). On it, consider a particle that travels northwards: from the equator (at latitude $\eta = 0$), to the north pole (at latitude $\eta = \pi/2$).

Assume that the particle moves really fast: it can reach the north pole in just one second. Still, before doing this, it must first make half the way, and reach latitude $\eta = \pi/4$. This should take half a second. After doing this, it must still do one more quarter of the way, and reach latitude $\eta = 3\pi/8$. This should take another quarter of a second, and so on.

So, we have a paradox here: the particle must carry out infinitely many steps, each shorter than the previous ones. This is an infinite list. How could the particle ever complete the entire way, and reach the north pole in a finite number of steps? This is Zeno's paradox.

### 11.7.2 Are You Afraid of Infinity?

Fortunately, by now, we're no longer afraid of infinity. After all, we can now use an infinite power series to calculate the total time that the particle needs:

$$\frac{1}{2} + \frac{1}{4} + \frac{1}{8} + \frac{1}{16} + \cdots = 1.$$

Thus, the total time is indeed one second, as expected.

### 11.7.3 Towards a Black Hole

But what if the particle slows down? For instance, assume that, at each step, the particle gets twice as slow. In this case, the total time would be as large as

$$1 + 1 + 1 + 1 + \cdots = \infty,$$

and the particle would never reach!

When could this happen? When our hemisphere models a more complicated process: a particle traveling towards a black hole. Later on, we'll discuss this a little more, including an interesting phenomena: in our eyes, the particle can never arrive!

### 11.7.4 The Nonphysical Event Horizon

The black hole is surrounded by a (theoretical) sphere: the event horizon. The particle can only approach it from the outside, but never arrive. To reach it physically, it needs an infinite time.

Why? Because, as the particle approaches the event horizon, its time gets slower and slower. This is gravitational time dilation, to be discussed later.

This is only in our eyes. In its own self-system, the particle feels no change at all: its time keeps ticking as usual. Thus, in its own self system, it'll eventually pass the event horizon smoothly, and fall into the black hole. After all, the event horizon is not physical, but only mathematical.

### 11.7.5 Zeno's Paradox Comes True!

We, on the other hand, who look at the particle from here, see it slowing down, and never reaching the event horizon. Zeno's paradox comes true!

This also works the other way around: no particle can ever leave the event horizon and travel to our eyes in any finite time.

### 11.7.6 The Dark Event Horizon

Even a photon can never leave the event horizon and reach our eyes. This is why the event horizon looks dark.

Although it is as fast as light, the photon has not enough time to complete the entire journey. This is why, from our perspective, it remains stuck in the event horizon forever.

In its own self system, on the other hand, its time ticks as uniformly as ever. From its own point of view, it advances towards us, as fast as light. After all, it has no idea that it can never leave the event horizon in any finite time.

### 11.7.7 Singularity at the Event Horizon: Time Stops!

To us, the event horizon makes a singularity. On the way towards it, time gets slower and slower. For us, at the event horizon itself, time stops!

To make things a little more "plausible," let's see how this could help in science fiction.

### 11.7.8 Honey, I Shrunk the Kids!

In this movie (by Disney's studios), there is an absent-minded scientist who builds a new machine that can create a new environment with the same effects as in a very high speed: time slows down, and length shrinks.

Unfortunately, by mistake, the machine starts working with no control, and shrinks the kids! They were thrown into a strange part of spacetime, where both time and space shrink by the same factor. Below, we'll illustrate this geometrically.

There, the kids are in a totally different system, completely detached from here and now. In their jump to the new system, they feel or see no change at all. After all, their time and length keep the same ratio as before. This way, they still have the same speed of light. We, on the other hand, see them shrinking, and their time slowing down.

These systems are inherently disjoint, and could never unite or even meet. Still, in the movie, they do interact, for the sake of more drama and fun... This trick was already used in *The Wonderful Adventure of Nils Holgersson* by Selma Lagerlioff.

Let's discuss spacetime a little more, and use geometry to understand it better.

## 11.8 Spacetime in Special Relativity

### 11.8.1 Our Coordinate System

In special relativity, spacetime is particularly easy to draw [15, 61]. After all, everything is linear: there is no gravity, nor any other force, nor any curvature.

For simplicity, assume that our space is one-dimensional. In our system, this makes the $x$-axis. Together with the time axis on top, we have the familiar $x$-$t$ system.

For even more simplicity, assume also that $x$ is measured in light-seconds. This way, light travels at speed 1: one light-second per second. In spacetime, this draws an oblique line of angle $45°$ (Figure 11.10).

This is nice and symmetric: space and time take equal status. This way, a light ray could split spacetime obliquely and evenly. This works well not only in our coordinate system but also in any other (inertial) coordinate system, moving at a constant velocity with respect to ours.

### 11.8.2 Traveling Coordinate System

Now, consider a spaceship, traveling at a constant speed $v$ away from us. (Again, $v$ is measured in light-seconds per second.) Clearly, $v$ can never exceed the speed of light:

$$|v| < 1.$$

In the spaceship, you carry your own clock and ruler. In them, you can measure your own time $t'$ and length $x'$. These make two new axes: the new $x'$-$t'$ coordinate system. In it, light still has the same speed: 1. Therefore, light must still draw an oblique line in between.

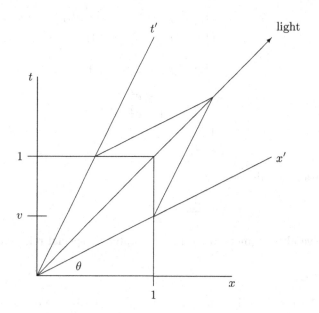

**Fig. 11.10.** Spacetime, illustrated in two different coordinate systems. The $x$-$t$ axes span our own system here. The $x'$-$t'$ axes, on the other hand, span another system, traveling rightwards at the constant speed $v$ (away from us). In both systems, light travels at the same speed, drawing an oblique line in spacetime.

### 11.8.3 Area in the New Coordinates

Look again at Figure 11.10. In it, what is the angle between the $x$- and $x'$-axes? This is $\theta$, satisfying

$$\tan(\theta) = v.$$

Thanks to symmetry, this is also the angle between the $t$- and $t'$-axes. After all, in special relativity, time and space take equal status.

Now, the $x'$-$t'$ axes make a new diamond. What is its lower-left angle? This is just

$$\frac{\pi}{2} - 2\theta.$$

This is also the upper-right angle.

How long is the edge of the diamond? Thanks to Pythagoras theorem, the answer is

$$\sqrt{1 + v^2}.$$

Thanks to Euclidean geometry, we are now ready to calculate the area of the entire diamond:

$$\text{area of diamond} = \left(1+v^2\right)\sin\left(\frac{\pi}{2}-2\theta\right)$$
$$= \left(1+v^2\right)\cos(2\theta)$$
$$= \left(1+v^2\right)\left(\cos^2(\theta)-\sin^2(\theta)\right)$$
$$= \left(1+v^2\right)\left(1-2\sin^2(\theta)\right)$$
$$= \left(1+v^2\right)\left(1-\frac{2v^2}{1+v^2}\right)$$
$$= \left(1+v^2\right)\frac{1-v^2}{1+v^2}$$
$$= 1-v^2$$
$$= \gamma^{-2},$$

where $\gamma$ is a new parameter, defined by

$$\gamma \equiv \frac{1}{\sqrt{1-v^2}} \geq 1.$$

Is this familiar? We've already met this $\gamma$ in the chapter on special relativity.

### 11.8.4 Time Dilation

Thus, to have area 1 (like the original square in Figure 11.10), the diamond must get bigger. For this purpose, the $t'$-axis must stretch by factor $\gamma$: from here, each second measured in the spaceship looks as long as $\gamma$ seconds (Figure 11.11). This is called time dilation. Is this familiar? We've already seen this before. Here, however, it is more vivid: not only algebraic, but also geometrical.

### 11.8.5 Length Contraction

For the sake of symmetry, the $x'$-axis must stretch by the same factor: $\gamma$. This means that, from here, your ruler (that you carry on the spaceship) looks $\gamma$ times as short. Why?

Well, assume that your ruler is as long as a light-second, so it ends at $x' = 1$. Now, in the spaceship, the ruler remains static, ending at $x' = 1$ all the time, and drawing the oblique line

$$x' \equiv 1$$

(the short oblique line in Figure 11.11). What slope does it have? Well, it is parallel to the $t'$-axis, so it makes angle $\theta$ with the $t$-axis. Thanks to triangle similarity, it hits the $x$-axis at

$$\gamma - \frac{(\gamma v)^2}{\gamma} = \gamma\left(1-\frac{(\gamma v)^2}{\gamma^2}\right) = \gamma\left(1-v^2\right) = \gamma\cdot\gamma^{-2} = \gamma^{-1}.$$

This is why, from here, the ruler indeed looks $\gamma$ times as short. This is called length contraction.

Isn't this familiar? We've already seen it before. This is indeed special relativity: no gravity is assumed. What happens in general relativity? Well, gravity is now allowed. As a result, we may have the same effects as above. After all, at a place of high gravity, objects may fall at a very high speed.

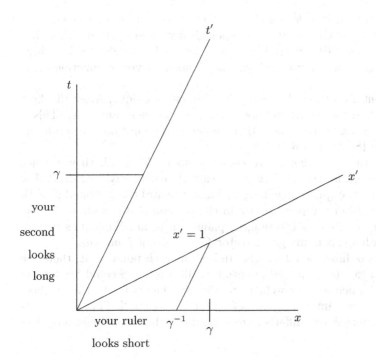

**Fig. 11.11.** The stretched $x'$-$t'$ system of the spaceship. From here, one second of their clock looks as long as $\gamma \geq 1$ seconds. This is time dilation, illustrated by the short horizontal line. Furthermore, from here, their meter looks as short as $\gamma^{-1} \leq 1$ meters. This is indeed length contraction, illustrated by the short oblique line.

### 11.8.6 Singularity

What happens when the spaceship is nearly as fast as light:

$$v \to 1?$$

In this limit case, the above diamond gets degenerate: too narrow, with no width at all. This is a singularity: space and time shrink to zero. Time is too lazy to move, so every motion must be degenerate: too small to notice at all. Energy and momentum, on the other hand, get infinite. This looks like complete chaos: time stops!

In spatial relativity, this is only theoretical: a limit case. In general relativity, on the other hand, this may really happen at the center of a black hole. There, only the laws of quantum mechanics could save us.

## 11.9 Spacetime in General Relativity

### 11.9.1 From the Sphere to Spacetime

To understand spacetime better, let's go back to our sphere for another look. What is so special about the sphere? Well, it is topologically different from $\mathbb{R}^2$: it has two "sides:" the inner, and the outer. This is why we need not only two but three vector functions to generate (represent) just any (smooth) vector function on it (Section 11.6.6).

Still, this representation has a drawback: it is no longer unique. After all, three basis functions are often too many, and not linearly independent any more. This is why they no longer form a basis. In fact, they no longer *span* but only *generate* the given vector function (Section 11.6.6).

Vector fields, on the other hand, are easier to span. After all, they are not defined globally in the entire sphere, but only locally, in each individual chart. For example, each point on the sphere may have its local (private) chart around it, with two local axes issuing from the point: $\partial/\partial x$ in the horizontal direction, and $\partial/\partial y$ in the perpendicular direction in the tangent plane. These are Riemann's normal coordinates that can help span any given vector *field* (but not *function*).

The sphere is a two-dimensional surface. Indeed, at each point on it, there are just two coordinates: say, latitude and longitude. Still, the same could be done in three (or even four) dimensions as well. In particular, thanks to set theory (Sections 11.3.9–11.3.11), a maximal atlas exists for spacetime as well. This way, spacetime makes a four-dimensional differentiable manifold: the time dimension, and three spatial dimensions.

### 11.9.2 Curvature

In special relativity, spacetime is completely flat and linear, with no curvature at all. The sphere, on the other hand, has a constant positive curvature. This is why it is nice and symmetric. Unlike both, in general relativity, spacetime may be completely nonlinear and uneven, with a nonconstant curvature that may change sharply from event to event. Fortunately, in each individual event, in just any coordinate system around it, the curvature can still be written in terms of energy and momentum.

### 11.9.3 Spacetime and Its Tangent Spaces

How to linearize our sphere? Envelope it with tangent planes. Likewise, spacetime can be enveloped with tangent spaces. Still, because spacetime is four-dimensional, these spaces are four-dimensional as well.

Around each event in spacetime, in its own private chart, we could then design four new vector fields, to span the tangent space. These are often named $\partial/\partial x^0$, $\partial/\partial x^1$, $\partial/\partial x^2$, and $\partial/\partial x^3$. These could then span just any given vector field, locally in each tangent space.

### 11.9.4 Basis of Local Tangent Vector Fields

Around each individual event in spacetime, these four would then be tangent to the original level sets in the original chart, which define the four local coordinates. In particular, if the original event has coordinates $(0, 0, 0, 0)$, then these are Riemann's normal coordinates, issuing from it.

If spacetime is linear and flat, as in special relativity, then these four could be viewed not only as vector fields but also as vector functions, defined globally and smoothly, everywhere in spacetime. These could then serve as a basis, to span just any given vector function. Indeed, in this case, spacetime is just $\mathbb{R}^4$, and the (constant) basis functions are just the standard unit vectors:

$$\frac{\partial}{\partial x^0} = \begin{pmatrix} 1 \\ 0 \\ 0 \\ 0 \end{pmatrix}, \quad \frac{\partial}{\partial x^1} = \begin{pmatrix} 0 \\ 1 \\ 0 \\ 0 \end{pmatrix}, \quad \frac{\partial}{\partial x^2} = \begin{pmatrix} 0 \\ 0 \\ 1 \\ 0 \end{pmatrix}, \quad \frac{\partial}{\partial x^3} = \begin{pmatrix} 0 \\ 0 \\ 0 \\ 1 \end{pmatrix}.$$

In general relativity, on the other hand, spacetime is not flat any more.

### 11.9.5 Basis vs. Generating System

If, on the other hand, spacetime has a more complex geometry, as in general relativity, then four are not enough any more. To generate just any given vector function, one needs five elementary vector functions. After all, as we'll see later, Riemann's normal coordinates are only local, not global.

Fortunately, here we are more interested in vector fields, defined only locally, not globally. To span these, we only need four tangent vector fields, to form a basis, and local coordinates.

## 11.10 Spacetime and Gravity

### 11.10.1 Curved Geometry in Spacetime

What is our sphere? It is a curved two-dimensional manifold. Furthermore, at each point on it, the curvature is a positive constant, which depends on the radius only. Spacetime, on the other hand, is much more complicated: it is a four-dimensional manifold. Indeed, it has three spatial dimensions, plus time on top.

In spacetime, the curvature is not only spatial but also temporal. In other words, it stems not only from the nonlinear gravity in space, but also from its effect on time. Indeed, gravity makes time nonuniform and nonlinear: slow near a massive star, and quick elsewhere.

How curved is spacetime, and where? In other words, what determines the curvature in spacetime?

### 11.10.2 Metric and Shortest Path

Thanks to Einstein, energy and momentum are not only physical but also geometrical: sources of mass, gravity, curvature, and symmetry. They determine the true metric in spacetime. This way, around each event in spacetime, you can imagine spacetime as a four-dimensional hypercube in the $x^0$-$x^1$-$x^2$-$x^3$ coordinates (picked arbitrarily in advance). Now, instead of stretching spacetime to have its true curvature, just define its metric, which tells us the true distance between two events: the length of the shortest path between them.

### 11.10.3 It Looks Curved, But it is Straight

Still, because spacetime is curved, such a path may no longer look straight in the usual sense. After all, the time axis is no longer uniform: the time scale may change from place to place. Near a massive star, for example, time gets slower.

This is indeed relativity: time is no longer absolute, but only relative to the place where it is being measured. If this place contains a lot of matter, then time is slower there.

### 11.10.4 Gravitational Time Dilation

This is called gravitational time dilation. To understand this better, consider an atomic clock that uses some radioactive process to measure time. Assume that the clock is placed near a massive star. To us humans, a second of this clock seems slower: it may take a few seconds of our time, here on the Earth.

Assume that one second has passed there. During this time, their light made one light-second there. To us humans, this looks slow. After all, we think that a few seconds passed, so light should make a few light-seconds! Once we realize that their clock runs slow, we accept this: their light should indeed make one light-second only, without violating Einstein's Law.

### 11.10.5 Gravitational Redshift

The light not only travels at a constant speed but also oscillates like a wave, at a constant frequency: constant number of cycles per second. In each cycle, it covers a certain number of kilometers. This is its wave-length, which specifies its color.

Still, here on the Earth, our second is quicker. Therefore, upon arriving on Earth, the light is less frequent: per Earth-second, it makes less cycles. As a result, it changes color: it becomes less blue, and more red.

This is called gravitational redshift: we see the light redder than it originally was upon leaving the star in the first place. In other words, the star looks redder than it really is. A very massive star may even look so red that it could hardly be seen at all.

## 11.11  Light Can Curve!

### 11.11.1  Light Near a Massive Star

What happens to a light ray that passes by a massive star? Well, as it gets closer and closer to the star, its time gets slower, so it makes a shorter distance than before. So, it must turn a little, and draw an arc around the star.

Around a star as massive and dense as a black hole, the light ray could even orbit forever, with no escape. This happens at the event horizon. If it gets any closer to the black hole, then it would spiral, and eventually fall right into the black hole. In either case, we'd never see it any more.

This is indeed why it is called black hole: it looks dark: no light could ever escape from it, and get to our eyes. This could also be explained in terms of gravitational redshift: it gets so red that it can no longer be seen at all.

### 11.11.2  "Straight" Line

In our eyes, this kind of spiral doesn't look straight at all. In terms of the curved geometry in spacetime, on the other hand, this is as straight as ever. Indeed, in spacetime, to be straight means to follow a "valley" where time is as slow as possible. Indeed, in a slow-time zone, the light ray follows a path that is highly stretched-out and straight (as in a degenerate diamond in Figure 11.11). In terms of its own self system, on the other hand, the light ray is static, making a vertical line, along its own self-time-axis, in its own self-spacetime.

From Earth, the light ray may seem curved. From its own perspective, on the other hand, it is vertical. After all, its own private time (measured in its own clock) remains as fast as before. Their time looks slow only from here, not from there.

### 11.11.3  Light Near a Black Hole

A black hole is a special kind of star — so massive and dense that even light can't escape from its powerful gravity. Still, a light ray that approaches a black hole would "feel" nothing unusual. Why? Because its own private time still ticks at the same rate as before.

After all, the light ray remains in a free fall, feeling no new force at all. Although gravity acts upon it quite strongly, this affects it only in the eyes of an observer who watches it from here. From its own perspective, on the other hand, the light ray feels much calmer: no force or acceleration or any change to its time rate.

### 11.11.4  Einstein's Happy Thought

This is Einstein's happy thought (or equivalence principle), which started general relativity in the first place. This could also be viewed geometrically: at each individual event, spacetime is nearly the same as its tangent space, which is as flat as in special relativity. Therefore, at this particular event, the photon feels as at a constant velocity, which doesn't accelerate at all. At the next event, on the other hand, there will be yet another tangent space. After all, spacetime is a legitimate differentiable manifold: Riemann's normal coordinates will remain valid. This reasoning could continue on and on forever.

## 11.12  Metric in Spacetime

### 11.12.1  Spacetime — a Four-Dimensional Manifold

Where could a motion take place? In classical mechanics, it takes place in space. In general relativity, on the other hand, it takes place in spacetime. Together with the time dimension, this is a new four-dimensional differentiable manifold. Each individual point in spacetime is called an event: a four-dimensional vector, specifying not only the spatial location but also the time.

What coordinates are needed in spacetime? Pick four, as you like: say, $t$, $x$, $y$, and $z$. Fortunately, we can measure them in our lab. For the sake of uniformity, they are often denoted by an upper index:

$$x^0 \equiv t$$
$$x^1 \equiv x$$
$$x^2 \equiv y$$
$$x^3 \equiv z.$$

Here, the superscript 0, 1, 2, or 3 has nothing to do with power: it just indexes a coordinate in spacetime. This upper index is often denoted by a small Greek letter, say $\alpha = 0, 1, 2, 3$.

### 11.12.2  Four-Dimensional Vector

Likewise, a four-dimensional row vector contains four numbers in a row:

$$v \equiv \left(v^0, v^1, v^2, v^3\right).$$

Again, the upper index is often denoted by a small Greek letter:

$$v \equiv (v^\alpha)_{0 \leq \alpha \leq 3}.$$

Again, $\alpha$ is no power: it just indexes the four components in the vector. Let's extend this into a square matrix.

### 11.12.3  Square Matrix

The above vector contains four numbers (components), listed one by one in a row. A $4 \times 4$ matrix, on the other hand, is a table that contains 16 numbers (elements), listed row by row:

$$A \equiv (a_{\mu\nu})_{0 \leq \mu,\nu \leq 3} = \begin{pmatrix} a_{00} & a_{01} & a_{02} & a_{03} \\ a_{10} & a_{11} & a_{12} & a_{13} \\ a_{20} & a_{21} & a_{22} & a_{23} \\ a_{30} & a_{31} & a_{32} & a_{33} \end{pmatrix}.$$

(Here, $a_{\mu\nu}$ means $a_{\mu,\nu}$ — the comma is often dropped.) In $A$, we indeed have 16 elements of the form $a_{\mu\nu}$, indexed by two lower indices: $\mu, \nu = 0, 1, 2, 3$.

An important matrix is the metric:

$$g \equiv (g_{\mu\nu})_{0 \le \mu,\nu \le 3} = \begin{pmatrix} g_{00} & g_{01} & g_{02} & g_{03} \\ g_{10} & g_{11} & g_{12} & g_{13} \\ g_{20} & g_{21} & g_{22} & g_{23} \\ g_{30} & g_{31} & g_{32} & g_{33} \end{pmatrix}.$$

This is not a constant matrix. On the contrary: it depends on the particular event under consideration, and may change from event to event in spacetime.

The metric $g$ encapsulates important geometrical information. At each individual event in spacetime, it tells us how spacetime "stretches," and in what direction. Unfortunately, the true metric is not yet available. To uncover it, one must solve Einstein's equations (at least numerically). This requires a new algebraic tool: tensor.

## 11.13  Tensor: Extended Matrix

### 11.13.1  Tensor

The tensor may be viewed as an extension of the matrix: it can use more indices, upper or lower. For example, the $4 \times 4 \times 4 \times 4 \times 4$ tensor

$$T \equiv \left( T^{\alpha\beta}_{\gamma\delta\epsilon} \right)_{0 \le \alpha,\beta,\gamma,\delta,\epsilon \le 3}$$

contains as many as

$$4^5 = 1024$$

numbers (entries), indexed by two upper indices ($\alpha$ and $\beta$), and three lower indices ($\gamma$, $\delta$, and $\epsilon$).

Let's go ahead and "play" with tensors algebraically. Two tensors with the same number of indices could be easily added to each other, entry by entry. Still, there is a trickier algebraic operation: contraction.

### 11.13.2  Contraction

In the original tensor $T$, take one upper index, say $\beta$, and one lower index, say $\delta$. Rename both $\beta$ and $\delta$ by one and the same index, say $\rho$. Then, sum over $\rho = 0, 1, 2, 3$:

$$\sum_{\rho=0}^{3} T^{\alpha\rho}_{\gamma\rho\epsilon}.$$

This gives a new tensor: the contracted tensor. In it, $\rho$ is just a dummy index, summed over $\rho = 0, 1, 2, 3$. As such, $\rho$ is no longer free: you can't specify it any more. The only free indices are now $\alpha$, $\gamma$, and $\epsilon$: you can still specify them as you like. As a result, the new (contracted) tensor is smaller than the original one: it contains

$$4^3 = 64$$

entries only.

### 11.13.3 Einstein Summation Convention

Einstein's summation convention makes our lives easier. It says: drop the '$\sum$' sign! This gives the new shorthand:

$$T^{\alpha\rho}_{\gamma\rho\epsilon} \equiv \sum_{\rho=0}^{3} T^{\alpha\rho}_{\gamma\rho\epsilon}.$$

On the left-hand side, $\rho$ appears not only as an upper but also as a lower index. This tells us that $\rho$ is no longer free to specify. On the contrary: it is just a dummy index, summed over $\rho = 0, 1, 2, 3$, as in the right-hand side. This shorthand is used often in general relativity.

## 11.14   Tensors and Trees

### 11.14.1   Naive Implementation

In the above, we've considered a tensor $T$ with five indices. This, however, is not the maximum: a tensor could be yet bigger, and use as many indices as you like. How to implement a general tensor, with arbitrarily many indices?

The naive approach is geometrical: to implement a vector, use a row of four components. Besides, store one more bit: 1, to tell us that the index is upper, not lower.

Next, to implement a matrix, use four rows, to store a total of 16 elements. Besides, store two bits: 0 and 0, to tell us that both indices are lower, not upper.

Next, introduce more and more dimensions, recursively and geometrically. This way, to implement the above tensor $T$, use a $4 \times 4 \times 4 \times 4 \times 4$ hypercube, containing

$$4^5 = 1024$$

entries, as required. Besides, store five more bits: 1, 1, 0, 0, and 0, to tell us that $T$ has two upper indices, followed by three lower indices.

In practice, however, the computer must "know" the dimension in advance, to allocate sufficient memory. Therefore, recursion doesn't work any more: to have a bigger tensor with one more index, one must start from scratch, and implement new algebraic operations all over again. Is there a more efficient way?

To save work, better implement all possible tensors in one go, with as many indices as you like. For this purpose, better use a truly recursive structure: a multilevel hierarchy, or a tree.

### 11.14.2   Quaternary Tree

In discrete math, we've already seen a perfect binary tree (Figure 7.9). Let's extend this into a quaternary tree. For this purpose, make the following change: from each node, issue not only two but four branches.

How to do this formally? As always, use mathematical induction. A one-level quaternary tree is just a dangling node. It could be used to store a degenerate "tensor," with no index at all: a single scalar (number).

Now, for $n = 1, 2, 3, \ldots$, assume that we already know how to define an $n$-level quaternary tree. (This is the induction hypothesis.) At its bottom, it contains $4^{n-1}$ leaves, ready to store $4^{n-1}$ numbers (entries).

### 11.14.3  Implementing a Vector

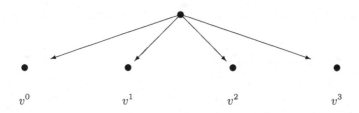

**Fig. 11.12.** A four-dimensional vector, implemented as a two-level tree. There are four leaves, ready to store the four components: $v^0$, $v^1$, $v^2$, and $v^3$.

For $n = 2$, for example, this is a two-level tree, with only four leaves, ready to store the components $v^0$, $v^1$, $v^2$, and $v^3$ in a given vector $v$ (Figure 11.12). Besides, store one more bit: 1, to tell us that the index is upper, not lower.

### 11.14.4  Implementing a Matrix

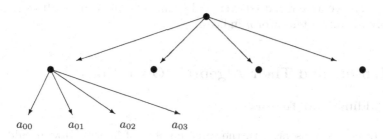

**Fig. 11.13.** A $4 \times 4$ matrix, implemented as a three-level tree. To reach $a_{03}$, pick the leftmost subtree. In it, pick the rightmost leaf. To reach $a_{30}$, on the other hand, work the other way around: pick the rightmost subtree. In it, pick the leftmost leaf.

For $n = 3$, on the other hand, this is already a three-level tree, with 16 leaves, ready to store the elements $a_{\mu\nu}$ in a given matrix $A$. In this tree, how to reach the element $a_{03}$, for instance? Look at Figure 11.13: pick the leftmost subtree. In it, pick the rightmost leaf. It indeed contains $a_{03}$, as required.

How to reach a more general element of the form $a_{\mu\nu}$ (for some $0 \leq \mu, \nu \leq 3$)? Look at Figure 11.14: pick the $\mu$th subtree. In it, pick the $\nu$th leaf. It indeed contains $a_{\mu\nu}$, as required.

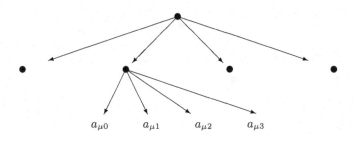

**Fig. 11.14.** To reach a general element of the form $a_{\mu\nu}$, pick the $\mu$th subtree. In it, pick the $\nu$th leaf.

Finally, store two more bits: 0 and 0, to tell us that both indices are lower, not upper. Next, let's extend the above: define an $(n+1)$-level quaternary tree, ready to store a tensor of $n$ indices.

### 11.14.5  The Induction Step

Thanks to the induction hypothesis, we can now go ahead and define a new $(n+1)$-level quaternary tree as well: place a new node at the head, and issue four branches from it. At the end of each branch, stick a new $n$-level quaternary subtree. This way, the new $(n+1)$-level tree is indeed even and symmetric, as required. It contains $4^n$ leaves, ready to store a tensor of $n$ indices.

## 11.15  Tensors and Their Algebraic Operations

### 11.15.1  Adding Two Tensors

Let $R$ and $Q$ be two tensors, with the same number of indices. How to add them to each other, entry by entry? Recursively, of course! After all, they are already implemented as quaternary trees, with the same number of levels. How many? If one, then this is easy: both $R$ and $Q$ are just scalars, easily added to each other. If, on the other hand, these are multilevel trees, then add them recursively, subtree by subtree.

### 11.15.2  Multiplying by a Scalar

Let $Q$ be a given tensor, with as many indices as you like. Let $r$ be a given scalar. How to multiply $r$ times $Q$, entry by entry? Recursively, of course! Indeed, if $Q$ has no index at all, then it is just a scalar, easily multiplied by $r$. If, on the other hand, $Q$ is implemented as a multilevel tree, then multiply it recursively, subtree by subtree.

### 11.15.3  Tensor Product

Thanks to the above algorithm, we can now calculate the (outer) product of two tensors. Let $R$ be a given tensor, with $k \geq 0$ indices (and $4^k$ entries). Let $Q$ be yet another tensor, with $l \geq 0$ indices (and $4^l$ entries). The tensor product $RQ$ is a new (bigger) tensor, with $k + l$ indices (and $4^{k+l}$ entries): each new entry is obtained as the product of an entry from $R$ times an entry from $Q$.

How to calculate $RQ$? Recursively, of course! Indeed, if $R$ has no index at all, then it is just a scalar, easy to multiply $Q$ (Section 11.15.2). If, on the other hand, $R$ is a bigger tensor, then it is implemented as a multilevel tree, with four subtrees. Each of them can now multiply $Q$ recursively. This completes the calculation of the new tensor product $RQ$, as required.

## 11.16  Contraction

### 11.16.1  Substitute by a Subtree

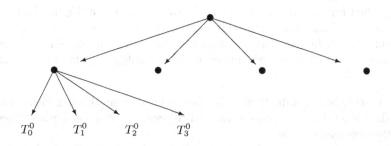

**Fig. 11.15.** In the algorithm Substitute$(q, 1)$, the original tree is replaced by its own $q$th subtree. If $q = 0$, for instance, then the original tree is replaced by its leftmost subtree.

To help contract a tensor, we need some preparation work. In particular, we need to be able to reduce the number of levels in our tree. In other words, we want to be able to "drop" one level from our tree.

For a start, we want to be able to replace our original tree by one of its own subtrees: say, the $q$th subtree (for some fixed $0 \leq q \leq 3$). This is illustrated in Figure 11.15.

More than that: we may also need to replace not the entire tree but only a few subtrees in it. For example, we may need to modify the $l$th level (for some $l \geq 1$), and replace all subtrees in it by their own $q$th subsubtree. This is illustrated in Figure 11.16.

How to do this? Recursively, of course! Indeed, this is done by the following algorithm:

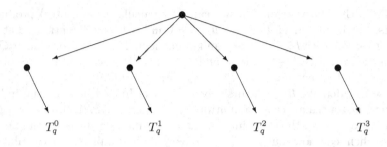

**Fig. 11.16.** In the algorithm Substitute($q, 2$), on the other hand, each subtree is replaced by its own $q$th leaf. For example, the leftmost subtree is replaced by $T_q^0$. In summary, the original tensor $(T_\nu^\mu)$ is replaced by the new vector $(T_q^\mu)$, where $q$ is fixed, not free.

Substitute($q, l$):

1. Start from the head.
2. If $l = 1$, then replace the entire tree by its own $q$th subtree (Figure 11.15).
3. If, on the other hand, $l > 1$, then don't replace the entire tree. Instead, scan the four subtrees, one by one. To each subtree, apply Substitute($q, l-1$) recursively. For example, if $l = 2$, then each subtree would be replaced by its own $q$th subtree (Figure 11.16).

Thanks to mathematical induction, it is easy to prove that this algorithm indeed works: at the $l$th level, all subtrees have been replaced by their own $q$th subtree, as required. (See exercises below.)

What happens to the original tensor? Well, its $l$th index is now no longer free, but fixed: it must be the same as $q$. In other words, one index is gone. This will help contract the original tensor.

### 11.16.2  Contract the First and Second Indices

Thanks to the above algorithm, we can now contract a given tensor. For this purpose, let's contract two indices with each other: the $k$th index with the $l$th index (for some $l > k \geq 1$).

Let's start from a simple case: $k = 1$ and $l = 2$. This means that we contract the first and second indices with each other.

### 11.16.3  A Three-Level Tree

For instance, assume that our original tensor $T$ has just two indices: an upper index, followed by a lower index:

$$T \equiv \left( T_\delta^\beta \right)_{0 \leq \beta, \delta \leq 3}.$$

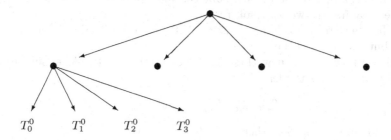

**Fig. 11.17.** To the leftmost subtree, apply Substitute(0, 1), to replace it by its own leftmost leaf: the scalar $T_0^0$.

Thanks to Einstein's summation convention, its contraction is just a scalar:

$$T_\rho^\rho \equiv T_0^0 + T_1^1 + T_2^2 + T_3^3.$$

How to calculate this number? Use the previous algorithm:

Contract($k = 1, l = 2$):

1. Start from the head of the original three-level tree.
2. For $i = 0, 1, 2, 3$, look at the $i$th subtree, and apply to it Substitute($i, 1$), to replace it by its $i$th leaf.
3. This produces a new two-level tree, with the leaves

$$T_0^0, \ T_1^1, \ T_2^2, \quad \text{and} \quad T_3^3$$

(Figure 11.17).
4. Sum them up into one new leaf. This gives the required scalar: the contracted tensor.
5. Finally, use this new leaf to replace the original three-level tree.

Let's extend this algorithm further.

### 11.16.4  A Four-Level Tree

Let's extend this to a slightly bigger tensor, with one more (lower) index:

$$T \equiv \left(T_{\delta\epsilon}^\beta\right)_{0 \leq \beta, \delta, \epsilon \leq 3}.$$

Now, contract it as follows: replace both $\beta$ and $\delta$ by the dummy index $\rho$, and sum over $\rho = 0, 1, 2, 3$:

$$T_{\rho\epsilon}^\rho \equiv T_{0\epsilon}^0 + T_{1\epsilon}^1 + T_{2\epsilon}^2 + T_{3\epsilon}^3.$$

This leaves just one free index: $\epsilon$.

How to calculate this? As before:

Contract($k = 1, l = 2$):

1. Start from the head of the original four-level tree.
2. For $i = 0, 1, 2, 3$, look at the $i$th subtree, and apply to it Substitute($i, 1$), to replace it by its $i$th two-level subtree.
3. What happens to the original tensor? Well, its second index is not free any more, but the same as the first index: $i$.
4. Indeed, we now have a new three-level tree, with four two-level subtrees. Its leftmost subtree has the leaves

$$T_{00}^0, \ T_{01}^0, \ T_{02}^0, \quad \text{and} \quad T_{03}^0.$$

Its second subtree, on the other hand, has the leaves

$$T_{10}^1, \ T_{11}^1, \ T_{12}^1, \quad \text{and} \quad T_{13}^1.$$

Its third subtree, on the other hand, has the leaves

$$T_{20}^2, \ T_{21}^2, \ T_{22}^2, \quad \text{and} \quad T_{23}^2.$$

Finally, its rightmost subtree has the leaves

$$T_{30}^3, \ T_{31}^3, \ T_{32}^3, \quad \text{and} \quad T_{33}^3.$$

5. Add these four subtrees to each other (Section 11.15.1).
6. This produces the required two-level tree: the contracted tensor. For example, its leftmost leaf becomes now

$$T_{00}^0 + T_{10}^1 + T_{20}^2 + T_{30}^3,$$

as required.
7. Finally, use this new two-level tree to replace the original four-level tree.

We can now go ahead and introduce many more indices at the end of the original tensor: the same algorithm still works.

### 11.16.5  Contract with the First Index

Let's extend the above algorithm to a yet more difficult case: contract the $k$th and $l$th indices with each other, for $k = 1$ and some $l \geq 2$. This means that the first index is now being contracted with any other index (not necessarily the second one).

For example, $T$ could now use even four indices: say, one upper, and three lower:

$$T \equiv \left( T_{\gamma\delta\epsilon}^\beta \right)_{0 \leq \beta, \gamma, \delta, \epsilon \leq 3}.$$

Thanks to Einstein's summation convention, the contracted tensor is

$$T_{\gamma\rho\epsilon}^\rho \equiv \sum_{\rho=0}^{3} T_{\gamma\rho\epsilon}^\rho.$$

(In this example, $l = 3$: the first index is contracted with the third one.)

How to calculate this? As before:

Contract($k = 1, l \geq 2$):

1. Start from the head of our original tree.
2. For $i = 0, 1, 2, 3$, look at the $i$th subtree, and apply to it Substitute$(i, l - 1)$.
3. What happens to the original tensor? Well, its $l$th index is now dummy: it is always the same as the first index: $i$.
4. From the resulting tree, pick the four subtrees, one by one, and add them to each other (Section 11.15.1).
5. This sum of subtrees is indeed the contracted tensor, as required.
6. Use it to replace the original tree.

We can now go ahead and introduce many more indices at the end of the original tensor: the same algorithm still works.

### 11.16.6  Contract in General

Let's extend the above algorithm to the most general case: contract the $k$th and $l$th indices with each other, for $k \geq 1$ and $l > k$. For example, $T$ could now have even five indices: say, two upper, and three lower:

$$T \equiv \left( T^{\alpha\beta}_{\gamma\delta\epsilon} \right)_{0 \leq \alpha,\beta,\gamma,\delta,\epsilon \leq 3}.$$

Thanks to Einstein's summation convention, the contracted tensor is

$$T^{\alpha\rho}_{\gamma\rho\epsilon} \equiv \sum_{\rho=0}^{3} T^{\alpha\rho}_{\gamma\rho\epsilon}.$$

(In this example, $k = 2$ and $l = 4$, so the second index is contracted with the fourth index.)

How to calculate this? Recursively, of course:

Contract$(k \geq 1, l > k)$:

1. If $k = 1$, then use the previous version (Section 11.16.5).
2. If, on the other hand, $k > 1$, then scan the four subtrees, one by one. To each subtree, apply Contract$(k - 1, l - 1)$ recursively.

We can now go ahead and introduce many more indices at the end of the original tensor: the same algorithm still works.

## 11.17  Symmetric Matrix and Tensor

### 11.17.1  Symmetric Matrix

So far, we've used discrete math for a numerical purpose: to calculate a tensor, and its algebraic operations, with as many indices as you like. Next, let's use discrete math for yet another purpose. This will lead to an important physical consequence: Einstein equivalence principle.

To do this, let's start from a symmetric matrix. Consider a $4 \times 4$ matrix:

$$A \equiv (a_{\mu\nu})_{0 \leq \mu, \nu \leq 3} = \begin{pmatrix} a_{00} & a_{01} & a_{02} & a_{03} \\ a_{10} & a_{11} & a_{12} & a_{13} \\ a_{20} & a_{21} & a_{22} & a_{23} \\ a_{30} & a_{31} & a_{32} & a_{33} \end{pmatrix}.$$

Its main diagonal contains just four elements:

$$\begin{pmatrix} a_{00} & & & \\ & a_{11} & & \\ & & a_{22} & \\ & & & a_{33} \end{pmatrix}.$$

In a symmetric matrix, the main diagonal acts like a mirror: each element below it is mirrored by an element above it:

$$a_{\mu\nu} = a_{\nu\mu}, \quad 0 \leq \mu, \nu \leq 3.$$

For example, the following elements are the same:

$$\begin{pmatrix} & & a_{02} & \\ & & & \\ a_{20} & & & \\ & & & \end{pmatrix}.$$

For yet another example, the following elements are the same:

$$\begin{pmatrix} & & & a_{03} \\ & & & \\ & & & \\ a_{30} & & & \end{pmatrix},$$

and so on.

## 11.17.2   Degrees of Freedom

Thus, there is no need to store all 16 elements. It is sufficient to store ten elements only:

$$(a_{\mu\nu})_{0 \leq \mu \leq \nu \leq 3} = \begin{pmatrix} a_{00} & a_{01} & a_{02} & a_{03} \\ & a_{11} & a_{12} & a_{13} \\ & & a_{22} & a_{23} \\ & & & a_{33} \end{pmatrix}.$$

This is the upper triangular part of $A$. It contains four diagonals: the main diagonal, and three more diagonals above it. Below it, on the other hand, we have three more diagonals:

$$(a_{\mu\nu})_{0 \leq \nu < \mu \leq 3} = \begin{pmatrix} & & & \\ a_{10} & & & \\ a_{20} & a_{21} & & \\ a_{30} & a_{31} & a_{32} & \end{pmatrix}.$$

This is the *strictly* lower triangular part of $A$. There is no need to store it. After all, it is mirrored by the upper triangular part, which has already been stored.

So, to define the symmetric matrix $A$, we have only ten degrees of freedom: we are free to pick ten elements only. By doing this, we actually specify all 16 elements, as required.

### 11.17.3  Counting in a Discrete Triangle

In the above, we've counted the degrees of freedom in a symmetric $4 \times 4$ matrix: ten. Let's check this. For this purpose, let's pretend we don't know the answer as yet. This will give us a more general approach, easy to extend to more complicated cases.

In our symmetric matrix, how many degrees of freedom are there? In other words, in the upper triangular part, how many elements are there? Let's mirror each such element by a point in a discrete triangle of size 3 (Figure 9.5).

More precisely, in the upper triangular part, each element has a pair of indices:

$$\mu \leq \nu.$$

How to count such pairs? In Chapter 9, Section 9.2.4, set

$$k \equiv 2$$
$$t_1 \equiv \mu$$
$$t_2 \equiv \nu$$
$$m \equiv 3.$$

Thus, the total number of such pairs is

$$\binom{m+k}{k} = \binom{3+2}{2} = \frac{5 \cdot 4}{2} = 10.$$

This is no surprise. After all, we could also count diagonal by diagonal:

$$4 + 3 + 2 + 1 = 10.$$

As a matter of fact, this is just a special case of the formula in Chapter 9, Section 9.3.6.

So, what's the big deal? To see this, let's move on to a more complicated case.

### 11.17.4  Symmetric Tensor

Let's extend the above to a tensor of three indices:

$$(T_{\alpha\beta\gamma})_{0 \leq \alpha, \beta, \gamma \leq 3} \, .$$

(For simplicity, assume that they are lower.) This way, $T$ contains

$$4^3 = 64$$

entries. Still, in a symmetric tensor, not all of them are free to specify.

We say that $T$ is symmetric if interchanging indices has no effect. For example, interchanging $\alpha$ and $\gamma$ has no effect:

$$T_{\alpha\beta\gamma} = T_{\gamma\beta\alpha}.$$

Likewise, interchanging $\alpha$ and $\beta$ (or $\beta$ and $\gamma$) has no effect either. So, there is actually no need to specify all 64 entries. It is sufficient to specify only those indexed with

$$\alpha \leq \beta \leq \gamma.$$

### 11.17.5   Counting in a Discrete Tetrahedron

How many such triplets are there? Let's mirror each such triplet by a point in a discrete tetrahedron of size 3 (Figure 9.6).

More precisely, each triplet of the form

$$\alpha \leq \beta \leq \gamma$$

could be renamed as

$$t_1 \leq t_2 \leq t_3.$$

For this purpose, in Chapter 9, Section 9.2.4, set

$$k \equiv 3$$
$$t_1 \equiv \alpha$$
$$t_2 \equiv \beta$$
$$t_3 \equiv \gamma$$
$$m \equiv 3.$$

### 11.17.6   Degrees of Freedom

Thus, the total number of such triplets is

$$\binom{m+k}{k} = \binom{3+3}{3} = \frac{6 \cdot 5 \cdot 4}{6} = 20.$$

In summary, the symmetric tensor $T$ has just 20 degrees of freedom: it is sufficient to specify just 20 independent entries, and the rest are specified automatically as well. Let's use this observation in physics.

## 11.18   Geometry and Physics

### 11.18.1   Is the Earth Flat?

In the ancient day, it was believed that the Earth was flat. After all, the land looks flat: to reach a distant place, just walk straight towards it.

Nowadays, on the other hand, we know better: the Earth is more like a ball, a little "fat" at the equator. Why does it look flat? Because this "ball" is very big relative to us.

### 11.18.2   Tangent Planes

In fact, at each individual place, the surface of the Earth could be approximated by a tangent plane (Figure 11.18). We humans may then think that we live on this plane, not on the round surface.

This picture uses two spatial dimensions. After all, the face of the Earth is two-dimensional: nearly a round sphere, with a (constant) positive curvature. In

**Fig. 11.18.** The Earth, and its tangent plane at the north pole.

particular, it is also smooth: at each individual point on it, there is a (unique) tangent plane, perpendicular to the radius.

These tangent planes are different from each other: they may have a different slope. Together, they envelope the entire Earth all over (Figures 11.7–11.8).

Let's extend this to four dimensions as well. This way, we can use the above geometrical insight in spacetime as well.

## 11.19  Riemann Normal Coordinates in Spacetime

### 11.19.1  Nearly Constant Metric

Recall that the metric is a $4 \times 4$ matrix:

$$g \equiv (g_{\mu\nu})_{0 \leq \mu,\nu \leq 3} = \begin{pmatrix} g_{00} & g_{01} & g_{02} & g_{03} \\ g_{10} & g_{11} & g_{12} & g_{13} \\ g_{20} & g_{21} & g_{22} & g_{23} \\ g_{30} & g_{31} & g_{32} & g_{33} \end{pmatrix}.$$

This is not a constant matrix: it depends on the event under consideration, and may change from event to event in spacetime.

The metric mustn't depend on the coordinates that happen to be used. After all, the metric is physical: it tells us how spacetime really curves. As such, it can never depend on the coordinates, which are just mathematical, not physical. Still, in each system of coordinates that describe spacetime, the metric may be written in a different way.

In the standard coordinates

$$x^0 \equiv t$$
$$x^1 \equiv x$$
$$x^2 \equiv y$$
$$x^3 \equiv z,$$

the metric takes the above form: $g$. In other coordinates, on the other hand, it may take a different form. Later on, we'll see how to transform from one coordinate system to another. In this context, discrete math will prove most useful.

As a matter of fact, there are no standard or favorite coordinates. After all, the coordinates are just mathematical (nonphysical) parameters, to help model spacetime.

Thus, we could actually adopt just any system of coordinates. In particular, we could adopt a new system of coordinates, in which the metric is nearly constant: its partial derivatives vanish. This way, the metric is nearly as simple as

$$\begin{pmatrix} -1 & & & \\ & 1 & & \\ & & 1 & \\ & & & 1 \end{pmatrix}.$$

This diagonal metric comes from special relativity, where no gravity is assumed at all.

### 11.19.2  No Self System!

These new coordinates are Riemann's normal coordinates (Sections 11.3.13 and 11.5.1). They are similar to the self coordinates, used often in special relativity. Still, there is a major difference.

In fact, Riemann's normal coordinates are only local, not global: each individual event in spacetime has its own private normal coordinates to span its own tangent space. In terms of these local coordinates, the metric is nearly constant only around this particular event, not around others. These different coordinate systems can never be "tied" together to form one global coordinate system.

### 11.19.3  Tangent Hyperboloids

What does this mean geometrically? Well, at each individual event in spacetime, there is a tangent four-dimensional space (Section 11.9.3). This means that spacetime is indeed smooth.

In this tangent space, gravity has no effect whatsoever. As a matter of fact, things are as simple as in special relativity. Geometrically, this means that the tangent space could split into disjoint (three-dimensional) hyperboloids. Later on, we'll discuss the physical meaning of this.

### 11.19.4  Constant Metric — No Curvature

In Riemann normal coordinates, the metric is nearly constant. Later on, we'll see how to design these coordinates, not globally, but only locally: around a particular event in spacetime.

Is it possible to design a coordinate system in which the metric is *exactly* constant? No! Indeed, if the metric were constant even in a tiny spot in spacetime, then there would be no curvature there at all, and no gravity either, as in special relativity. Spacetime would then be completely flat and linear, not only in this spot, but also everywhere. Indeed, just look around you. Could the Earth be flat only in a little area around you? Of course not: if it were, then it would be flat everywhere as well, which is untrue.

## 11.20  Transformation of Coordinates

### 11.20.1  Physics — Independent of Coordinates

In physics, the laws mustn't depend on the coordinates. In my lab, I may have my own time and space. You, on the other hand, may measure your own time and space in your own lab. Still, both of us must see the same nature, with the same laws.

Spacetime is real: physical, not just mathematical. Therefore, it must have the same geometry and curvature, regardless of the coordinates that are used to observe and measure them. After all, the coordinates are just mathematical artifacts, with no physical meaning whatsoever. Indeed, they are just relative, not absolute. For example, in my lab, I measure my own private coordinates, which seem static. From my lab, I see your coordinate system as moving (as in special relativity in Figure 11.10), or even accelerating (as in general relativity). Still, nothing is absolute: from your perspective, your system is static, while mine is moving (or even accelerating).

### 11.20.2  Metric — Dynamic Variables

So, the coordinates are nonphysical: they only help describe nature. What is physical? Spacetime! This is a four-dimensional manifold, whose geometry and curvature are encapsulated in the metric. Thus, the metric is physical and real: it is an integral part of spacetime, determining its shape. Still, in each coordinate system, the metric may take a different mathematical face.

The metric is a symmetric $4 \times 4$ matrix. As such, it contains ten degrees of freedom (Section 11.17.2). They are not constant: they depend on the particular event under consideration, and may change from event to event in spacetime. This is why they are called dynamic variables: they are functions of four independent variables — $x^0$, $x^1$, $x^2$, and $x^3$:

$$g_{\mu\nu} \equiv g_{\mu\nu}\left(x^0, x^1, x^2, x^3\right), \quad 0 \le \mu \le \nu \le 3.$$

This is how the (nonphysical) independent variables $x^0$, $x^1$, $x^2$, and $x^3$ help describe the metric, and indeed the real physics in spacetime. Still, this is not a must: we could equally well use any other system of independent variables. The same metric would then be written in a new way: as a composite function of the new coordinates.

### 11.20.3  New Coordinates

How to transform? Suppose that we want to use new coordinates:

$$y^0 \equiv y^0\left(x^0, x^1, x^2, x^3\right)$$
$$y^1 \equiv y^1\left(x^0, x^1, x^2, x^3\right)$$
$$y^2 \equiv y^2\left(x^0, x^1, x^2, x^3\right)$$
$$y^3 \equiv y^3\left(x^0, x^1, x^2, x^3\right).$$

(These round parentheses are often dropped for short.) This transforms the old coordinates into new ones. We assume that this transformation is invertible: the new coordinates could be transformed back to the old ones smoothly (Sections 11.3.9–11.3.11).

### 11.20.4  Partial Derivatives

We can now go ahead and calculate the partial derivative of any old coordinate with respect to a new one. For example, the old time $x^0 = t$ could be differentiated with respect to the new "time" $y^0$:

$$\frac{\partial x^0}{\partial y^0} \equiv \frac{\partial x^0}{\partial y^0} \left( y^0, y^1, y^2, y^3 \right).$$

(These round parentheses are often dropped for short.) To obtain this partial derivative, hold $y^1$, $y^2$, and $y^3$ fixed, and differentiate with respect to $y^0$ only. Let's go ahead and use this.

### 11.20.5  Transformation of the Metric

In terms of the new coordinates, how does the metric look like? For this purpose, one could write the original metric $g$ as a composite function of the new coordinates. This, however, could still be nonlinear and complicated. Is there a more direct way?

In the new coordinates, the metric takes a new mathematical face: it is no longer called $g$, but $h$. This is a new symmetric $4 \times 4$ matrix, written as a function of the new coordinates.

To have $h$ explicitly, let's index it in new (lower) indices — $\alpha$ and $\beta$. The new matrix $h$ is then obtained from the formula

$$h_{\alpha\beta} \equiv \frac{\partial x^\mu}{\partial y^\alpha} \cdot \frac{\partial x^\nu}{\partial y^\beta} \cdot g_{\mu\nu}, \quad 0 \le \alpha \le \beta \le 3.$$

Here, Einstein's summation convention has been used twice. This way, $\mu$ and $\nu$ are now just dummy indices, summed over $\mu, \nu = 0, 1, 2, 3$. Only $\alpha$ and $\beta$ are free. Furthermore, they could interchange with each other, with no effect whatsoever. This proves that $h$ is indeed a symmetric $4 \times 4$ matrix as well, and contains only ten degrees of freedom, as required (Section 11.17.2).

## 11.21  Riemann Normal Coordinates and Einstein's Happy Thought

### 11.21.1  Nearly Constant Metric

Consider a particular event in spacetime. To describe it, use the new coordinates. Around the event, draw a small spot (a little four-dimensional domain in spacetime). Could the metric be constant there? For example, could $h$ have the constant form

$$h = \begin{pmatrix} -1 & & & \\ & 1 & & \\ & & 1 & \\ & & & 1 \end{pmatrix}$$

there? In other words, if I "lived" in the spot, could I feel no gravity at all, as in special relativity?

Of course not! We've already discussed this, both physically and geometrically. Let's support this algebraically as well.

Indeed, let's try and design our new coordinates in such a way that $h$ indeed has this form. After all, we haven't specified the new coordinates as yet.

In our spot, let's focus on our original event. At this event, could we force $h$ to have this simple form? Easy: in Section 11.20.5, the symmetric matrix $h$ is defined in just ten equations. These are not too many. Indeed, to satisfy them, we have freedom to pick as many as 16 unspecified "unknowns:"

$$\frac{\partial x^{\mu}}{\partial y^{\alpha}}, \quad 0 \leq \alpha, \mu \leq 3.$$

Why are these called "unknowns?" Because we haven't defined the new coordinates as yet, so the partial derivatives with respect to them are still unspecified. Later on, we'll define the new coordinates cleverly, with partial derivatives that make $h$ constant, as in the above diagonal matrix.

This is good: more unknowns than equations [33]. Thus, it is easy to specify the unknowns properly, to make $h$ constant, as in the above diagonal form.

This is done at our event. Still, this doesn't guarantee that $h$ has the same form in the entire spot. After all, $h$ may change from event to event. Still, we could force $h$ to be *nearly* constant in some little spot. For this purpose, $h$ must have zero partial derivatives at our original event. To achieve this, at our event, differentiate the ten equations

$$h_{\alpha\beta} \equiv \frac{\partial x^{\mu}}{\partial y^{\alpha}} \cdot \frac{\partial x^{\nu}}{\partial y^{\beta}} \cdot g_{\mu\nu}, \quad 0 \leq \alpha \leq \beta \leq 3,$$

with respect to $y^{\delta}$ ($0 \leq \delta \leq 3$). This gives 40 new equations.

How many unknowns are there? Well, we now have new unknowns: the *second* partial derivative of an old coordinate, with respect to two new coordinates:

$$\frac{\partial^{2} x^{\mu}}{\partial y^{\alpha} \partial y^{\delta}}, \quad 0 \leq \mu \leq 3, \ 0 \leq \alpha \leq \delta \leq 3.$$

Why are these called "unknowns?" Because we haven't defined our new coordinates completely as yet. After all, we've only determined how the old coordinates differentiate with respect to them. This only determines the first partial derivatives, not the second ones!

Now, these new unknowns are symmetric in terms of $\alpha$ and $\delta$. Indeed, these indices could interchange, with no effect whatsoever. So, these indices could be picked in ten different ways. On top of that, $\mu$ could be picked in four different ways. So, in total, we have 40 unknowns, just enough to solve our 40 equations.

## 11.21.2   Tangent Space

By picking our new coordinates cleverly, we can therefore make sure that these new unknowns are indeed as required to force $h$ to have zero second partial derivatives at our original event. This is good: in a little neighborhood around our event, $h$ hardly changes, and nearly keeps the same diagonal form.

### 11.21.3  Einstein's Happy Thought

These are Riemann normal coordinates. In them, spacetime is smooth: well-approximated by a flat linear tangent space. Thus, at the original event, things are as in special relativity: you'd feel as in a free fall, with no force or acceleration at all, as in Einstein's happy thought.

### 11.21.4  Riemann Normal Coordinates: Local and Implicit

There is no need to calculate Riemann normal coordinates explicitly: they can remain theoretical and implicit. Furthermore, they are not global, but only local. Indeed, in the above algorithm, they have been calculated separately for each individual event in spacetime. Thus, they depend on the event under consideration, and are defined only in a small neighborhood around it.

### 11.21.5  Is There a Flat Spot in Spacetime?

By now, we've got what we wanted: in terms of Riemann normal coordinates, $h$ has zero first partial derivatives at our event, and is therefore nearly constant in a small neighborhood around it. Still, could we have more? Could we force $h$ to have zero *second* partial derivatives as well? Not any more! Indeed, let's differentiate the above ten equations twice: not only with respect to $y^\delta$ but also with respect to $y^\eta$. Fortunately, in partial differentiation, order doesn't matter. Therefore, we only need to consider ten cases: $0 \leq \delta \leq \eta \leq 3$. So, in total, we have 100 equations to solve.

How many unknowns are there? Well, the new unknowns are now of the form

$$\frac{\partial^3 x^\mu}{\partial y^\alpha \partial y^\delta \partial y^\eta}, \quad 0 \leq \mu \leq 3, \; 0 \leq \alpha \leq \delta \leq \eta \leq 3.$$

(After all, in partial differentiation, order doesn't matter.) So, we actually have here four symmetric tensors, or a total of 80 unknowns (Section 11.17.6).

This is not enough! After all, 100 is more than 80. Therefore, in general, we can't force $h$ to be constant, not even in a tiny neighborhood of our original event.

### 11.21.6  Spacetime: Smooth, but Never Flat

This is no surprise: spacetime behaves just like our original example: the sphere (Section 11.18.1). Indeed, we already know a lot about the sphere: it is completely round, with positive (constant) curvature everywhere. As such, it contains no flat area at all. Still, it is smooth: each point on it has a (unique) tangent plane. In principle, spacetime is the same: smooth everywhere, but flat nowhere.

## 11.22  Special vs. General Relativity

### 11.22.1  Flat Spacetime — Constant Metric

In general, 80 unknowns are not enough to solve 100 equations. Still, in one special case, they are. When? When some of these equations are duplicate copies of others. In other words, they are not entirely independent of each other.

In this case, one could keep differentiating the above ten equations (where $h$ is defined) more and more, use discrete math to count the equations, and make sure that they are not entirely independent of each other, so there are enough unknowns to solve them. This way, at our original event, all partial derivatives of $h$ vanish, even the high-order ones. This means that $h$ must be constant in some (tiny) spot around our event. What does this mean geometrically? No curvature at all in the entire spot!

Could spacetime contain such a spot? As discussed above, this would lead to a rather strange spacetime: completely flat. This could happen only theoretically in special relativity (Figure 11.10), but not in real physics: in general relativity, gravity is felt everywhere, and spacetime is curved everywhere, not flat.

### 11.22.2 Einstein Equivalence Principle

Thanks to Riemann normal coordinates, spacetime is smooth. Indeed, locally, spacetime is nearly flat: at each individual event, it has a tangent space.

This is a good geometrical property: in terms of Riemann normal coordinates, spacetime is nearly flat. More precisely, it is nearly hyperbolic: the metric is nearly as simple as

$$\begin{pmatrix} -1 & & & \\ & 1 & & \\ & & 1 & \\ & & & 1 \end{pmatrix}.$$

This is indeed the metric used in special relativity, where no gravity is assumed at all.

What does this mean physically? Well, if you "live" at this event, then you'd feel no force at all! After all, around this event, there is a (tiny) spot where the metric is nearly as in special relativity. Relative to the other events in the spot, you'd have nearly the same metric: there is hardly any change.

To you, the spot is just too big to feel any curvature. Just like you think the Earth is flat, you'd think the spot is flat as well. Physically, you are actually in a free fall.

At the event, the new system of Riemann normal coordinates encapsulates all acceleration due to gravity. This is why you feel no force at all: you are carried effortless by the new coordinates. After all, you could "live" in the new coordinates, and ignore how they were formed in the first place. Just relax, enjoy your new coordinates, and disregard how spacetime really curves or changes around you.

After all, you have no other coordinates to compare to. On the contrary: to you, Riemann normal coordinates are most natural and realistic. In fact, to you, they are as straight and linear as ever. This is why you feel as in a free fall. Why transform? Better stick to these lovely coordinates!

Now, there is nothing special about our original event. The same could be done in each and every event in spacetime. This way, around the next event, design a new tiny spot, with slightly different Riemann normal coordinates. They are still local: with respect to them, the new event lies at the origin: $(0,0,0,0)$. This could be done in each and every event in spacetime.

This is indeed Einstein equivalence principle (or happy thought): your local coordinates could take on the job of encapsulating gravity. In terms of them, you'd then feel no gravity at all!

Historically, things happened the other way around. In the beginning, Einstein used a thought experiment to discover his equivalence principle. Later, he used differential geometry to develop the complete mathematical theory of general relativity and gravity.

## 11.23  Anti-Symmetric Matrix and Tensor

### 11.23.1  Anti-Symmetric Matrix

Assume now that our matrix $A$ is not symmetric but anti-symmetric. This means that interchanging indices picks a minus sign:

$$a_{\mu\nu} = -a_{\nu\mu}, \quad 0 \le \mu, \nu \le 3.$$

For this reason, each main-diagonal element must vanish. After all, it is the same as the minus of itself. Thus, it is sufficient to store the *strictly* upper triangular part:

$$(a_{\mu\nu})_{0 \le \mu < \nu \le 3} = \begin{pmatrix} a_{01} \; a_{02} \; a_{03} \\ \phantom{a_{01}} a_{12} \; a_{13} \\ \phantom{a_{01} \; a_{02}} a_{23} \end{pmatrix}.$$

The rest of the elements, on the other hand, don't have to be stored. After all, we already know what they are: on the main diagonal, they vanish. Below it, on the other hand, they are anti-mirrored by the (strictly) upper triangular part.

### 11.23.2  How Many Degrees of Freedom?

How many degrees of freedom are there? In other words, how many elements are there in the *strictly* upper triangular part? Well, each such element is indexed by

$$\mu < \nu.$$

How many such pairs are there? To count, let's use a general method, easy to extend to more complicated cases as well. In Chapter 9, Section 9.2.3, set

$$k \equiv 2$$
$$s_1 \equiv \mu + 1$$
$$s_2 \equiv \nu + 1$$
$$n \equiv 3 + 1 = 4.$$

This way,

$$1 \le s_1 < s_2 \le n,$$

as required. Thus, the total number of such pairs is

$$\binom{n}{k} = \binom{4}{2} = \frac{4 \cdot 3}{2} = 6.$$

Indeed, in the *strictly* upper triangular part, there are exactly six elements:

$$a_{01}, \; a_{02}, \; a_{03}, \; a_{12}, \; a_{13}, \quad \text{and} \quad a_{23}.$$

### 11.23.3   Anti-Symmetric Tensor

Let's extend the above to a tensor of three indices:

$$(T_{\alpha\beta\gamma})_{0 \le \alpha,\beta,\gamma \le 3}.$$

(For simplicity, assume that all indices are lower.) We say that $T$ is anti-symmetric if interchanging two indices picks a minus sign. For example, interchanging $\alpha$ and $\gamma$ leads to

$$T_{\alpha\beta\gamma} = -T_{\gamma\beta\alpha},$$

and so on.

So, many entries must vanish: an entry that has two identical indices must be the minus of itself, or just zero. As a matter of fact, it is sufficient to specify those entries indexed with

$$\alpha < \beta < \gamma.$$

### 11.23.4   How Many Degrees of Freedom?

How many such triplets are there? In Chapter 9, Section 9.2.3, set

$$k \equiv 3$$
$$s_1 \equiv \alpha + 1$$
$$s_2 \equiv \beta + 1$$
$$s_3 \equiv \gamma + 1$$
$$n \equiv 3 + 1 = 4.$$

This way,

$$1 \le s_1 < s_2 < s_3 \le n,$$

as required. Thus, the total number of such triplets is

$$\binom{n}{k} = \binom{4}{3} = 4.$$

In summary, the anti-symmetric tensor $T$ has just four degrees of freedom:

$$T_{012}, \; T_{013}, \; T_{023}, \quad \text{and} \quad T_{123}.$$

## 11.24   Towards Bundles and Lie Groups

### 11.24.1   Heraclitus: Dynamics and Relativity

To carry out an experiment in physics, you need a lab, with four coordinates: your own (proper) time, plus three spatial coordinates. This is indeed relativity: your lab is not static, but dynamic: it travels with the entire Earth around the sun, which travels around the center of the milky way. In fact, there is no such thing as "static:" a position is only relative, not absolute. As Heraclitus said, everything is moving and flowing.

## 11.24.2  Local Experiment in the Lab

Furthermore, your lab is only local: it focuses on a particular event, here and now, and on its immediate neighborhood in spacetime.

In your lab, how is the event specified? This is done in terms of your own coordinates: your (proper) time, and your position. Still, the result of your experiment mustn't depend on the coordinates, which are just mathematical, not physical. This is why tensors are so useful: they have *a-priori* rules that tell us how to transform them from one coordinate system to another. Fortunately, the results of the experiment could often be placed in a tensor, and transform in a compatible way. This way, although they may change numerically, they still have the same physical (and geometrical) meaning.

## 11.24.3  Metric and Shortest Path in Spacetime

In general relativity, the dynamic variables are placed in a symmetric $4\times4$ tensor: the metric (Section 11.20.2). This has not only physical but also geometrical meaning: it tells us what is the shortest path between two events in spacetime. The integration along the path is carried out with respect to the proper time, measured in a dynamic clock, carried along the path. To discover this optimal path, one must use the Euler–Lagrange equations [33].

The metric is then designed locally to fit these equations. Thanks to the metric, one can then define the curvature, to tell us how curved spacetime is at each individual event. Again, this is a local and nonconstant property: it depends on the particular event under consideration, and may change from event to event in spacetime.

Furthermore, thanks to the metric, one can also define the covariant derivative, to help differentiate along the curved spacetime. This helps introduce Einstein's equations, which tell us the whole story of general relativity: how the metric is related to the energy and the momentum throughout spacetime [4, 5, 11, 42, 15, 33]. This is how geometry is used fully to explain physics!

## 11.24.4  High-Dimensional Bundle

In [44], on the other hand, there is a yet deeper interpretation: not only local but also global. For this purpose, embed spacetime in a high-dimensional manifold: the principal bundle, mirrored by the associated bundle.

How to do this? To each individual event in spacetime, attach a complete Lie group, to model all possible (smooth) coordinate systems. This group is also mirrored by a yet simpler structure: a Lie algebra (discussed later in the book).

Still, this new mathematical structure may seem a bit artificial and unnecessary. After all, it is designed to fit the physics we already know... So, don't expect it to help in the practical calculations.

Moreover, to a physicist, this approach may look like fancy mathematics. After all, we are soon going to project the high-dimensional bundle back to spacetime, and filter out the entire Lie group, as if it had never been. This way, all different coordinate systems are lumped and united again, and considered as one and the same. So, why was all this necessary? Why not stick to good old spacetime?

### 11.24.5 Vertical Projection back to Spacetime

Still, the new bundle may give us a lot of geometrical intuition: onto each individual event in spacetime, project vertically, along the "vertical" Lie group. The remaining (horizontal) component, on the other hand, aligns with the original spacetime. Along it, one can then define the curvature, to tell us how curved spacetime is at this particular event.

Furthermore, one can also differentiate in the horizontal direction: this is the covariant derivative along the curved spacetime. Upon projecting back to spacetime, we also obtain a useful field: the Young-Mills field.

### 11.24.6 Gauge Transformation and the Homology Group

A shift shouldn't affect differentiation. If you change the initial time, or the location of the origin, then this should have no effect on any partial derivative with respect to time or space. In other words, the coordinate system should be insensitive to any shift by a constant vector. Such a shift is often called a gauge transformation. Why? Because it helps look at things from yet another angle. After all, coordinates like time and space could be viewed as observables: they could be measured in your system, and tell you something about physics (as observed from this perspective).

Together, these shifts make a new group: the homology group. As a matter of fact, our Lie group is far bigger: it contains not only shifts but also many other (compatible) transformations of coordinates. This is indeed a more general kind of gauge transformation. Why? Because new coordinates are just a new way to look at things: physics (and indeed geometry) remains the same.

Geometrically, how does such a transformation look like? Just "jump" vertically in the bundle, and reach a new element from the Lie group: a new coordinate system. Physically, nothing is changed. This is why you are still above the same event in spacetime. Still, now you look at things from another perspective: new time, and new spatial coordinates. This is why you may get new results in your experiments. Still, this is just a numerical change, not physical or geometrical.

### 11.24.7 Could Bundles Really Help?

Although they give no computational gain, bundles may still help interpret space-time topologically better. Indeed, once the differential details are shed off, the topology takes a simple algebraic face: the holonomy group. Furthermore, thanks to the underlying Lie algebra, physics gets as simple as ever. For example, the Schur lemma helps introduce angular momentum and spin. Moreover, Weyl groups and Dynkin diagrams help sort Lie algebras, and model interesting physical phenomena.

Later on, in quantum mechanics, we'll look at the position and momentum operators [6, 8, 21, 41, 47]. Thanks to bundles, one could even reverse-engineer the Young-Mills field, change the coordinates properly, with no effect on physics or geometry at all.

Later on, we'll also introduce Hamiltonian mechanics. In this vein, we'll use the Poisson brackets to design a new Lie algebra, and obtain Noether's theorem for free. This may help understand conservation laws from a geometrical point of view: pick a conserved observable, and mirror it by a kinematic symmetry of the Hamiltonian: an isometric direction, in which the metric remains unchanged.

## 11.25   Exercises: Trees and Tensors

### 11.25.1   Mathematical Induction in a Tree

1. Let $v$ be a four-dimensional vector. What tree could be used to store it? Hint: a two-level tree, with four leaves, to store the components $v^0$, $v^1$, $v^2$, and $v^3$ (Figure 11.12).
2. How to store the information that the index is upper, not lower? Hint: for this purpose, store one more bit: 1.
3. Let $A$ be a $4 \times 4$ matrix. What tree could be used to store it? Hint: a three-level tree, with four subtrees, each being a two-level tree as above.
4. How to reach the element $a_{03}$? Hint: see Figure 11.13.
5. On the other hand, how to reach the element $a_{30}$?
6. In general, for given $0 \le \mu, \nu \le 3$, how to reach the element $a_{\mu,\nu}$ (or $a_{\mu\nu}$, with no comma)? Hint: see Figure 11.14.
7. How to store the information that both $\mu$ and $\nu$ are lower, not upper? Hint: for this purpose, store two bits: 0 and 0.
8. Let $T$ be a big tensor, with five indices (Section 11.13.1). What tree could be used to store it? Hint: a six-level quaternary tree.
9. How to store the information that $T$ has two upper indices, followed by three lower indices? Hint: for this purpose, store five bits: 1, 1, 0, 0, and 0.
10. In this $T$, can it be allowed to contract the first and second indices with each other? Hint: no! They are both upper.
11. Furthermore, can it be allowed to contract the third and fourth indices with each other? Hint: no! They are both lower.
12. On the other hand, can it be allowed to contract the second and fourth indices with each other? Hint: yes! One is upper, and the other is lower, as required.
13. Use mathematical induction to show that the recursive algorithm in Section 11.15.1 indeed works, and adds two tensors (of the same number of indices) to each other. Hint: for $n = 0$, this is just a scalar addition. For $n = 0, 1, 2, 3, \ldots$, on the other hand, assume that the algorithm indeed works for $n$ indices (or $n+1$ levels). (This is the induction hypothesis.) Now, for $n + 2$ levels, the subtrees have $n + 1$ levels each, so they are indeed added properly.
14. Use mathematical induction to show that the recursive algorithm in Section 11.15.2 indeed works, and multiplies a tensor by a scalar. Hint: for $n = 0$, this is just scalar times scalar. For $n = 0, 1, 2, 3, \ldots$, on the other hand, assume that the algorithm indeed works for $n$ indices (or $n + 1$ levels). (This is the induction hypothesis.) Now, for $n + 2$ levels, each subtree has $n + 1$ levels, and can indeed be multiplied recursively.
15. Use mathematical induction on $k$ to show that the recursive algorithm in Section 11.15.3 indeed works, and calculates the tensor product $RQ$. Hint: for $k = 0$, this is just scalar-times-tensor, which we already know how to do. For $k = 0, 1, 2, 3, \ldots$, on the other hand, assume that the algorithm indeed works when $R$ has $k$ indices (or $k+1$ levels). (This is the induction hypothesis.) Now, when $R$ has $k + 2$ levels, its subtrees have $k + 1$ levels each, and can therefore multiply $Q$ recursively.
16. Use mathematical induction on $l$ to show that the recursive algorithm in Section 11.16.1 indeed works, and sets the $l$th index to a fixed instance: $q$.

Hint: for $l = 1$, the algorithm indeed picks the $q$th subtree, as required. For $l = 1, 2, 3, 4, \ldots$, on the other hand, assume that the algorithm indeed works for the $l$th index. (This is the induction hypothesis.) Thanks to this, the $(l + 1)$st index can be now set as well: scan the four subtrees, one by one, and set their $l$th index to $q$.

17. Show that the algorithm in Section 11.16.5 indeed works, and contracts the first index with the $l$th index ($l \geq 2$). Hint: first, it sets the $l$th index to have the same value as the first index: $i = 0$, 1, 2, or 3. Finally, it scans the four subtrees, and sums them up, as required.

18. Use mathematical induction on $k$ to show that the recursive algorithm in Section 11.16.6 indeed works, and contracts the $k$th index with the $l$th index ($l > k \geq 1$). Hint: for $k = 1$, see the previous exercise. For $k = 1, 2, 3, 4, \ldots$, on the other hand, assume that the algorithm indeed works for some $k$ (for every $l > k$). (This is the induction hypothesis.) Thanks to this, it also works well for $k + 1$ (for every $l + 1 > k + +1$). Indeed, it scans the four subtrees, one by one, and contracts their own $k$th and $l$th indices with each other, as required.

19. Use the above algorithm to contract the tensor $T$ as in Section 11.13.2. Hint: use $k = 2$ and $l = 4$ to contract the second and fourth indices with each other.

20. In the previous exercise, how does the recursion look like? Hint: in each subtree, the first and third indices contracts with each other, using the version in Section 11.16.5.

21. How many degrees of freedom are there in a $4 \times 4 \times 4 \times 4$ tensor? Hint: all entries are free to specify: $4^4 = 256$.

22. How many degrees of freedom are there in a $4 \times 4 \times 4 \times 4$ symmetric tensor? Hint: 35 (Section 11.17.6).

23. How many degrees of freedom are there in a $4 \times 4 \times 4 \times 4$ anti-symmetric tensor? Hint: just one (Section 11.23.4).

24. What is it? Hint: $T_{0123}$. Indeed, all the others follow from it by interchanging indices and picking a minus sign.

25. Consider a sphere. Could it contain a flat area?

26. Consider a smooth two-dimensional surface, such as an ellipsoid or a hyperboloid. Could it contain a flat area, with no curvature at all?

27. What is the curvature in an ellipsoid? Hint: positive everywhere.

28. What is the curvature in a hyperboloid? Hint: negative everywhere.

29. What is spacetime? Hint: a four-dimensional differentiable manifold.

30. Could spacetime contain a flat spot, with no curvature at all? Hint: if it did, then the entire spacetime would have to be flat. This is science fiction: the theoretical laws of special relativity override the realistic laws of general relativity.

31. What is a vector field on the Cartesian plane? Hint: see Figure 11.6.

32. Is it also a vector function? Hint: yes, it maps $\mathbb{R}^2$ into $\mathbb{R}^2$. Indeed, to each individual point, assign one arrow.

33. What is a vector field on the sphere?

34. Is it also a function? Hint: no! The arrows are not in the same plane any more. Indeed, to each individual point on the sphere, an arrow is assigned in a different tangent plane (or local chart).

35. What are the advantages and disadvantages of our generating system? Hint: on one hand, it helps represent just any (smooth) vector function on the sphere. On the other hand, this representation is not unique.

36. In the new coordinates in Section 11.20.5, show that the metric $h$ is still symmetric:

$$h_{\alpha\beta} = h_{\beta\alpha}, \quad 0 \leq \alpha, \beta \leq 3.$$

Hint: thanks to Einstein summation convention, the dummy indices $\mu$ and $\nu$ could interchange, with no effect.

37. Use dynamic trees to implement tensors in C++. The solution can be found in Chapter 20 in the appendix.

# 12

# Zorn's Lemma
# in Quantum Mechanics

In classical mechanics, a particle has a well-defined position (or location): a specific point $(x, y, z)$, dependent on time only. Likewise, its momentum is well-defined as well: a concrete three-dimensional vector $p$, dependent on time as well. This is easy enough to understand.

In quantum mechanics, on the other hand, things are not so nice and simple any more. The particle has no deterministic position any more. On the contrary: at each given time, it could be at many possible locations, each at a certain probability. This is a random variable. Furthermore, it has no definite momentum either: at each given time, it may have many possible momenta, each at a certain probability. This is a superposition: all these momenta "exist" at the same time, and help form reality: no longer concrete, but only stochastic.

Still, how could reality be so ambiguous? What is reality in fact? How does reality exist uniquely in nature around us?

To make quantum mechanics more plausible (not only mathematically but also physically), we have the theory of parallel universes [9]. In this theory, all possible universes are formed, and get to exist. You could be in either of them. Still, you are highly likely to find yourself in the probable ones, not in the improbable ones. In this sense, you are not the only one: duplicate copies of yourself also live in each and every parallel universe. Only upon making an experiment (or a measurement, or an observation, like just looking around you) do you get to see and realize in which universe exactly you are.

Thus, reality is nothing but an endless sequence of observations, picking correct universes over and over again. Fortunately, Zorn's lemma makes sure that such a chain of universes indeed exists. This helps design a concrete chain of events, one by one, straight into the future, with no end or bound. In this chain, quantum mechanics gets its real face, just like classical mechanics before. In this sense, Zorn's lemma helps solve Zeno's paradox once and for all: time never stops, but always restarts.

## 12.1 Nondeterminism

### 12.1.1 Particle and Its Position and Momentum

Consider an individual particle. In quantum mechanics, things are nondeterministic. At each given time, we can't tell where exactly the particle is, but only where it *could* be, at some probability. This is a random variable. Likewise, at each given time, the particle *could* also travel at some speed and direction (at some probability). This is its (momentary) velocity: uncertain, not concrete. As a matter of fact, velocity reflects a more fundamental property: (momentary) momentum. This is mass times velocity. It is nondeterministic too: at each given time, we can't tell what exactly the momentum is, but only what it *could* be (at some probability). Thus, both position and momentum are stochastic: random variables. Only one of them could be measured precisely, but this is not advisable, because then the other one can never be measured any more, not even approximately.

### 12.1.2 Nondeterminism and Uncertainty

In theory, we could measure the position of the particle, and come up with an exact answer. Still, this is not advisable: this would change its wave function for good, with no trace. As a result, we'd never know what its momentum was. This is indeed the uncertainty principle.

### 12.1.3 Uncertainty: Position and Momentum

This also works the other way around: at each given time, we could measure the momentum of the particle, and come up with a definite answer. Again, this is not a good idea: it would change its wave function forever, with no return. As a result, we'd never know what the position was.

You can never escape nondeterminism: you can never measure both position and momentum accurately at the same time, but only one of them, leaving the other hidden and missing forever. This is not because you are ignorant, but because nature is stochastic: position and momentum are random variables, and their probability distributions are stored in the wave function. Thus, the true reality is encapsulated in the wave function, and we better be careful not to destroy or "collapse" it. For this reason, better take no measurement, until absolutely necessary.

## 12.2 Light Cones

### 12.2.1 Future Light Cone

In quantum mechanics, reality takes a stochastic face: the wave function tells us the probability to observe a certain position (or momentum). Still, in reality, we never see the wave function itself. Instead, we see a chain of events. How could this emerge and materialize from the mathematical probability distribution?

For simplicity, consider just one individual particle. Assume that, at the initial time $t_0$, we already know its position: $(x, y, z)$. In spacetime, this makes an event: $(x, y, z, t_0)$.

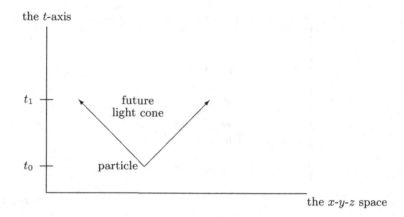

**Fig. 12.1.** Initially, the particle is at the former event $(x, y, z, t_0)$ in spacetime. Thanks to quantum mechanics, it could then travel in just any direction, not faster than light. Therefore, at $t_1 > t_0$, it must remain in the future light cone.

From now on, how could the particle travel? Well, it must never exceed the speed of light. Thus, it must remain within its light cone (Figure 12.1). In this cone, it could travel in just any direction (at some probability). Each such direction marks a time-like line in spacetime, suitable for motion. Any other direction that points outside the cone, on the other hand, marks a space-like line, unsuitable for motion.

### 12.2.2 Past Light Cone

Consider again a particle at some event in spacetime (Figure 12.2). Where did it come from? Well, it could come from just any event in its past light cone (at some probability). Indeed, in this cone, each line leading to the particle is time-like, thus suitable for a legitimate motion. Any space-like line, leading to the particle from *outside* the cone, on the other hand, is unsuitable and illegitimate: a motion along it would exceed the speed of light, which is nonphysical.

## 12.3 Universe: Set of Particles

### 12.3.1 Universe: Bounded Set in 3-D

A universe is a (bounded) collection of particles in the three-dimensional Cartesian space. In physics, we already know that it is discrete, and even finite (although huge). In the present discussion, on the other hand, we allow even an uncountable continuum, so long as it is still bounded in 3-D.

In your universe, consider one particle. What can you tell about it?

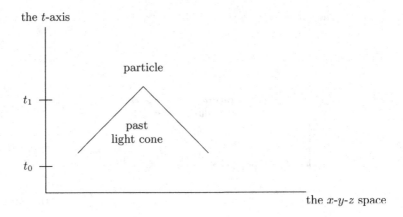

**Fig. 12.2.** The particle at $(x, y, z, t_1)$ could come from just any event in its past light cone (at some probability).

### 12.3.2 Knowledge vs. Entropy

In Chapter 8, uncertainty was given a new name: entropy. In quantum mechanics, the particle has two options:

- either its position is known exactly, with no entropy at all, and its momentum is completely uncertain, with maximum entropy;
- or the other way around: its momentum is known exactly, with zero entropy, but its position remains nondeterministic, at maximum entropy.

Besides, there are many possibilities in between, in which both position and momentum have some (nonzero but nonmaximal) uncertainty (or entropy).

### 12.3.3 Classical Universe

Our parallel universes, on the other hand, should better be not quantum-mechanical but classical: each particle has its own position and momentum (although unknown forever). After all, this is how we want to make quantum mechanics more plausible. In what (classical) universe are you? Well, it depends on you: if you look *where* the particle is, then you get to a universe where the particle is there, with some momentum. If, on the other hand, you look *how fast* the particle moves, then you get to a universe where the particle has that momentum, at some place. Finally, if you don't look at the particle at all, then you get to a universe where the particle is at such a place that could be reached from its previous place, in your previous universe, where you were before.

# 12.4 Universes and Their Order

### 12.4.1 Two Universes

Consider now two universes: $P$ and $Q$. When can we say that

$$P < Q?$$

In other words, how to define a (strict) partial order between universes?

### 12.4.2 Strict Partial Order

Let $P$ and $Q$ be two distinct universes. Assume that $Q$ could emerge from $P$. In other words, at one and the same time, each and every individual particle in $Q$ could be produced from some particle(s) in $P$ (at some positive probability). In this case, we can indeed write

$$P < Q.$$

Is this a legitimate (strict) partial order? Well, we've already proved this in a simpler model: tree of universes (Chapter 2, Section 2.12.2). Let's extend this to the present model as well.

Indeed, our order is clearly transitive. Is it also anti-symmetric? Well, could $P$ emerge from $Q$ back again? No! Indeed, since both are bounded, $Q$ could return to $P$ only by force, consuming energy, and increasing entropy. Thus, the result can never be $P$ any more.

In other words, the entire transition from $P$ to $Q$ and back must have contained a real physical non-adiabatic (irreversible) process, with not only a shock but also a rarefaction wave, increasing entropy (as in Chapter 9 in [45]). Therefore, it could never produce $P$ back again.

Furthermore, in our parallel universes, entropy increases in yet another sense: information is lost to other universes, created simultaneously at the same time. This will be discussed more below. (We disregard virtual or massless particles for now.)

Usually, anti-symmetry implies anti-reflexivity. Here, however, we assumed that $P \neq Q$, so we must also prove anti-reflexivity. For this, ask: could $P$ be absolutely static? No! Indeed, a static universe must have a temperature as low as the absolute zero: $0°K$, or $-273°C$. This is completely nonphysical, and could be excluded. Likewise, a singular (degenerate) universe of just one point is nonphysical as well, and could be excluded as well.

Thus, our partial order is not only transitive but also anti-symmetric and anti-reflexive, as required. This shows quite clearly how plausible the laws of thermodynamics really are, both physically and mathematically.

### 12.4.3 Traveling Universe

How could $Q$ emerge from $P$? Here is a simple example: to produce $Q$, $P$ could just travel at a constant speed, not exceeding the speed of light (Figures 12.3–12.4). In quantum mechanics, such a travel could take place at some (small) probability. We can therefore write:

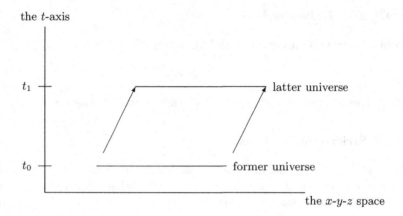

**Fig. 12.3.** In quantum mechanics, motion is nondeterministic. At some (small) probability, all particles could travel rightwards at the same speed. This way, the entire universe travels rightwards as a whole (not faster than light).

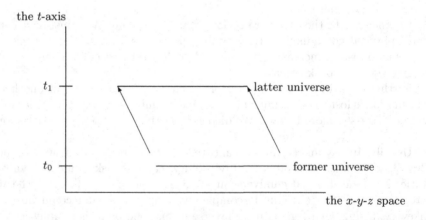

**Fig. 12.4.** If, on the other hand, all particles happen to travel leftward rather than rightward, then the entire universe travels leftward as well (not faster than light).

$$P < Q.$$

This is a very simple example: $P$ and $Q$ are equivalent to each other: they are essentially the same.

Could $P$ emerge from $Q$ back again? No! After all, this would require some force to change direction, in violation of conservation of momentum and energy. Once again, we get to see how plausible the laws of thermodynamics really are, not only physically but also mathematically.

Still, this is a rather fictitious example. After all, in what frame of references does this travel take place? Essentially, this is just a static universe, which we already excluded. Let's look at a more genuine example.

### 12.4.4 Expanding Universe

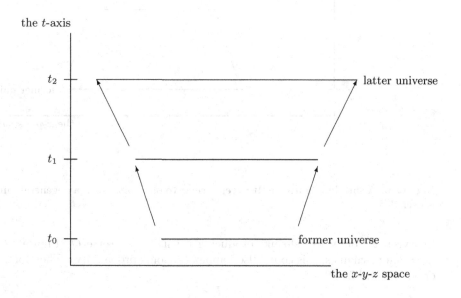

**Fig. 12.5.** An expanding universe.

Consider now a more interesting example: $P$ expands, producing a bigger universe $Q$ (Figure 12.5). In quantum mechanics, this could also happen (at some probability). Therefore, in our order,

$$P < Q.$$

This also works the other way around:

### 12.4.5 Shrinking Universe

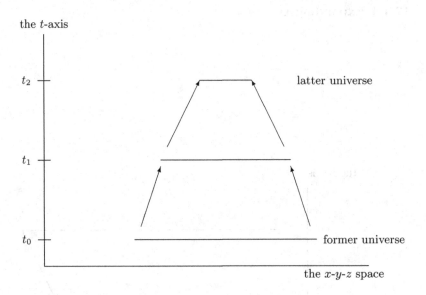

**Fig. 12.6.** A shrinking universe. It never shrinks to one point: such a degenerate universe is excluded.

Likewise, $P$ could also shrink, producing a smaller universe $Q$ (Figure 12.6). In quantum mechanics, this could also happen (at some probability). Therefore, in our order, we still have

$$P < Q.$$

Although $Q$ is smaller, it is still nonsingular: it contains more than one point. Therefore, it is not excluded. Let's move on to yet more complicated examples.

### 12.4.6 Split or Merge Particles

At some probability, some particles in $P$ could also split or merge, as in a Feynman diagram. (That is, if they have enough energy for this: see Chapter 19.) In this process, energy is still conserved: some of their original mass may transform into a new kinetic energy, allowing the new particles to fly away (or the other way around). Thus, we can still write

$$P < Q,$$

as required. Our partial order is now ready: we can now use Zorn's lemma to design reality. Still, before getting into this, let's see how our new partial order fits with the main phenomenon in quantum mechanics: interference.

## 12.5 Least Action and Lagrangian

### 12.5.1 Quantum-Mechanical Universes

So far, we mainly looked at classical universes. Next, let's focus on two quantum-mechanical universes: $P < Q$. This means that $Q$ could follow from $P$ not only classically but also quantum-mechanically: interference. Still, as before, entropy must increase. How come?

### 12.5.2 Straight Line in Spacetime

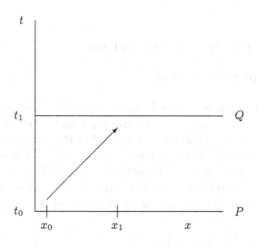

**Fig. 12.7.** No potential: the particle travels in a straight line in spacetime, from $x_0 \in P$ to $x_1 \in Q$, arriving at time $t_1 > t_0$.

At the initial time $t_0$, our particle is at the initial point $x_0 \in P$. At time $t_1 > t_0$, on the other hand, it arrives at a new point: $x_1 \in Q$. What path does it take in spacetime? This is the straight line from $(x_0, t_0)$ to $(x_1, t_1)$ (Figure 12.7). Why?

Because this line minimizes the Lagrangian, which (in the lack of any potential) is just the kinetic energy:

$$\text{Lagrangian} = \text{kinetic energy} = \frac{1}{2}\text{mass} \cdot \text{speed}^2.$$

### 12.5.3 No Potential: Minimal Kinetic Energy

More precisely, in spacetime, the straight line minimizes the action: the integral of the Lagrangian. Indeed, on the straight line, the kinetic energy is distributed uniformly and evenly. This is as efficient as ever. After all, doubling the speed for a short while would cost energy four times as large, unnecessarily. This is why runners are often advised to run moderately: not too fast, and not too slowly.

### 12.5.4 Uncertainty: Position and Momentum

At time $t_1$, our particle finally arrives at $x_1$. What happened to entropy? Could it decrease? Of course not. Still, could it remain the same? In other words, could this travel be adiabatic and reversible? No! After all, due to the uncertainty principle, information was lost forever. Indeed, at $x_1 \in Q$, we know nothing about momentum, and therefore have no idea where the particle came from. In fact, we lost all information about those parallel universes created simultaneously at $t_1$ as well. This means more entropy: $Q > P$, as required.

## 12.6 Path Integral and Least Action

### 12.6.1 Principle of Least Action

The above is just a simple example. In general, on the other hand, we also have a potential. This may make the optimal path not straight but curved in spacetime. In classical mechanics, this is indeed the path that the particle takes in spacetime. In quantum mechanics, on the other hand, the particle is highly likely to take this path, but still could (at some small probability) use other paths as well to get from $(x_0, t_0)$ to $(x_1, t_1)$ in spacetime. Each path may have a different probability to be used. To obtain the probability, we must first calculate the complex amplitude $a$:

$$
\begin{aligned}
a(\text{path}) &= \exp\left(-\frac{\sqrt{-1}}{\hbar}\text{action}\right) \\
&= \exp\left(-\frac{\sqrt{-1}}{\hbar}\int_{\text{on the path}} \text{Lagrangian } dt\right) \\
&= \exp\left(-\frac{\sqrt{-1}}{\hbar}\int_{\text{on the path}} (\text{kinetic minus potential energy}) \, dt\right),
\end{aligned}
$$

where $\hbar$ is Planck's constant.

### 12.6.2 Paths and Their Interference

In the above definition, the action depends on the particular path used in spacetime. To define it, the integral is taken along this path in spacetime, with respect to the time segment $dt$. (In relativity, on the other hand, $t$ means proper time, read from a clock carried with the particle itself.) Minimizing the action (in magnitude) means that the particle indeed picks the best path: it uses its potential energy best to approximate the kinetic energy that it needs, as efficiently and moderately as possible: not too early, and not too late. In classical mechanics, this is indeed the solution: the optimal path that the particle indeed chooses in spacetime.

In quantum mechanics, on the other hand, this is the most probable path, but not the only one. This path is most likely to be picked, because it suffers from little interference. Any other path, with a stronger action, on the other hand, is less likely: its complex amplitude oscillates more frequently, and may suffer from destructive interference with other paths that lie nearby. (See also Chapter 18, Sections 18.6.1–18.6.2 below.)

Thus, at $(x_0, t_0)$, $P$ splits into a lot of intermediate universes that lie between $P$ and $Q$, each uses a different math, to get to $Q$, as required. Upon arriving at $Q$, we'll never know which path was used in spacetime. After all, one particle is not enough to see an interference pattern: for this, we must have many.

## 12.7 Interference: Don't Detect!

### 12.7.1 Example: Double Slit

Let's look once again at a simple example, with no potential at all. In Figure 12.7, we've already seen a one-dimensional case. Here, on the other hand, we have one more dimension: the vertical $y$-coordinate, which is proportional to the time $t$. Thanks to no potential, the particle travels in straight lines, not only in the $(x, y)$-plane but also in the $(x, y, t)$-spacetime (Figure 12.8). For simplicity, we also pick $x_1 = x_0$. This is indeed the double-slit experiment.

In between $P$ and $Q$, there are two intermediate universes. We're equally likely to use either one to get to $Q$: either pass through the left slit, or through the right one. After that, we may well find ourselves at $(x_1, y_1) \in Q$, as required. In either case, entropy indeed increases: information is lost to parallel universes, created at $t_1$ as well. With just one particle, however, we can't see interference as yet. For this, we need many.

### 12.7.2 Interference: a Statistical Phenomenon

Consider now not only one but also many particles: say, photons sent together in one light pulse, or electrons sent one by one towards the slits. In either case, if we place a screen at $Q$, then we get to see interference on it: alternate strips of light, dark, light, dark, and so on. How come?

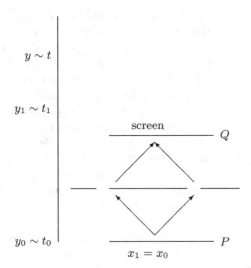

**Fig. 12.8.** Double slit: in spacetime, the particles travel from $(x_0, y_0, t_0)$ to $(x_1, y_1, t_1)$ (no potential — straight lines). In the $(x, y)$-plane, on the other hand, they travel through the left (or right) slit, superpose, and interfere with themselves, to print an interference pattern on the screen in $Q$.

### 12.7.3 Universe with an Interference Pattern

In quantum mechanics, each particle passes through both slits at the same time. This way, it superposes and interferes with itself. This is why it is highly likely to hit the screen in a strip of light: constructive interference. This is indeed a $Q$ we are likely to find ourselves in. Likewise, once many particles hit the screen, we're highly likely to find ourselves in a $Q$ with an interference pattern on its screen.

### 12.7.4 Entangle with the Detector: No Interference!

Better not place a detector by the slit to find out which slit was used. This is not a good idea: it destroys the interference completely! In fact, this way, there is no superposition at all. Instead, we're highly likely to find ourselves in a $Q$ to which half the particles arrived through the left, and half through the right slit. Instead of the above interesting $Q$, with an interference pattern on its screen, we're much more likely to find ourselves in a boring $Q$, with just two spots of light, and dark on the rest of the screen.

Why? Due to entanglement: to detect which slit was used, each particle must get entangled with the detector. This distinguishes between the two faces of the particle, and prevents them from superposing with one another. Now, the particle must make a decision: either pass through the left, or through the right slit, but not through both at the same time! This is why we're highly likely to find ourselves in a $Q$ with no interference on its screen any more.

### 12.7.5 Entanglement Entropy

Let's return to the more interesting (no detector) case, in which we are highly likely to find ourselves in a $Q$ with an interference pattern on its screen: alternate strips of light, dark, light, dark, and so on. To this $Q$, each particle had arrived as a superposition, interfering with itself on the way. Still, is $Q > P$? Well, $Q$ clearly follows from $P$. Furthermore, it has more entropy. Indeed, some information was lost to parallel universes, created at $t_1$ as well. This means that entropy increases, as required.

Moreover, to print the interference pattern onto the screen, some heat is needed. In thermodynamics, this means more entropy. In chemistry, on the other hand, this has a deeper interpretation: on the screen, molecules start to move fast, increasing entropy, as in Section 12.5.4. Better yet, in quantum mechanics, this has a yet more subtle interpretation: in $Q$, the particles get entangled with the screen. This introduces a new kind of entropy: entanglement entropy.

## 12.8 Using Zorn's Lemma

### 12.8.1 No Maximal Universe

Let $P$ be some initial universe. Look at those universes that could emerge from it (at some probability):

$$\{Q \mid Q > P\}.$$

In this set, is there a maximal universe? No! Indeed, in a maximal universe, let time tick forward. This way, you must always get the same universe again. But this is a static universe that was already excluded. Thus, no maximal universe could ever be found.

### 12.8.2 Chain with No Upper Bound

Thanks to Zorn's lemma, there is a chain of universes, emerging from one another, with no upper bound at all. This is a well-ordered chain: every subchain starts from its own initial universe, at its own initial time (Chapter 2, Section 2.13.4). In quantum mechanics, this could now serve as a concrete chain of real universes. Indeed, the chain makes a concrete history, starting from the initial universe, and developing from universe to universe, throughout the entire chain, endlessly and indefinitely.

In the chain, the universes are concrete and classical. This is how quantum mechanics stops being stochastic and mathematical, and gets more realistic and physical. In fact, the well-ordered chain tells us how the universe develops in time. This way, Zorn's lemma answers Zeno's paradox once and for all: in the well-ordered chain, time never stops, but always restarts.

### 12.8.3 Chain of Life

Better yet, let $L$ be a fixed chain of universes. These are the universes that you get to observe throughout life. How to extend it to an endless reality?

Easy: restrict the above discussion only to those chains that include $L$. Thanks to Zorn's lemma, there is a (well-ordered) chain with no upper bound, which includes $L$. This is what we wanted: an endless reality that extends $L$. This is indeed your own reality.

## 12.9 Exercises: Measurement and Result

### 12.9.1 What Do You Really Measure?

1. Make some experiment. Say, measure the magnetic field, or the position of a particle.
2. What is the result? Hint: look at your equipment. It must have a screen. Look at the answer, and write it down.
3. What is the meaning of this result?
4. Is this the position of the particle? Hint: not quite! This is only what the equipment *thinks* that the position is. After all, to calculate the answer, the equipment uses some inner process. During this process, there is a lot of room for inaccuracy and uncertainty. For example, who says that the electrons used to print the answer on the screen were generated in the right way? Who says that the inner mechanism really measures what it is supposed to measure?
5. How does your equipment work? Does it really tell you the truth?

# Part VI

Applications
in Cryptography
and Error Correction

# Applications in Cryptography and Error Correction

Set theory is the basis for mathematical logic, which is the basis for computer science (Figures 1.12–1.13). In Chapters 7–8, we've already seen an example of a useful logical structure: a binary tree. This is the basis for useful recursive algorithms, such as fast Fourier transform (FFT). Later on, we'll introduce FFT in three ways: recursively, nonrecursively, and even on a quantum computer. This will be useful in cryptography.

In this part, we design a practical coding-decoding algorithm, to help transfer secret messages over the internet. For this purpose, we need a very long prime number.

Recall that a prime number is a special kind of natural number: it can never be factored as a product of two other natural numbers. Now, assume that we already have a very long prime number, with a lot of digits. Thanks to it, we can implement a safe coding-decoding algorithm: the RSA key exchange.

To prove the RSA algorithm, we must first prove a fundamental theorem in number theory: Fermat's little theorem. For this purpose, we use the binomial formula, with modular arithmetic.

To help store long numbers, best use their binary representation as binary polynomials. This way, it is particularly easy to use them in modular arithmetic. To multiply very long numbers, best use FFT. Later on, we'll also combine ideas from FFT and number theory, and design a practical error-correction algorithm: the Reed-Solomon code, with an interesting application in computational biology.

To mirror FFT, we use a virtual binary tree. This is easily implemented recursively on a digital computer. Still, later on, we'll also do this nonrecursively, and even on a quantum computer.

Unfortunately, no quantum computer is currently available: this is still science fiction. Still, it may help in another aspect: to give more insight about quantum mechanics, including its potential and limits. Moreover, it may help understand the computational algorithm better, including its cost in terms of time and storage. This may lead to an efficient implementation on the traditional digital computer, saving a lot of memory, and leaving a lot of room for parallelism.

Still, the RSA algorithm contains an expensive part: to find a very long prime number in the first place. This is not an easy task. To avoid it, better compromise: instead of a prime number, use another number that is only *probably* prime. With it, our coding-decoding algorithm may still be safe enough for all practical purposes.

# 13

## Coding–Decoding:
## the RSA Key Exchange

In this chapter, we look at the integer numbers. On them, we define an equivalence relation. This leads to modular arithmetic. Thanks to this, we design a practical coding-decoding algorithm in cryptography: the RSA key exchange. To do this, we assume that we already have a very long prime number, containing a lot of binary digits.

Thanks to the binomial formula, we can then prove Fermat's little theorem, which helps prove correctness of the RSA algorithm. Finally, instead of a very long prime number (which is too hard to find), it is good enough to use a *probably* prime number. This still leads to a safe method to send secret messages on the internet.

## 13.1 Prime Numbers

### 13.1.1 Prime Numbers

What is a prime number? This is a special kind of natural number that cannot be divided by any other number (evenly, with no remainder). Here are some prime numbers:

$$2, \ 3, \ 5, \ 7, \ 11, \ 13, \ 17, \ 19, \ 23, \ 29, \ 31, \ 37, \ \ldots.$$

Indeed, 11 could be divided by neither 2 nor 3 nor 5 nor 7. Of course, 11 could be divided by 1 and 11 itself, but this doesn't count.

### 13.1.2 Prime Numbers and Their Density

Consider an elementary problem: find all the prime numbers between 2 and $N$, where $N$ is a given (large) number. How many are there? Quite many: as many as $N/\log N$ (times a constant that is independent of $N$). This is also denoted by

$$O\left(\frac{N}{\log N}\right).$$

Thus, the density of prime numbers is as large as

$$O\left(\frac{1}{\log N}\right).$$

Thus, if you pick a number between 2 and $N$ at random, then it has probability $O(1/\log N)$ to be prime. As $N$ increases, this probability approaches 0 quite slowly. Next, how to find these prime numbers in practice?

### 13.1.3 Finding Prime Numbers: a Naive Algorithm

The naive algorithm uses mathematical induction: assume that we have already found all the prime numbers between 2 and $N-1$. (This is the induction hypothesis.) Now, scan them one by one, and check whether they divide $N$ or not. If any of them does, then $N$ is nonprime. If none of them does, then $N$ is prime.

As a matter of fact, there is no need to scan them all. One could stop at the largest prime number $p_N$ for which

$$p_N^2 \leq N.$$

After all, if $N = pq$ for some prime numbers $p$ and $q$, then

$$\min\left(p^2, q^2\right) \leq pq = N,$$

so either $p$ or $q$ is scanned, and there is no need to scan the other.

### 13.1.4 The Sieve of Eratosthenes

In the induction step above, how to check whether a prime number $p$ divides $N$ or not? For this purpose, one could divide, and check whether there is any remainder or not. Even if both $N$ and $p$ are in their efficient binary representation, this may still cost too much: as many as

$$\log_2 N \log_2 p$$

bit operations. This is too slow.

To avoid this, better use a different approach: look at things the other way around — not from the perspective of $N$, but from the perspective of $p$. More specifically, once a prime number $p$ has been found, don't wait: use it immediately to find all its multiples, and dismiss them as nonprimes.

This is the sieve of Eratosthenes. It models the following algorithm. Scan the prime numbers one by one, starting from

$$p = 2, 3, 5, \ldots, p_N.$$

For each prime number $p$, drop all its multiples. After all, they are surely nonprime. In the end, this leaves only the prime numbers between 2 and $N$, as required.

## 13.2 Congruence: a Mathematical Equivalence Relation

### 13.2.1 Division with Remainder

Let's look at the integer numbers. Thanks to set theory, they enjoy all sorts of mathematical equivalence relations.

Let $n > m > 0$ be two natural numbers. What is the ratio between $n$ and $m$? By "ratio," we don't mean the simple fraction $n/m$, but only its integer part: the maximal integer number that doesn't exceed it:

$$\left\lfloor \frac{n}{m} \right\rfloor = \max\left\{ j \in \mathbb{Z} \mid j \leq \frac{n}{m} \right\},$$

where $\mathbb{Z}$ is the set of integer numbers. This way, we can now divide $n$ by $m$ with remainder (residual). This means representing $n$ in the following form:

$$n = km + l,$$

where

$$k = \left\lfloor \frac{n}{m} \right\rfloor,$$

and $l$ is the remainder (or residual) that is too small to be divided by $m$:

$$0 \leq l < m.$$

This $l$ has a new name: $n$ modulus $m$. To be recognized as one thing, it is often placed in parentheses:

$$l = (n \bmod m).$$

### 13.2.2 Congruence: Same Remainder

The above equation could also be written as

$$n \equiv l \bmod m.$$

This means that $n$ and $l$ are related to each other: they are *congruent* modulus $m$:

$$n = km + l$$
$$l = 0 \cdot m + l.$$

This means that both $n$ and $l$ are indistinguishable modulus $m$, and have the same remainder: $l$:

$$(n \bmod m) = l = (l \bmod m).$$

In other words, their difference $|n - l|$ could be divided by $m$ (evenly, with no remainder at all):

$$m \mid |n - l|.$$

This is indeed a mathematical equivalence relation. (Check!)

## 13.3 Greatest Common Divisor

### 13.3.1 Common Divisor

Consider now the special case in which there is no remainder at all:

$$(n \bmod m) = 0.$$

In this case, we say that $m$ divides $n$:

$$m \mid n.$$

In general, however, $m$ doesn't divide $n$. Still, $m$ may share a common divisor with $n$: a third number that divides both $n$ and $m$. For example, if both $n$ and $m$ are even, then 2 is a common divisor. Still, there could be a few common divisors. What is the maximal one?

### 13.3.2 The Euclidean Algorithm

How to calculate the greatest common divisor of $n$ and $m$? Well, there are two possibilities: if $m$ divides $n$, then $m$ itself is the greatest common divisor. If not, then

$$
\begin{aligned}
\mathrm{GCD}(n, m) &= \mathrm{GCD}(km + l, m) \\
&= \mathrm{GCD}(l, m) \\
&= \mathrm{GCD}(m, l) \\
&= \mathrm{GCD}(m, n \bmod m).
\end{aligned}
$$

So, instead of the original numbers

$$n > m,$$

we can now work with two smaller numbers:

$$m > l = (n \bmod m).$$

This leads to the (recursive) Euclidean algorithm:

$$
\mathrm{GCD}(n, m) = \begin{cases} m & \text{if } (n \bmod m) = 0 \\ \mathrm{GCD}(m, n \bmod m) & \text{if } (n \bmod m) > 0. \end{cases}
$$

How to prove this? By mathematical induction on $n = 2, 3, 4, \ldots$ (see below).

### 13.3.3 The Extended Euclidean Algorithm

Thanks to this kind of mathematical induction, one could also write the greatest common divisor as a linear combination:

$$\mathrm{GCD}(n, m) = an + bm,$$

for some integer numbers $a$ and $b$ (positive or negative or zero). Indeed, if $m$ divides $n$, then this is easy:

$$\mathrm{GCD}(n, m) = m = 0 \cdot n + 1 \cdot m,$$

as required. If not, then the induction hypothesis tells us that

$$\mathrm{GCD}(n, m) = \mathrm{GCD}(m, l) = \tilde{a} \cdot m + \tilde{b} \cdot l,$$

for some new integer numbers $\tilde{a}$ and $\tilde{b}$ (positive or negative or zero). Now, in this formula, substitute

$$l = n - km.$$

This gives

$$\begin{aligned}
\mathrm{GCD}(n, m) &= \mathrm{GCD}(m, l) \\
&= \tilde{a} \cdot m + \tilde{b} \cdot l \\
&= \tilde{a} \cdot m + \tilde{b}\,(n - km) \\
&= \tilde{b} \cdot n + \left(\tilde{a} - k \cdot \tilde{b}\right) m.
\end{aligned}$$

To have the desired linear combination, just define the new coefficients

$$a = \tilde{b}$$
$$b = \tilde{a} - k \cdot \tilde{b} = \tilde{a} - \left\lfloor \frac{n}{m} \right\rfloor \tilde{b}.$$

This is the extended Euclidean algorithm.

### 13.3.4 The Modular Extended Euclidean Algorithm

In practice, however, there is still a problem: the new coefficients $a$ and $b$ could still be negative. Worse, they could be too big, and even exceed $n$. How to avoid this? Easy: throughout the above algorithm (including the recursion), introduce just one change: take modulus $n$. This will force the new coefficients to remain moderate: between 0 and $n - 1$:

$$0 \le a, b < n.$$

Likewise, the induction hypothesis should inherit the same change as well, and confine its own coefficients to the same interval: $[0, n - 1]$:

$$0 \le \tilde{a}, \tilde{b} < n.$$

To make this change, we must first rephrase the original algorithm as follows:

$$a = \tilde{b}$$
$$s = \left\lfloor \frac{n}{m} \right\rfloor \tilde{b}$$
$$b = \tilde{a} - s.$$

This changes nothing. It is now time to make a real change. To make sure that both $a$ and $b$ are between 0 and $n - 1$, make just a tiny change: take modulus $n$:

$$a = \tilde{b}$$
$$s = \left(\left\lfloor \frac{n}{m} \right\rfloor \tilde{b}\right) \bmod n$$
$$b = \begin{cases} \tilde{a} - s & \text{if} \quad \tilde{a} \ge s \\ \tilde{a} + n - s & \text{if} \quad \tilde{a} < s. \end{cases}$$

Now, in the induction hypothesis, assume also that

$$0 \leq \tilde{a}, \tilde{b} < n.$$

Thanks to this, we also have

$$0 \leq a, s, b < n,$$

as required. Furthermore, modulus $n$, nothing has changed: the algorithm still does the same job. Thus, we obtained a new linear combination modulus $n$:

$$\text{GCD}(n, m) \equiv (an + bm) \bmod n.$$

Let's go ahead and use this in practice.

## 13.4 Modular Arithmetic

### 13.4.1 Coprime

Consider a natural number $n$. What is a coprime of $n$? This is another number $p$ (prime or not) that shares no common divisor with $n$:

$$\text{GCD}(n, p) = 1.$$

There may be a few legitimate coprimes. Which one to pick? Better pick a moderate one. For this purpose, start from an initial guess, say a small prime number:

$$p \leftarrow 5.$$

Does it divide $n$? If not, then pick it as our desired coprime. If, on the other hand, $p$ divides $n$, then we must keep looking:

$$p \leftarrow \text{the next prime number,}$$

and so on. In at most $(\log_2 n)$ guesses, we'll eventually find a new prime number $p$ that doesn't divide $n$ any more, and can therefore serve as its legitimate coprime.

### 13.4.2 Modular Multiplication

Let's consider yet another task. Let $n$ and $j$ be two natural numbers. How to calculate the product

$$nj \bmod m?$$

Unfortunately, both $n$ and $j$ could be very long, and contain many digits. Their product $nj$ could be even longer, and not easy to store on the computer or work with. How to avoid this?

Fortunately, both $n$ and $j$ could be written in the form in Section 13.2.2:

$$n = \left\lfloor \frac{n}{m} \right\rfloor m + (n \bmod m)$$

$$j = \left\lfloor \frac{j}{m} \right\rfloor m + (j \bmod m).$$

On the right-hand side, only the latter term is interesting. The former term, on the other hand, is a multiple of $m$, which is going to drop anyway. Therefore, instead of $nj$, better calculate

$$(n \bmod m)(j \bmod m) \bmod m.$$

In other words, the modulus operation could be carried out not only after but also before starting to multiply. This way, we only multiply moderate numbers, which never exceed $m - 1$.

### 13.4.3 Modular Power

The same idea could be used time and again to calculate a power like

$$n^k \bmod m.$$

Here, even for a moderate $k$, $n^k$ may be too long to store or use. Thanks to the above idea, we can now avoid this: before using $n^2$, take modulus $m$:

$$\text{power}(n, k, m) = \begin{cases} 1 & \text{if } k = 0 \\ n \bmod m & \text{if } k = 1 \\ n \cdot \text{power}(n^2 \bmod m, (k-1)/2, m) \bmod m & \text{if } k > 1 \text{ and } k \text{ is odd} \\ \text{power}(n^2 \bmod m, k/2, m) & \text{if } k > 1 \text{ and } k \text{ is even.} \end{cases}$$

Better yet, before starting the calculation, substitute

$$n \leftarrow (n \bmod m).$$

This avoids the large number $n^2$ in the first place. Instead, the recursion is applied to the moderate number

$$(n \bmod m)^2 \leq (m - 1)^2.$$

This way, throughout the entire calculation, all intermediate products are kept moderate, and never exceed $(m - 1)^2$. To prove this, use mathematical induction on $k$ (see exercises below).

Still, this seems a bit too pedantic. After all, the modulus operation could be costly too, so why apply it so often? Better apply it only when absolutely necessary: when detecting an inner (temporary) variable too long to store or use efficiently.

## 13.5 Modular Inverse: Set and Its Mapping

### 13.5.1 Modular Inverse

Let $p$ and $q$ be some given coprime numbers (prime or not):

$$\text{GCD}(p, q) = 1.$$

Another important task is to find the inverse of $q$ modulus $p$, denoted by

$$q^{-1} \bmod p.$$

What is this inverse? It is the unique solution $x$ of the equation

$$qx \equiv 1 \bmod p.$$

## 13.5.2 Set and Its Mapping

Does $x$ exist? Moreover, is it unique? To prove this, consider the set $S$, containing the integer numbers from 0 to $p - 1$:

$$S = \{0, 1, 2, \ldots, p - 1\}.$$

Define the new mapping

$$M : S \to S$$

by

$$M(s) = (qs \bmod p), \quad s \in S.$$

Our aim is to show that $M$ maps some number to 1. This number will then serve as the desired solution $x$. To show this, let's look at $M$, and study its properties.

## 13.5.3 One-to-One Mapping

Is $M$ one-to-one? To check on this, consider two numbers $a, b \in S$ (say, $a \geq b$). Assume that

$$M(a) = M(b).$$

This means that

$$q(a - b) \equiv 0 \bmod p.$$

In other words, $p$ divides $q(a - b)$:

$$p \mid q(a - b).$$

Since $p$ shares no common divisor with $q$, it must divide $a - b$:

$$a \equiv b \bmod p.$$

## 13.5.4 Mapping Onto the Set

Thus, $M$ is indeed one-to-one. As such, it preserves the total number of elements in $S$:

$$|M(S)| = |S| = p.$$

Thus, $M$ is a one-to-one mapping from $S$ onto $S$.

## 13.5.5 The Inverse Mapping

Thus, it has an inverse mapping:

$$M^{-1} : S \to S.$$

In particular, we can now define

$$x = M^{-1}(1).$$

In summary, the inverse of $q$ modulus $p$ exists uniquely:

$$(q^{-1} \bmod p) = M^{-1}(1).$$

### 13.5.6 Using the Extended Euclidean Algorithm

How to solve for $x$ in practice? For this purpose, assume that

$$q < p.$$

(Otherwise, just substitute

$$q \leftarrow q - p$$

time and again, until $q$ gets as small as $p$. After all, this makes no difference to $x$.) Now, use the modular extended Euclidean algorithm (Section 13.3.4). This way, we can now write

$$1 = \mathrm{GCD}(p, q) \equiv (ap + bq) \equiv bq \bmod p,$$

for some integer numbers $a$ and $b$ that lie in between 0 and $p - 1$. Obviously, the desired solution is just

$$x = b.$$

In summary, the inverse of $q$ modulus $p$ is now available explicitly:

$$\left(q^{-1} \bmod p\right) = b.$$

This will be useful below.

## 13.6 Group: Set with an Algebraic Operation

### 13.6.1 Set of Integer Numbers

What is a group? This is a set with an algebraic operation that enjoys a few desirable rules. What is the most common group? This is the set of integer numbers:

$$\mathbb{Z} = \{\ldots, -3, -2, -1, 0, 1, 2, 3, \ldots\}.$$

What is the algebraic operation? Not multiplication, but addition! Is this a legitimate group?

### 13.6.2 Rules in a Group

Well, let's check the required rules:

- For every two integer numbers, their sum is integer as well.
- Addition is associative.
- To every number, you could add 0, with no effect whatsoever.
- To each number, you could add its negative counterpart, and obtain 0, as required.

These are indeed the laws that makes it a legitimate group.

### 13.6.3 Subgroup of Even Numbers

From this group, we can now extract a subgroup. Say, the even numbers:

$$2\mathbb{Z} = \{\ldots, -6, -4, -2, 0, 2, 4, 6, \ldots\}.$$

Is this a legitimate subgroup? Well, let's check:

- For every two even numbers, their sum is even as well.
- For every even number, its negative counterpart is even as well.

These are the laws that make it a legitimate subgroup. Likewise, one could go ahead and design more and more subgroups:

$$3\mathbb{Z}, \ 4\mathbb{Z}, \ 5\mathbb{Z}, \ldots.$$

### 13.6.4 Quotient Group

Note that $\mathbb{Z}$ has an attractive property: it is Abelian (commutative). Thus, every subgroup is normal [26, 46]. Therefore, we can go ahead and divide, and obtain a new group: the quotient (or factor) group. For example, divide by $4\mathbb{Z}$, to obtain the new group

$$\mathbb{Z}_4 = \mathbb{Z}/(4\mathbb{Z}).$$

In this new group, every two integer numbers are considered as one and the same if their difference is a multiple of 4. For example,

$$3 \equiv 7 \equiv 11 \equiv 15 \equiv \cdots \mod 4$$

are all considered as one and the same. Thus, in $\mathbb{Z}_4$, we have periodicity modulus 4. Indeed, adding a multiple of 4 has no effect whatsoever: the number remains the same. Thus, we actually have only four numbers:

$$\mathbb{Z}_4 = \{0, 1, 2, 3\},$$

and the algebraic operation is addition modulus 4. This is indeed a useful group in its own right.

## 13.7 Drop Zero!

### 13.7.1 Nonzero Integers: Not a Group!

Next, let's drop 0, and look at the rest of the integer numbers:

$$\mathbb{Z} \setminus \{0\} = \{\ldots, -3, -2, -1, 1, 2, 3, \ldots\}.$$

Since 0 is missing, the algebraic operation can't be addition any more: it must be multiplication. Is this a legitimate group? Let's check:

- For every two nonzero numbers, their product is still nonzero.
- Multiplication is associative.
- Multiplying by 1 has no effect whatsoever.
- Whoops: unfortunately, the reciprocal is often noninteger!

## 13.7.2 Modular Multiplication

How to fix this? Let's try the following trick: instead of $\mathbb{Z}$, look at $\mathbb{Z}_4$, and drop 0 from it:

$$\mathbb{Z}_4 \setminus \{0\} = \{1, 2, 3\}.$$

Is this a legitimate group?

## 13.7.3 Modular Inverse

To check on this, let's pick some number, say 3. Does it have an inverse modulus 4? Fortunately, 3 is coprime to 4 (Section 13.5.1). Therefore, it does have an inverse modulus 4: 3 itself. Indeed,

$$3 \cdot 3 = 9 \equiv 1 \bmod 4.$$

In other words,

$$\left(3^{-1} \bmod 4\right) = 3.$$

So, 3 has an inverse modulus 4. But what about 2? Does it have an inverse modulus 4? Unfortunately not: 2 is not coprime to 4, and has no inverse modulus 4.

## 13.7.4 Multiplication Modulus a Prime Number

Thus, $\mathbb{Z}_4 \setminus \{0\}$ is *not* a legitimate group. Instead, better look at

$$\mathbb{Z}_5 \setminus \{0\} = \{1, 2, 3, 4\}.$$

Is this a legitimate group? It sure is. Why? Because, unlike 4, 5 is prime. Therefore, 1, 2, 3, and 4 are coprime to 5, and have an inverse modulus 5:

$$
\begin{aligned}
1 \cdot 1 &\equiv 1 \quad \bmod 5 \\
2 \cdot 3 &\equiv 1 \quad \bmod 5 \\
3 \cdot 2 &\equiv 1 \quad \bmod 5 \\
4 \cdot 4 &\equiv 1 \quad \bmod 5.
\end{aligned}
$$

This will be useful later.

# 13.8 The RSA Algorithm: Coding

## 13.8.1 The Coding Part

We are now ready to introduce the RSA algorithm for coding-decoding [40]. Assume that the original message has already been translated into a long natural number (containing thousands of digits, or even more). Before it is sent over the internet, it must be coded safely. How to do this? Let

$$p < q$$

be two long prime numbers that are kept secret. Their product

$$n = pq,$$

on the other hand, is not secret at all: it is available to the public. Define also

$$\phi = (p - 1)(q - 1).$$

Although $n$ is public, $\phi$ is not. In fact, since both $p$ and $q$ are secret, $p-1$ and $q-1$ are secret as well, so their product can't be calculated by anyone.

Furthermore, let $e \geq 5$ be a moderate coprime of $\phi$. Unlike $\phi$, $e$ is not secret: it is placed in the public domain.

Now, let

$$m < p$$

be our message: a natural number (smaller than $p$) that should be sent secretly to the decoder. For this purpose, $m$ itself must never be sent: someone could intercept and read it! To avoid this, only a coded version should be sent.

Fortunately, like everyone, the sender has access to both $n$ and $e$, and can use them to prepare the coded message — a modular power of the original message:

$$c = (m^e \bmod n).$$

Now, send $c$ to the decoder. Don't worry: nobody can decode it, but only the legitimate decoder, who has access to $\phi$. For this purpose, the decoder needs to calculate

$$d = \left(e^{-1} \bmod \phi\right).$$

Fortunately, only the decoder can calculate this. After all, nobody else has access to $\phi$. Later on, $d$ will be used to uncover the original message $m$.

This is the coding part. To introduce the decoding part, we need a fundamental theorem in number theory: Fermat's little theorem.

## 13.9 Fermat's Little Theorem

### 13.9.1 Binomial Coefficients

To help prove Fermat's little theorem, we need a simple lemma. Let $p$ be a prime number. For any natural number $0 < i < p$, consider the binomial coefficient

$$\binom{p}{i} = \frac{p!}{i!(p-i)!}.$$

In this expression, where could $p$ be found? Only in the numerator, not in the denominator! In other words,

$$p \mid \binom{p}{i}.$$

This will be useful below.

### 13.9.2 Fermat's Little Theorem

We are now ready to introduce Fermat's little theorem. Let $p$ be a prime number. For any natural number $a$, the theorem tells us an interesting fact: the $p$th power has no effect:

$$a^p \equiv a \bmod p.$$

This was proved by Euler:

### 13.9.3 Euler's Proof

The proof is by mathematical induction on $a$. Indeed, for $a = 1$, we clearly have

$$a^p = 1^p = 1.$$

We are now ready for the induction step. For some $a \geq 1$, assume that

$$a^p \equiv a \bmod p.$$

This is the induction hypothesis. Let's use it in Newton's binomial formula. Indeed, thanks to the above lemma, most binomial coefficients vanish (modulus $p$):

$$(a + 1)^p = \sum_{i=0}^{p} \binom{p}{i} a^i \equiv a^p + 1 \equiv a + 1 \bmod p,$$

as asserted. This completes the induction step. This is Euler's proof.

### 13.9.4 Corollary

So far, we proved that

$$p \mid a^p - a = (a^{p-1} - 1)a.$$

On the right-hand side, we have a product of two factors: $a^{p-1} - 1$ times $a$. Now, we already know that $p$ is prime. Therefore, $p$ must divide either the former factor, $a^{p-1} - 1$, or the latter factor, $a$.

Does $p$ divide $a$? Let's assume not. In other words, assume that $a$ is not a multiple of $p$:

$$p \nmid a.$$

In other words, assume that $a$ is coprime to $p$:

$$\mathrm{GCD}(a, p) = 1.$$

In this case, $p$ must divide the former factor:

$$p \mid a^{p-1} - 1.$$

In other words,

$$a^{p-1} \equiv 1 \bmod p.$$

In particular, this holds whenever $a < p$. After all, in this case, $a$ is indeed coprime to $p$, as required: This will be useful below.

### 13.9.5 The Order of $a$

This leads to yet another corollary. Again, assume that $a$ is not a multiple of $p$:

$$GCD(a, p) = 1.$$

Let $q$ be the minimal natural number for which

$$a^q \equiv 1 \bmod p.$$

($q$ is called the order of $a$, and is denoted by $q = o_p(a)$.) Then $q$ divides $p - 1$:

$$q \mid p - 1.$$

Indeed, from the definition of $q$ as minimal,

$$q \leq p - 1.$$

Let's show that $q$ indeed divides $p-1$. To see this, divide $p-1$ by $q$ (with remainder):

$$p - 1 = lq + r,$$

where $l$ is some natural number, and $r$ is the remainder:

$$0 \leq r < q.$$

This way,

$$a^r = a^r 1^l \equiv a^r a^{ql} = a^{lq+r} = a^{p-1} \equiv 1 \bmod p.$$

This is in violation of the very definition of $q$ as minimal, unless

$$r = 0.$$

This proves that

$$q \mid p - 1,$$

as asserted.

## 13.10   The RSA Algorithm: Decoding

### 13.10.1   The Decoding Part

We are now ready to introduce the decoding part. For this purpose, we use the same notations as in Section 13.8.1: $m$ is the original message, $p$ and $q$ are (private) prime numbers, etc.

Since $m < p < q$, $m$ is coprime to both $p$ and $q$. Likewise, every power of $m$ is also coprime to both $p$ and $q$, and can therefore play the role of $a$ in Section 13.9.4. Thus,

$$m^\phi = \left(m^{p-1}\right)^{q-1} = 1 + Jq$$
$$m^\phi = \left(m^{q-1}\right)^{p-1} = 1 + Lp$$

for some nonnegative integers $J$ and $L$, satisfying

$$Jq = Lp.$$

Thus,

$$p \mid Lp = Jq.$$

Since both $p$ and $q$ are primes, $p$ must divide $J$ as well:

$$J = J'p$$

(for some new nonnegative integer $J'$). Thus,

$$Jq = (J'p)\, q = J'(pq) = J'n.$$

This means that

$$m^\phi \equiv 1 \bmod n.$$

So, using Fermat's little theorem twice, we managed to get rid of both $p$ and $q$, and obtain a new formula modulus $n$.

We are now ready to decode the coded message $c$ (Section 13.8.1). Fortunately, only the decoder has access to $d$, and can use it in a new modular power:

$$c^d \equiv (m^e)^d = m^{ed} = m^{1+K\phi} \equiv m \bmod n$$

(for some new nonnegative integer $K$). This is indeed decoding: by calculating the modular power $c^d \bmod n$, the decoder uncovers the original message $m$ secretly, as required.

Is this a good code? This depends on yet another question: are $p$ and $q$ safe? Well, to discover them, a hacker has a rather difficult task: to factorize $n$ as $n = pq$. This is not easy: if $p$ and $q$ are picked big enough, then $n$ would be even bigger, and too hard to factorize in an acceptable time. No hacker of a right mind would even attempt to do this: they'd probably find something else to do, and leave us in peace.

## 13.11 Probably Prime Numbers

### 13.11.1 Probably Prime Numbers

Thus, to have a safe coding-decoding algorithm, we need two very long prime numbers: $p$ and $q$. How to find such? This is too difficult. Instead, let them be only *probably* prime. This is still good enough. After all, no hacker of a right mind would waste his time on a code that is probably unbreakable. And what about those of a twisted mind? Let's leave them alone.

To find a probably prime number, we must be able to test: given a natural number, is it prime or not?

### 13.11.2  The Naive Test

Given a natural number $n$, is it prime or not? Consider a naive test: pick some integer number $a$ in between

$$1 < a < n.$$

Now, if

$$a^{n-1} \equiv 1 \bmod n,$$

then the test is passed. Otherwise, the test fails.

How reliable is this test? Well, thanks to Fermat's little theorem, if $n$ happens to be prime, then the test must be passed:

$$n \text{ is prime} \quad \Rightarrow \quad \text{the test is passed.}$$

So, in this direction, the test never lies: if $n$ is prime, then we are going to know this. We'll never miss any prime number.

Still, what about the other way around? Is it also true that

$$\text{the test is passed} \quad \Rightarrow \quad n \text{ is prime?}$$

Unfortunately not. There are nonprime numbers $n$ that do pass the test for too many $a$'s. Such $a$'s are called liars.

So, there are nonprime numbers with too many liars. This makes the test unreliable and unsafe. It is too easy to pass, and could mislead us to believe that a nonprime number is prime.

### 13.11.3  The Stronger Rabin–Miller Test

Fortunately, the Rabin–Miller test is safer [35, 39]. This test applies to any odd number $n > 3$. This way, $n - 1$ is even, so it makes sense to filter out powers of 2 from it: factorize $n - 1$ in the form

$$n - 1 = 2^r s,$$

where $r > 0$ is maximal, so $s$ is odd. Now, pick some integer number $a$ in between

$$1 < a < n - 1.$$

Next, look at the following factorization:

$$
\begin{aligned}
a^{n-1} - 1 \;=\;& a^{2^r s} - 1 \\
=\;& \left(a^{2^{r-1}s} - 1\right)\left(a^{2^{r-1}s} + 1\right) \\
=\;& \left(a^{2^{r-2}s} - 1\right)\left(a^{2^{r-2}s} + 1\right)\left(a^{2^{r-1}s} + 1\right) \\
=\;& \left(a^{2^{r-3}s} - 1\right)\left(a^{2^{r-3}s} + 1\right)\left(a^{2^{r-2}s} + 1\right)\left(a^{2^{r-1}s} + 1\right) \\
=\;& \cdots = (a^s - 1)(a^s + 1)\left(a^{2s} + 1\right)\left(a^{2^2 s} + 1\right)\cdots\left(a^{2^{r-1}s} + 1\right).
\end{aligned}
$$

We say that the test is passed if $n$ divides any factor on the right-hand side. In other words, the test is passed if either

$$a^s \equiv 1 \bmod n$$

or

$$a^{2^j s} \equiv (n-1) \bmod n$$

for any $0 \le j < r$. Otherwise, the test fails.

How reliable is this test? Well, if $n$ happens to be prime, then Fermat's little theorem guarantees that $n$ divides the left-hand side in the above factorization. Therefore, $n$ must also divide at least one factor on the right-hand side. Thus, the test is passed:

$$n \text{ is prime} \quad \Rightarrow \quad \text{the test is passed.}$$

This is good: we are never going to miss any prime number. If $n$ is prime, then we are going to know this.

Still, what about the other way around? Assume now that the test is passed. This means that $n$ divides at least one factor on the right-hand side. Therefore, $n$ also divides the left-hand side. Thus, this test is stronger (less easy) than the naive test:

$$\text{the test is passed} \quad \Rightarrow \quad \text{the naive test is passed.}$$

Still, is it also true that

$$\text{the test is passed} \quad \Rightarrow \quad n \text{ is prime?}$$

Not quite. Still, the situation is now better than before. Why? Because the Rabin–Miller test is often *strictly* stronger.

The naive test is unreliable: there are nonprime $n$'s that do divide $a^{n-1} - 1$ for too many liar $a$'s. Fortunately, such an $n$ may *not* pass the (stronger) Rabin–Miller test. In fact, for most $a$'s, such an $n$ may fail to divide *any* of the factors on the right-hand side. For example, such an $n$ could be a product of two factors (or their subfactors). In such a case, the Rabin–Miller test is *strictly* stronger, and tells us the truth: $n$ is nonprime! In this sense, it is better and safer.

Still, the Rabin–Miller test is not absolute: there may be a nonprime $n$ that does pass it for a few liar $a$'s. Fortunately, there are not too many liars: at most 25%. Thus, if $a$ is picked at random from the interval $[2, n-2]$, then it is not very likely to be a liar: the probability for this is as low as $1/4$.

To improve on this, repeat the test $k$ independent times. (This could be done in parallel.) At each time, pick a new $a$ at random. Only if all tests are passed is $n$ declared as probably prime. After all, a nonprime number is highly unlikely to pass $k$ independent tests. Indeed, the probability to pick $k$ independent liars is as low as

$$\left(\frac{1}{4}\right)^k = 4^{-k}.$$

For $k = 10$, for instance, this is as small as

$$4^{-10} = 2^{-20} = 1024^{-2} < 10^{-6}.$$

### 13.11.4  How to Break the Code?

Suppose that we want to use the RSA code. For this, we've already picked a probably prime number $p$. Now, we need one more: another probably prime number $q > p$. How to pick it?

Better be careful: since the product $n = pq$ is nonprime, it should better *not* pass the naive test for too many liar $a$'s. If it did, then any hacker could easily break our RSA code.

Indeed, the hacker could pick such an $a$, and use it to test $n$ with the Rabin–Miller test. After all, although $a$ is a liar for the naive test, it is probably *not* a liar for the Rabin–Miller test: with this $a$, $n$ doesn't pass the Rabin–Miller test. How could this happen? We already know: on the right-hand side, there must be a factor that is divided only by $p$, but not by $q$. This is not good: the hacker could just take this factor, and calculate its GCD with $n$. This would uncover $p$, and break our RSA code!

How to avoid this? Better pick $q$ in such a way that $n = pq$ has only a few liars, even with respect to the naive test. This way, no hacker of a right mind would even attempt to break your code. And what about those of a wrong mind? Let them work forever, searching for that special $a$ that is a liar only for the naive test, but not for the Rabin–Miller test.

## 13.12  Exercises

### 13.12.1  Modular Power

1. Show that the relation
$$l \equiv n \bmod m$$
   (Sections 13.2.1–13.2.2) is indeed a mathematical equivalence relation between $l$ and $n$. Hint: see Chapter 1, Section 1.6.1.
2. Prove that the sieve of Eratosthenes indeed works, and finds all the prime numbers between 2 and $N$.
3. How to calculate $n^k \bmod m$ efficiently?
4. The naive method uses two stages: first, calculate $n^k$. Then, take the modulus with $m$. What's wrong with this? Hint: see below.
5. In this method, what kind of numbers are used? How large could they be? Could they be too long to store on the computer? Could they be too long to work with?
6. How to avoid this?
7. Look at the formula at the end of Section 13.4.3. Does this avoid the above problem? How?
8. Does this use short numbers only? How short?
9. Does this still give the correct answer? Hint: use mathematical induction on $k = 1, 2, 3, \ldots$.
10. Before using this formula, substitute
$$n \leftarrow (n \bmod m).$$
    This way, throughout the entire calculation, are there short numbers only? How short?

11. More specifically, throughout the entire calculation, is there any (intermediate) product as large as $m^2$? Why?

12. Prove that there is none. Hint: use mathematical induction on $k = 1, 2, 3, \ldots$. Note that the inner recursion is applied to

$$(n \bmod m)^2 \bmod m < m,$$

and can therefore benefit from the induction hypothesis.

13. Still, is this worthwhile? After all, the modulus operation may be costly as well. How to avoid using it too often? Hint: use it only for (intermediate) numbers that are too long to work with.

14. Prove that the Euclidean algorithm (Section 13.3.2) indeed works, and finds the greatest common divisor. Hint: use mathematical induction on $n = 2, 3, 4, \ldots$.

15. Prove that the extended Euclidean algorithm (Section 13.3.3) indeed works, and helps write the greatest common divisor as a linear combination. Hint: use mathematical induction on $n = 2, 3, 4, \ldots$.

16. In this linear combination, could the coefficients be negative? How to avoid this? Hint: see Section 13.3.4.

17. Look at the algorithm in Section 13.4.1. Does it work? Why?

18. Prove that it indeed finds a small coprime, as required.

## 13.12.2  Probably Prime Numbers

1. Consider a prime number $n$. Must it pass the naive test in Section 13.11.2? Hint: this follows from Fermat's little theorem.

2. Consider an odd prime number $n > 3$. Must it pass the Rabin–Miller test? Hint: this follows from Fermat's little theorem and the factorization in Section 13.11.3.

3. Consider a number $n$ that passes the naive test. Is it prime? Hint: not necessarily: $a$ could be a liar.

4. Assume that this $n$ is also odd and greater than 3. Must it pass the Rabin–Miller test as well? Hint: not necessarily: it could be a product of two factors (or their subfactors) from the factorization in Section 13.11.3.

5. Consider an odd number $n > 3$ that passes the Rabin–Miller test. Must it pass the naive test as well? Hint: this follows from this factorization.

6. Must it be prime?

7. Is it more likely to be prime?

8. How likely is it to be prime? Hint: at probability 3/4 or even more. After all, $a$ was picked at random from the interval $[2, n-2]$. In this interval, at most 25% are liars for the Rabin–Miller test, and the rest are not.

9. How to increase this probability even more? Hint: repeat the test ten times. In each time, pick a new $a$ at random. Only if all tests are passed is $n$ declared as probably prime. After all, it is highly unlikely that $n$ was nonprime, and that all these independent $a$'s were liars.

10. Would any hacker ever attempt to break an RSA code that uses probably prime numbers? Why? Hint: the code is probably unbreakable, so no hacker would waste time on it.

11. Implement the RSA algorithm on your computer, with probably prime numbers. Hint: the solution can be found in Chapter 5 in [45].

# 14

# Fast Fourier Transform: a Virtual Binary Tree

In Chapters 7–8, we've already designed a binary tree, and used it to model a computer code. Here, we use the same logic to design a practical algorithm in numerical analysis and cryptography: fast Fourier transform (FFT).

In cryptography, we often use very long natural numbers, with thousands of digits, or even more. How to store them on the computer? Best use their binary representation as binary polynomials. After all, in this form, they are particularly easy to multiply. To implement this efficiently, one could use fast Fourier transform (FFT).

To apply the discrete Fourier transform to a given vector, best use FFT (introduced by Cooley and Tukey). To introduce FFT in a simple way, best use a new algebraic tool: a polynomial, sampled evenly on the unit circle in the complex plane. This way, it is particularly easy to see how the recursive calls make a virtual binary tree. This helps estimate the complexity (total cost), and see how low it indeed is.

Thanks to FFT, we can now calculate convolution fast. This could help multiply long natural numbers, in four stages:

- Apply FFT.
- Multiply digit by digit.
- Apply inverse FFT, which is nothing but FFT itself.
- Finally, push digits ahead, if necessary.

This is easy enough to implement on the digital computer. Later on, we'll also discuss a (theoretical) quantum algorithm.

## 14.1 The Discrete Fourier Transform

### 14.1.1 Vector and Its Polynomial

Consider the $n$-dimensional complex vector

$$f \equiv (f_0, f_1, f_2, \ldots, f_{n-1})^t \in \mathbb{C}^n,$$

with $n$ complex components. Let's use these components as coefficients in a new polynomial:

$$f(z) \equiv f_0 + f_1 z + f_2 z^2 + \cdots + f_{n-1} z^{n-1} = \sum_{i=0}^{n-1} f_i z^i,$$

where $z$ is the complex variable.

### 14.1.2 Roots of Unity

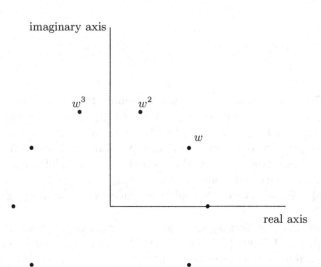

**Fig. 14.1.** In the complex plane $\mathbb{C}$, the roots of unity are $w, w^2, w^3, \ldots, w^n = 1$.

How to transform $f$? Just sample $f()$ at the roots of unity (Figure 14.1). These are $n$ distinct points, distributed evenly on the unit circle, in the complex plane. The rightmost point is the complex number 1. The next point above it is

$$w \equiv \exp\left(\frac{2\pi\sqrt{-1}}{n}\right) = \cos\left(\frac{2\pi}{n}\right) + \sqrt{-1}\sin\left(\frac{2\pi}{n}\right).$$

Why is this a root of unity? Because its $n$th power is 1:

$$w^n = 1.$$

What is so good about $w$? Well, its powers

$$\{w, w^2, w^3, \ldots, w^n = 1\}$$

are roots of unity in their own right. Indeed, for $j = 1, 2, 3, \ldots$, look at $w^j$, and calculate its $n$th power:

$$(w^j)^n = w^{jn} = w^{nj} = (w^n)^j = 1^j = 1.$$

### 14.1.3 The Discrete Fourier Transform

What does the discrete Fourier transform do? It transforms the original vector of coefficients

$$f = (f_0, \ f_1, \ f_2, \ \ldots, \ f_{n-1})^t$$

to the new vector of values of $f()$, sampled at the above roots of unity:

$$\left(f(1), \ f(w), \ f\left(w^2\right), \ f\left(w^3\right), \ \ldots, \ f\left(w^{n-1}\right)\right)^t.$$

How to calculate this fast, in one go?

## 14.2 FFT: a Virtual Binary Tree

### 14.2.1 FFT

How to apply the discrete Fourier transform? For this purpose, we need to sample the original polynomial $f(z)$ at $n$ distinct complex numbers:

$$z \in \left\{w, \ w^2, \ w^3, \ \ldots, w^n = 1\right\}.$$

How to do this? Break $f()$ into two polynomials:

$$f(z) = u(z^2) + z \cdot v(z^2),$$

where the new polynomials $u()$ and $v()$ are obtained from $f()$: $u()$ contains the even powers:

$$u(z) = f_0 + f_2 z + f_4 z^2 + f_6 z^3 + \cdots = \sum_{0 \le i < n, \ i \ \text{is even}} f_i z^{i/2},$$

and $v()$ contains the odd powers:

$$v(z) = f_1 + f_3 z + f_5 z^2 + f_7 z^3 + \cdots = \sum_{0 \le i < n, \ i \ \text{is odd}} f_i z^{(i-1)/2}.$$

What's so good about this new decomposition? Well, the new polynomials $u()$ and $v()$ are sampled at $z^2$, not $z$. This could be done recursively.

### 14.2.2 Virtual Binary Tree

For this purpose, assume that $n$ is even. In this case, one could take $w$, and raise it to power $n/2$:

$$w^{n/2} = \exp\left(\frac{2\pi\sqrt{-1}}{n} \cdot \frac{n}{2}\right) = \exp(\pi\sqrt{-1}) = -1.$$

This way, the roots of unity split into two subsets — the upper, mirrored by the lower:

$$z \in \left\{ w, \ w^2, \ w^3, \ \ldots, \ w^{n/2-1}, \ w^{n/2}, \ w^{n/2+1}, \ w^{n/2+2}, \ \ldots, \ w^n = 1 \right\}$$
$$= \left\{ w, \ w^2, \ w^3, \ \ldots, \ w^{n/2-1}, \ -1, \ -w, \ -w^2, \ -w^3, \ \ldots, \ 1 \right\}.$$

Here, the former $n/2$ points are on the upper semicircle in the complex plane. These are followed by $n/2$ more points that are only slightly different: they pick a minus sign, and mirror from below, on the lower semicircle (Figure 14.1). Each subset could now be placed in its own subtree in a new (virtual) binary tree.

Upon taking the square, the minus sign drops, with no effect whatsoever. Therefore, $u()$ and $v()$ must be sampled at $n/2$ points only:

$$z^2 \in \left\{ w^2, \ w^4, \ w^6, \ \ldots, \ w^{n-2}, \ 1 \right\}.$$

How to do this? Recursively, of course! For this purpose, assume that $n$ is a power of 2:

$$n = 2^k,$$

for some fixed integer number $k \geq 0$. This way, recursion is indeed possible. If, on the other hand, $n$ is not a power of 2, then extend $f$ with a few dummy components (tailing zeroes).

Thanks to recursion, we can sample $u()$ and $v()$ fast as well. (This is the induction hypothesis, to be made precise later.) This helps sample $f()$ too, as required. This completes the entire algorithm: FFT. Let's calculate its complexity (total cost).

### 14.2.3 Complexity

What is the total cost? Not too much: only $n \log_2 n$ complex multiplications, and $n \log_2 n$ complex additions. To prove this, use mathematical induction on $k = 0, 1, 2, 3, \ldots$, or $n = 1, 2, 4, 8, \ldots$. For $n = 1$, there is no work at all. This agrees with the above formula:

$$1 \cdot \log_2 1 = 1 \cdot 0 = 0.$$

Assume now that the above cost is correct for $n/2$. (This is the induction hypothesis.) Thanks to this, we already know the cost of the recursion used to sample $u$ (or $v$) at the above $n/2$ points:

$$\frac{n}{2} \log_2 \left( \frac{n}{2} \right) = \frac{n}{2} \left( \log_2 n - 1 \right)$$

multiplications, and the same number of additions. Now, to sample $f()$, there are $n$ more multiplications to calculate $z$ and $zv(z^2)$. Furthermore, to add $u(z^2)$ and $zv(z^2)$, there are $n$ more additions (or subtractions). Together with the recursion, this makes a total of

$$2 \frac{n}{2} \left( \log_2 n - 1 \right) + n = n (\log_2 n - 1) + n = n \log_2 n$$

multiplications (and the same number of additions), as asserted.

## 14.3 The Inverse Transform

### 14.3.1 Matrix Form

The discrete Fourier transform could also be written in a matrix form (see Chapter 1 in [46]):

$$f \to Wf,$$

where $W$ is an $n \times n$ complex matrix, whose elements are roots of unity in their own right:

$$W \equiv \left(w^{ij}\right)_{0 \leq i,j < n}.$$

This way, $W$ is the same as its own transpose:

$$W = W^t.$$

### 14.3.2 The Inverse Matrix

Furthermore, once normalized by $\sqrt{n}$, $W$ becomes unitary. Therefore,

$$\sqrt{n} \cdot W^{-1} = \left(\frac{1}{\sqrt{n}}W\right)^{-1} = \left(\frac{1}{\sqrt{n}}W\right)^{\dagger} = \frac{1}{\sqrt{n}}\bar{W}^t = \frac{1}{\sqrt{n}}\bar{W}.$$

Thus, to invert $W$, just take its complex conjugate, and divide by $n$:

$$W^{-1} = \frac{1}{n}\bar{W}.$$

### 14.3.3 The Inverse Transform

Thus, the inverse transform looks like this: for a given $n$-dimensional vector $g$,

$$g \to W^{-1}g = \frac{1}{n}\bar{W}g.$$

This could be carried out in two stages:

- Apply FFT, with just one change: $\bar{w}$ rather than $w$.
- Divide by $n$.

Better yet, this could also be carried out in four stages:

- Take the complex conjugate of $g$. This produces $\bar{g}$.
- Apply FFT. This produces $W\bar{g}$.
- Take the complex conjugate of $W\bar{g}$. This produces $\bar{W}g$.
- Finally, divide by $n$.

## 14.4 Convolution

### 14.4.1 Convolution

Why is FFT so useful? Well, consider two polynomials: $f()$ and $g()$. How to calculate the product $fg$, including its new coefficients? Well, we could use a direct approach (convolution), but this would be rather expensive. Better use FFT:

- Introduce a few tailing zeroes, to make both $f$ and $g$ have $n = 2^k$ coefficients.
- Use FFT to sample both $f()$ and $g()$ at the roots of unity (Figure 14.1).
- Multiply $f()$ by $g()$ pointwise at these points.
- Use the inverse FFT to transform back, and obtain the coefficients of $fg$, as required.
- We are not done yet: if $f()$ and $g()$ are binary polynomials that represent long natural numbers, then we may have some more work to do. After all, $fg$ may still contain coefficients (digits) that are not only 0 or 1 but also 2 or 3 or 4.... How to fix this? Push them ahead! For this purpose, start from the unit digit: the least significant digit. If it exceeds 1, then it must contribute to the more significant digit as well. For example, if it is 3, then reduce it to 1, and add 1 to the more significant digit. Then, advance to the more significant digit, and do the same, and so on.
- Finally, drop leading zeroes, if any.

This, however, is not the only application of FFT: it has many more applications in numerical analysis and signal processing.

## 14.5 Exercises: Complexity

### 14.5.1 FFT and Its Complexity

1. Consider two long natural numbers. What is the best way to store them? Hint: as binary polynomials, with binary coefficients (digits): 0 or 1.
2. Multiply them by each other, and obtain a new polynomial: their product or convolution.
3. How does this look like? Is this a legitimate binary polynomial as well? What is its length? Hint: about the same as the sum of lengths of the original natural numbers.
4. What coefficients (digits) are there in it? Hint: unfortunately, the new polynomial may contain coefficients that are not only 0 or 1 but also 2 or 3 or 4....
5. How to fix this? Hint: push digits ahead (Section 14.4.1).
6. What is the extra cost of this? Hint: like the length of the product — the sum of lengths of the original natural numbers.
7. Assume that the convolution uses a naive algorithm: a standard vertical multiplication. What is its cost (complexity)? Hint: $O(nm)$ bit operations, where $n$ and $m$ are the lengths of the original natural numbers.
8. Assume, for instance, that the original natural numbers contain as many as $n = m = 10^{10}$ binary digits. In this case, how big is the above cost? Hint: $nm = 10^{20}$.

9. To improve on this, use FFT (Sections 14.2.3–14.4.1). What is the cost? Is it less than before? Hint: FFT costs $2n \log_2 n$ complex operations, each as expensive as $10^3$ bit operations. This is still worthwhile:

$$10^3 \cdot 10 \cdot 10 \cdot 10^{10} \ll 10^{20}.$$

10. Implement FFT on your computer, and use it to carry out convolution.
11. Use this to multiply long natural numbers, stored as binary polynomials. In the end, don't forget to push digits ahead, as in Section 14.4.1. Hint: the solution can be found in Chapter 5 in [45].

# 15

## Error Correction:
## The Reed-Solomon Code

Thanks to set theory, we have mathematical equivalence relations. These lead to modular arithmetic (Chapter 13). Let's use this not only in cryptography but also in error correction [62]. In this context, coding means something else: no longer a method to encrypt a secret message, but a method to allow the receiver to correct a few errors that may have been introduced on the way.

In FFT, we have already seen a Vandermonde matrix. In the context of modular arithmetic, this takes a new face: a new generator matrix. This is the basis for linear coding. Thanks to the new code, our message gets more stable: even if contaminated with errors, it is still as valuable as before. In fact, the receiver can easily fix the errors, and receiver the original message. Moreover, the receiver can even fix a burst of errors, containing many bugged bits, one by one in a row.

## 15.1 Linear Code and Primitive Number

### 15.1.1 Finite Set of Numbers

Let $q$ be a (rather large) prime number. We have already seen how $q$ can be used in modular arithmetic: add, multiply, and divide modulus $q$. This is also useful to correct errors in a message. In fact, the message is a long list of bits, which could be interpreted as a long natural number, (in its binary representation). Still, why use base 2? Instead, better write it in base $q$. In other words, let's break it into $k$ blocks, each being a number between 0 and $q - 1$.

For this purpose, consider the set of integer numbers from 0 to $q - 1$:

$$Q = \{0, 1, 2, \ldots, q - 1\}.$$

### 15.1.2 The Original Message: a Vector

Our original message takes now the form of a $k$-dimensional vector, with components in $Q$:

$$f = \begin{pmatrix} f_0 \\ f_1 \\ f_2 \\ \vdots \\ f_{k-1} \end{pmatrix} \in Q^k.$$

### 15.1.3 Linear Code

We want to send $f$: a column vector, whose components are numbers in $Q$. How to do this? From $f$ we design our coded message: $Gf$, where $G$ is an $n \times k$ matrix: the generator matrix.

By picking $n$ and $G$ cleverly, we have our new linear code. Thanks to it, we can now form $Gf$ (a new $n$-dimensional column vector), and send it to the receiver. Thanks to this, the receiver will be able to uncover $f$ even if $Gf$ is contaminated with a few bugs in a few components.

### 15.1.4 Primitive Number

How to design $G$? It is well-known that there is a primitive number $\gamma \in Q$ that generates $Q$. In other words, $\gamma$ and its powers make the nonzero numbers in $Q$:

$$Q \setminus \{0\} = \{1, 2, 3, \ldots, q - 1\} = \{\gamma^0, \gamma^1, \gamma^2, \ldots, \gamma^{q-2}\}.$$

The latter set contains the same $q - 1$ numbers, probably in a different ordering. These numbers are now written as powers of $\gamma$:

$$\gamma^0 = 1,$$
$$\gamma^1 = \gamma,$$
$$\gamma^2,$$
$$\vdots$$
$$\gamma^{q-2}.$$

There is no need to include the next power, which coincides with 1:

$$\gamma^{q-1} \equiv 1 \mod q,$$

as follows from Fermat's little theorem (Chapter 13, Section 13.9.4).

## 15.2 The Generator Matrix

### 15.2.1 Vandermonde Matrix

Let's use our primitive number $\gamma$ and its powers to design a new $(q - 1) \times (q - 1)$ matrix:

$$A = \left(\gamma^{ij}\right)_{0 \le i,j < q-1}.$$

This is a Vandermonde matrix. It has a nice property: each and every submatrix in it is nonsingular. This will be useful below.

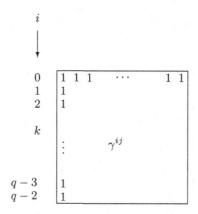

**Fig. 15.1.** The $(q-1) \times (q-1)$ matrix $A$. It contains 1's on the first row and column. The other elements are higher powers of the primitive number $\gamma$.

### 15.2.2 Relation to FFT

Is this familiar? Well, we've already seen a Vandermonde matrix in FFT (Chapter 14, Section 14.3.1). Here, however, there are a few changes: $q-1$ plays the role of $n$, and $\gamma$ comes instead of $w$. Still, in principle, things are the same. Like $w$, $\gamma$ could also be viewed as a root of unity (Figure 14.1). The only difference is that $w$ is in the complex plane, with complex arithmetic, whereas $\gamma$ is in a finite field of $q$ integer numbers, with modular arithmetic.

### 15.2.3 Picking Parameters

To detect and correct errors, we need to assume that

$$k < q - 1.$$

This puts a lower bound on $q$: it can't be as small as you like any more. Later on, we'll see how to get around this.

Furthermore, let $n$ be an intermediate number:

$$k < n < q.$$

For error detection, we'd like $n$ to be as large as possible: $n = q - 1$. This will allow us to detect as many errors as possible. We'll come back to this later.

### 15.2.4 The Generator Matrix

In $A$, look at an $n \times k$ rectangular submatrix: $G$. In Figure 15.2, for example, $G$ is the upper-left block. Still, this is not a must. In general, $G$ could be made of any other $n$ rows of $A$, and any $k$ (consecutive) columns.

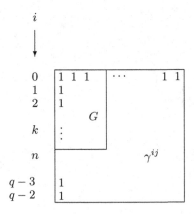

**Fig. 15.2.** The generator matrix $G$ is a rectangular submatrix of $A$, with $n$ rows, and $k$ columns.

### 15.2.5 The Linear Mapping

$G$ is the generator matrix, which makes the new linear mapping

$$G : Q^k \to Q^n.$$

### 15.2.6 The Coding Step

This will help carry out the coding step:

$$f \to Gf.$$

Instead of sending $f$, better send $Gf$. This is a bit more expensive: it requires sending $n > k$ components. Still, this is worthwhile: it allows detecting and even correcting a few errors. This way, even if $Gf$ arrives with a few bugged components, $f$ may still be recovered safely.

## 15.3 The Reed-Solomon Code

### 15.3.1 Coding

Thus, coding is quite simple: just calculate $Gf$. This is a linear operation. This is why the code is often called a linear code. The result $Gf$ is now ready to be sent to the receiver.

### 15.3.2 Vector and Polynomial

The components of $f$ could also be used to design a polynomial of degree $k-1$ (or less). We denote this new polynomial by the same letter $f$, followed by round parentheses:

$$f(x) = f_0 + f_1 x + f_2 x^2 + \cdots + f_{k-1} x^{k-1}.$$

This way, the components of $Gf$ could also be viewed as the values of $f()$ at $n$ distinct numbers:

$$Gf = \begin{pmatrix} f\left(\gamma^0\right) \\ f\left(\gamma^1\right) \\ f\left(\gamma^2\right) \\ \vdots \\ f\left(\gamma^{n-1}\right) \end{pmatrix}.$$

## 15.4 Error Detection

### 15.4.1 The Parity-Check Matrix

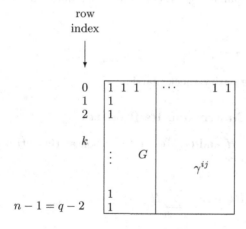

**Fig. 15.3.** A generator matrix $G$ with the maximal possible number of rows: $n = q - 1$.

Consider a special case, in which $n$ is maximal:

$$n = q - 1.$$

This way, $G$ occupies the left part of $A$ (Figure 15.3). In this case, it makes sense to define yet another interesting matrix in $A$: the rectangular $(n-k) \times n$ matrix $H$, which occupies the upper part of $A$ (not including the first row):

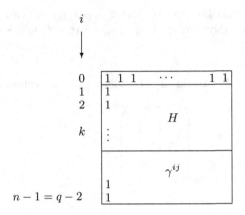

**Fig. 15.4.** The parity-check matrix $H$ is a rectangular submatrix of $A$, with $n - k$ rows, and $n$ columns, where $n$ is maximal: $n = q - 1$. ($H$ doesn't contain the top row of $A$, which contains 1's only.)

$$H = \left(\gamma^{ij}\right)_{1 \leq i \leq n-k,\ 0 \leq j < n}$$

(Figure 15.4). $H$ is called the parity-check matrix.

### 15.4.2 The Parity-Check Matrix and Its Properties

What is the relation between $H$ and $G$? To see this, look at the $m$th column of $G$ ($0 \leq m < k$), and apply $H$ to it:

$$
\begin{aligned}
(HG)_{i,m} &= \sum_{j=0}^{n-1} H_{i,j} G_{j,m} \\
&= \sum_{j=0}^{q-2} \gamma^{ij} \gamma^{jm} \\
&= \sum_{j=0}^{q-2} \gamma^{(i+m)j} \\
&= \sum_{j=0}^{q-2} \left(\gamma^{i+m}\right)^{j} \\
&= \frac{1 - \gamma^{(i+m)(q-1)}}{1 - \gamma^{i+m}} \\
&\equiv \frac{1 - 1}{1 - \gamma^{i+m}}
\end{aligned}
$$

$$\equiv 0 \mod q.$$

(Since $0 < i + m < n$, the above denominator never vanishes, so one can sum the power series as in Chapter 7, Section 7.5.2.)

### 15.4.3 Orthogonality in Modular Arithmetic

Thus, the columns of $G$ (and their linear combinations) are "orthogonal" to the rows of $H$ (and their linear combinations), in terms of modular arithmetic. Any other column vector that is not a linear combination of the columns of $G$, on the other hand, can never be orthogonal to *all* the rows of $H$.

## 15.5 The Detection Test

### 15.5.1 The Received Message

Thanks to this property, the receiver can now check for errors. This is done as follows. Denote the received message by

$$w = \begin{pmatrix} w_0 \\ w_1 \\ w_2 \\ \vdots \\ w_{n-1} \end{pmatrix}.$$

If we happen to know that this is the same as $Gf$, then we are done: just pick any $k$ rows from $G$, form a $k \times k$ nonsingular submatrix, and solve for $f$, as required.

### 15.5.2 Detecting an Error

But we are not yet there: we don't know whether $w = Gf$ or not. In fact, there may be some errors, or bugged components, for which

$$w_i \neq (Gf)_i.$$

Is there any? To check on this, just apply $H$ to $w$. What should the result be? Well, no errors means that

$$w = Gf.$$

For this to be true, we must have

$$Hw = H(Gf) = (HG)f \equiv (0)f \equiv 0 \mod q$$

(the $k$-dimensional zero vector, in terms of modular arithmetic):

### 15.5.3 The Test

Thus, the receiver can do the following test: take the received message $w$, and apply $H$ to it (from the left). If

$$Hw \not\equiv 0 \mod q,$$

then there must be a bug: $w$ must contain an error in at least one component.

### 15.5.4 No Error!

If, on the other hand,

$$Hw \equiv \mathbf{0} \mod q,$$

then there is still room for uncertainty. If we happen to know from some other source that there are at most $n - k$ errors (so $w$ contains at least $k$ correct components), then there is no error at all! After all, since $w$ is a linear combination of the columns of $G$, we could pick $k$ rows from $G$, and solve for $f$ uniquely. If, on the other hand, we're not so lucky to have this information in advance, then there could be as many as $n - k + 1$ errors (or even more). In this (rare) case, the receiver can never solve for $f$ uniquely and safely any more: the received message is too unreliable.

## 15.6 Decoding: Error Correction

### 15.6.1 Decoding

Assume now that the above test fails:

$$Hw \not\equiv \mathbf{0} \mod q,$$

This indicates that the received message $w$ contains an error in at least one component. Alternatively, assume that the test can't be carried out easily: say, $n < q - 1$, so the parity-check matrix is not easily available any more.

Still, assume that the receiver heard some good news from some other source: the received message $w$ contains no more than $(n - k)/2$ bugged components. How to recover $f$? How to decode?

### 15.6.2 The Berlekamp-Welch Algorithm

For this purpose, let's design two new polynomials (with coefficients and independent variable in $Q$, and modular arithmetic). The first polynomial, $e(x)$, will have a degree no more than $(n - k)/2$. Furthermore, its leading coefficient (the coefficient of the highest power) will be 1. This way, $e(x)$ will have no more than $(n - k)/2$ unknown coefficients. The second polynomial, $b(x)$, will be of degree no more than $k - 1 + (n - k)/2$. This way, it will have no more than $k + (n - k)/2$ unknown coefficients. Together, $e(x)$ and $b(x)$ will have no more than

$$\frac{n - k}{2} + k + \frac{n - k}{2} = k + n - k = n$$

unknown coefficients. How to find them?

### 15.6.3 The Linear System of Equations

For this purpose, for $0 \leq i < n$, solve the following equations (in modular arithmetic):

$$b\left(\gamma^i\right) \equiv w_i e\left(\gamma^i\right) \mod q.$$

Is there a solution?

### 15.6.4 A Specific Solution

Well, let's design one. Let $\tilde{e}(x)$ be the product of factors of the form $x - \gamma^i$, for those $i$'s that are in error:

$$w_i \not\equiv (Gf)_i \mod q.$$

This way, $\tilde{e}(x)$ indeed has a degree no more that $(n-k)/2$, as required. Furthermore, $\tilde{e}(x)$ also has leading coefficient 1, as required. Now, use this $\tilde{e}(x)$ to define $\tilde{b}(x)$:

$$\tilde{b}(x) = f(x)\tilde{e}(x)$$

(where $f$ is our original message). This way, $\tilde{b}(x)$ has degree no more than $k - 1 + (n - k)/2$, as required. These new polynomials indeed solve our equations:

$$\tilde{b}\left(\gamma^i\right) = f\left(\gamma^i\right)\tilde{e}\left(\gamma^i\right) \equiv w_i\tilde{e}\left(\gamma^i\right) \mod q.$$

Indeed, for those $i$'s not in error, both sides are the same. For those $i$'s that are in error, on the other hand, both sides vanish, and are again the same.

### 15.6.5 The General Solution

Assume now that we already solved our system of $n$ equations, and obtained some polynomials $e(x)$ and $b(x)$. How to use them to uncover $f(x)$? For this purpose, look at the polynomial

$$b(x)\tilde{e}(x) - \tilde{b}(x)e(x).$$

On one hand, it has degree no more than

$$\frac{n-k}{2} + k - 1 + \frac{n-k}{2} = k - 1 + n - k = n - 1.$$

On the other hand, for every $0 \le i < n$, it vanishes at $\gamma^i$, no matter whether the $i$th component is in error or not:

$$b\left(\gamma^i\right)\tilde{e}\left(\gamma^i\right) - \tilde{b}\left(\gamma^i\right)e\left(\gamma^i\right) \equiv w_i e\left(\gamma^i\right)\tilde{e}\left(\gamma^i\right) - w_i\tilde{e}\left(\gamma^i\right)e\left(\gamma^i\right) \equiv 0 \mod q.$$

Thus, it must be the zero polynomial that vanishes everywhere in $Q$:

$$b(x)\tilde{e}(x) - \tilde{b}(x)e(x) = 0,$$

or

$$b(x)\tilde{e}(x) = \tilde{b}(x)e(x).$$

Upon dividing by $\tilde{e}(x)$, we obtain

$$b(x) = \frac{\tilde{b}(x)}{\tilde{e}(x)}e(x) = f(x)e(x).$$

Finally, once this is divided by $e(x)$ (in terms of polynomial arithmetic, as in the exercises below), we obtain

$$\frac{b()}{e()} = f().$$

This gives us the unknown polynomial $f()$, coefficient by coefficient. These are the unknown components of the original message $f$, as required.

## 15.7 A Multilevel Algorithm

### 15.7.1 Picking Parameters in Practice

In practice, the original message is often very big. For example, it could be a big file, containing as many as $10^{16}$ bits. This could be viewed as a long natural number (in its binary representation), containing as many as $10^{16}$ binary digits. To transfer this, $q$ should be as large as

$$q \doteq 2^{(10^8)},$$

and $k$ as large as

$$k \doteq 10^8.$$

Indeed, this way,

$$q^k \doteq \left(2^{(10^8)}\right)^{10^8} = 2^{((10^8)^2)} = 2^{(10^{16})},$$

as required.

Next, how to pick $n$? Pick

$$n \doteq \frac{5}{4}k.$$

This way,

$$\frac{n-k}{2} \doteq \frac{1}{8}k.$$

This way, even if as many as 10% of our $n$ components are bugged, we can still fix them.

### 15.7.2 Recursion

But each component is still very big, and contains as many as $10^8$ bits. This could be viewed as a long natural number, with as many as $2^{(10^8)}$ binary digits. How to transfer it safely, with error correction? Recursively, of course! This should be carried out $k$ times, for each and every component. This way, we can also fix inner errors within an individual component, if there are any.

This makes a multilevel algorithm. In each level, one can pick a different $n$, depending on the percentage of errors that one wants to fix.

### 15.7.3 Burst of Consecutive Bugged Bits

In the bottom level, the innermost component will be moderate: as small as 100 bits or so. At this level, no need to call recursion any more. After all, errors often come in a burst, with many bugged bits, one by one in a row. Thus, there is no point to try and fix an individual bit. After all, if there is an error, it probably spoils all those 100 bits. Better fix them together, as one component.

## 15.8 Towards Polynomial Arithmetic

### 15.8.1 Replace Numbers by Polynomials

Better yet, instead of numbers, use polynomials. In this context, instead of standard arithmetic, use polynomial arithmetic (as in the exercises below). For example, use binary polynomials, whose coefficients are either 0 or 1 (with modulus-2 arithmetic).

### 15.8.2 Prime Binary Polynomial

Thus, $q$ will be replaced by a new prime polynomial $q(x)$. For example, to store (and transfer) a component with $m$ bits (for some even $m$, say $m = 10^8$), define $q(x)$ as

$$q(x) = 1 + x + x^2 + x^3 + \cdots + x^m.$$

Why is $q()$ prime? Because it has no roots! Indeed, neither 0 nor 1 are roots:

$$q(0) = 1 + 0 + 0 + 0 + \cdots = 1 \not\equiv 0 \quad \mod \quad 2$$
$$q(1) = 1 + 1 + 1 + 1 + 1 + \cdots = m + 1 \equiv 1 \not\equiv 0 \quad \mod \quad 2.$$

(If, on the other hand, $m$ is odd, then drop $x^{m-1}$ from the definition of $q(x)$.) Thus, $q()$ is irreducible, or prime. Thanks to this property, we can now use polynomial arithmetic, modulus $q()$.

Likewise, replace the primitive number $\gamma$ by a new primitive polynomial. In both coding and decoding, use polynomial arithmetic rather than standard modular arithmetic. This is quite efficient: in its binary form, each polynomial can be easily stored on the computer. This approach is based on Galois theory of finite fields.

## 15.9 Exercises: Polynomial Arithmetic

### 15.9.1 Natural Numbers and Polynomials

1. Consider a natural number $n$. Consider its binary representation. What is the rightmost (least significant) binary digit?
2. Assume that $n$ is even. What is its least significant binary digit? Hint: 0.
3. Assume that $n$ is odd. What is its least significant binary digit? Hint: 1.
4. No matter whether $n$ is odd or even, what is its least significant binary digit? Hint: $n$ mod 2.
5. Divide $n$ by 2 with remainder. What is the remainder? Hint: $n$ mod 2.
6. How is the remainder related to the least significant binary digit? Hint: they are the same.
7. What is the next (more significant) binary digit? Hint: divide $n$ by 2 with remainder. Disregard the remainder. Look at the result: $\lfloor n/2 \rfloor$. Find its least significant binary digit.
8. What are the rest of the binary digits in $n$? Hint: repeat the above recursively: drop the rightmost digit time and again.
9. How is this related to Horner's algorithm (Chapter 5 in [45])? Hint: both use the same recursion: divide by $x$ (or 2) with remainder, to shift a list of numbers time and again.
10. What is the difference? Hint: in the above, we shift the (unknown) binary digits, to uncover them one by one. In Horner's algorithm, on the other hand, we work the other way around: shift the (well-known) coefficients, and use them one by one, to uncover the (unknown) value.
11. In the above exercises, introduce a little change: base 3 rather than 2.
12. Pick an arbitrarily long natural number at random. Look at its representation in base 3. How likely is it to contain only the digits 0 and 2, but never the digit 1? Hint: the probability for this is as low as zero (Chapter 5).

13. How to add two (arbitrarily long) natural numbers to each other? Hint: write the bigger number in the top row, add the smaller in the next row below. Start from the rightmost (least significant) digit. Add digit by digit, all the way leftwards. As your teacher always said, don't forget the carry!

14. How to add two polynomials to each other? Hint: likewise: monomial by monomial.

15. How to subtract two natural numbers from each other? Hint: assume that the former is bigger. (Otherwise, interchange their roles, and attach a minus sign to the result.) Write it in the top row, and the other number in the next row below. Start from the rightmost digit. Work vertically, digit by digit, all the way leftwards.

16. How to subtract two polynomials from each other? Hint: likewise, monomial by monomial.

17. How to multiply two natural numbers by each other? Hint: vertically, leftwards, digit by digit.

18. Is there a better way? Hint: try FFT (Chapter 14).

19. How to multiply two polynomials by each other? Hint: likewise, monomial by monomial.

20. Is there a better way?

21. How to divide two natural numbers by each other (with remainder)? Hint: vertically, rightwards, digit by digit (assuming that the latter is nonzero).

22. How to divide two polynomials by each other (with remainder)? Hint: likewise, monomial by monomial.

23. After the division, look at the remainder. What is its meaning? Hint: the remainder is just the result of the modulus operation: the former modulus the latter.

24. Use the above to write the Euclidean algorithm for polynomials as well, and find their greatest common divisor: a new polynomial that divides them both (evenly).

25. Write the extended Euclidean algorithm (Chapter 13, Section 13.3.3) for polynomials as well: write their greatest common divisor as their "sum," with some "coefficients," which could be polynomials in their own right.

# 16

# Application
# in Computational Biology

Set theory is the key to mathematical logic and Boolean algebra (Figures 1.12–1.13). Thanks to this, we can now use the code developed above in an interesting application in computational biology: pool testing. This helps minimize the number of expensive tests, and save a lot. This is quite useful in practice.

## 16.1 Distinguishing Matrix

### 16.1.1 How to Distinguish Column from Column?

To develop our application in computational biology [62], we need a new algebra: Boolean algebra. Fortunately, set theory gives us the elementary logic required for this.

In what follows, we keep using the same parameters $n$, $k$, and $q$, as in Chapter 15. Let $C$ be a matrix with elements in $Q$: integer numbers between 0 and $q-1$. $C$ will be a distinguishing matrix. What does this mean? To see this, let's write $C$ column by column:

$$C = \left( c^{(1)} \mid c^{(2)} \mid c^{(3)} \mid \cdots \right).$$

Now, let's focus on $c^{(j)}$: the $j$th column in $C$ (for just any $j$). Assume that $c^{(j)}$ is not the zero vector:

$$c^{(j)} \neq \mathbf{0}.$$

In other words, $c^{(j)}$ contains at least one nonzero component. Now, thanks to the fact that $C$ is a distinguishing matrix, $c^{(j)}$ can be distinguished from any other column $c^{(l)}$: there is a row $i$ at which $c^{(l)}$ vanishes, but $c^{(j)}$ does not:

$$c_i^{(l)} = 0 \neq c_i^{(j)}.$$

### 16.1.2 How to Distinguish Column from Set of Columns?

Moreover, $c^{(j)}$ can be distinguished not only from one particular column but also from any set

$$S = \left\{ c^{()}, \ c^{()}, \ c^{()}, \ \cdots \right\}$$

that contains no more than $n/(k-1)$ columns from $C$. In other words, there is a row $i$ (for some new $i$) at which all columns in $S$ vanish, but $c^{(j)}$ does not:

$$c_i^{(j)} \neq 0, \quad \text{but} \quad c_i^{(l)} = 0 \quad \text{for every} \quad c^{(l)} \in S.$$

If this is true for any column $j$, then we say that $C$ is a distinguishing matrix. Let's design such a matrix.

### 16.1.3 The Kautz-Singleton Construction

For this purpose, let's use the Reed-Solomon code (Chapter 15). Recall that our original message is a $k$-dimensional vector, whose components are integer numbers between 0 and $q-1$:

$$f \in Q^k.$$

How many possible $f$'s could be designed? Clearly, their total number is

$$|Q^k| = q^k.$$

So, the total number of nonzero vectors in $Q^k$ is

$$|Q^k \setminus \{\mathbf{0}\}| = q^k - 1.$$

Now, for each such $f$, the Reed-Solomon code designs a coded message — a new $n$-dimensional (nonzero) vector:

$$Gf \in Q^n \setminus \{\mathbf{0}\}.$$

Let's list all these column vectors, one by one in a row. This makes our $C$: a new $n \times (q^k - 1)$ matrix. Is this a distinguishing matrix?

### 16.1.4 Rows of Agreement

Is $C$ a distinguishing matrix? To check on this, let $c^{(j)}$ be its $j$th column, as before. Thus, $c^{(j)}$ must be of the form

$$c^{(j)} = Gf^{(j)}$$

for some nonzero message $f^{(j)} \in Q^k \setminus \{\mathbf{0}\}$. Let $S$ contain no more than $n/(k-1)$ columns from $C$, other from $c^{(j)}$:

$$c^{(j)} \notin S.$$

Now, pick some column from $S$:

$$c^{(l)} \in S.$$

At how many rows $m$ could $c^{(j)}$ and $c^{(l)}$ agree, and satisfy

$$c_m^{(j)} = c_m^{(l)}?$$

At most $k-1$. After all, if there were $k$ (or more) such rows, then the polynomial

$$f^{(j)}(x) - f^{(l)}(x)$$

would have as many as $k$ roots, and would have to vanish identically.

Likewise, the same logic holds for all columns in $S$: each may have at most $k-1$ components that agree with their counterparts in $c^{(j)}$. In total, the columns in $S$ can have no more than

$$\frac{n}{k-1}(k-1) = n$$

components that agree with their counterparts in $c^{(j)}$.

### 16.1.5 Row of Disagreement

Assume now that $n$ is not divided by $k-1$:

$$(k-1) \nmid n.$$

In this case, $n/(k-1)$ is not an integer. This leads to a stricter estimate: the columns in $S$ must have strictly less than

$$\frac{n}{k-1}(k-1) = n$$

components that agree with their counterparts in $c^{(j)}$. So, there must be at least one row at which all columns in $S$ disagree with $c^{(j)}$. This is the key to designing our distinguishing matrix in its final form.

### 16.1.6 The Final Form

To obtain our distinguishing matrix in its final form, replace each element in $C$ by a new $q$-dimensional column vector. More specifically, each number in $Q$ is replaced by the corresponding standard unit vector of dimension $q$:

$$0 \leftarrow \begin{pmatrix} 1 \\ 0 \\ 0 \\ 0 \\ 0 \\ \vdots \end{pmatrix}, \quad 1 \leftarrow \begin{pmatrix} 0 \\ 1 \\ 0 \\ 0 \\ 0 \\ \vdots \end{pmatrix}, \quad 2 \leftarrow \begin{pmatrix} 0 \\ 0 \\ 1 \\ 0 \\ 0 \\ \vdots \end{pmatrix},$$

and so on. This makes $C$ binary: its elements are either 0 or 1. Furthermore, it is now $q$ times as big: not just $n \times (q^k - 1)$, but $(nq) \times (q^k - 1)$. In its final form, it is indeed a distinguishing matrix, as required.

## 16.2 Application: Pool Testing

### 16.2.1 Pooling

The distinguishing matrix designed above is quite practical: it could help testing a big population, and find those individuals who are ill with a certain disease. For this, each one needs to give blood, which is then tested for the disease. Unfortunately, the test is often quite expensive. So, testing each and every individual is too expensive. How to avoid this?

Fortunately, the number of ill people is often quite small: much smaller than the total number of people in the entire population. Under these circumstances, there is a trick: pooling (or pool testing). In this approach, the blood drawn from a few people is mixed, and poured into a common pool.

A pool is a big tube, in which many people can put their blood samples. Initially, the pool is completely empty. Then, people give their blood into it, one by one. In the end, some blood is drawn from the pool, and tested for the disease. If it turns out to be clean (uncontaminated, or negative), then these people must be healthy. If, on the other hand, it comes contaminated, then at least one of them is ill. Who? We remain puzzled. Fortunately, there is a trick to it.

Thanks to the trick, we save a lot: the expensive test is applied only to the pool, not to the individual people. Fortunately, the total number of pools is often quite small: much smaller than the total number of people. So, we indeed save a lot: instead of testing all people, we only test our pools.

Still, we are not done yet. After all, our results only tell us which pool is contaminated, and which is not. From this, how to uncover the ill people themselves? For this, we can use our distinguishing matrix.

### 16.2.2 Minimizing the Number of Expensive Tests

Assume that our population contains as many as $q^k - 1$ people. In it, assume that there are no more than $n/(k-1)$ ill people. How to find them? For this, we need only $nq$ pools. Often,

$$nq \ll q^k - 1.$$

Thus, we save quite a bit.

### 16.2.3 Designing Pools from the Distinguishing Matrix

In the context of pool testing, what is the meaning of our distinguishing matrix $C$? Well, let's design our pools cleverly, to give $C$ a lot of sense. For this, look at the $l$th person ($1 \leq l < q^k$). Does he/she give blood to the $i$th pool ($1 \leq i \leq nq$)? Well, this depends on the $(i, l)$th element in $C$:

$$c_i^{(l)} = 1 \quad \text{if and only if} \quad \text{person } l \text{ gives blood to pool } i$$
$$c_i^{(l)} = 0 \quad \text{if and only if} \quad \text{person } l \text{ doesn't give blood to pool } i.$$

(Recall that $C$ is a binary matrix, whose elements are either 0 or 1.) Let's use this to form our new system of equations.

## 16.3 The System of Equations

### 16.3.1 The Column Vectors

Let $g$ be the $(q^k - 1)$-dimensional binary vector that indicates the ill people: for $1 \leq l < q^k$, the $l$th person is ill if and only if

$$g_l = 1.$$

Otherwise,

$$g_l = 0$$

(because $g$ is a binary vector). Likewise, let $h$ be the $(nq)$-dimensional binary vector that indicates the contaminated pools: for $1 \leq i \leq nq$, the $i$th pool is contaminated if and only if

$$h_i = 1.$$

Otherwise,

$$h_i = 0$$

(because $h$ is a binary vector). What is the relation between $g$ and $h$? We are now ready to write this as a system of equations:

$$Cg = h.$$

Still, this is not in standard arithmetic, but in terms of a new arithmetic: Boolean algebra.

### 16.3.2 Boolean Algebra

In binary arithmetic, we take modulus 2. For example,

$$1 + 1 \equiv 0 \mod 2.$$

In Boolean algebra, on the other hand, there is a little change:

$$1 + 1 = 1$$

(Figure 1.12). This way, '+' stands for the "or" operation ('$\vee$' in Figure 7.8). Multiplication, on the other hand, remains the same as in binary arithmetic. This way, it stands for the "and" operation ('$\wedge$' in Figure 7.8). With these observations at hand, look at the $i$th equation in our system:

$$(Cg)_i = \sum_{l=1}^{q^k-1} c_i^{(l)} g_l = h_i.$$

Be aware: this equation is not in standard arithmetic, but in terms of Boolean algebra. Indeed, how could we possibly have

$$h_i = 1$$

(the $i$th pool gets contaminated)? Only if, for some $1 \leq l < q^k$,

$$c_i^{(l)} = g_l = 1$$

(the $l$th person is ill, and gives blood to the $i$th pool). In other words, to have a contaminated pool, at least one ill person must give blood to it. This is indeed what we want from a plausible pooling technique.

In pooling, however, we have information about $h$ only, not $g$. How to solve for $g$? How to uncover the individual ill people?

### 16.3.3 Finding the Ill People

To uncover $g$, use the following relation:

ill person $\Rightarrow$ all pools that contain his/her blood get contaminated

healthy person $\Rightarrow$ not all pools that contain his/her blood get contaminated.

Indeed, the former claim is clearly true. To prove the latter, assume that the $j$th person is healthy. Since there are no more than $n/(k-1)$ ill people, their columns can be placed in the set $S$ in Section 16.1.2. Since $C$ is a distinguishing matrix, the $i$th row distinguishes $c^{(j)}$ from $S$:

$$c_i^{(j)} = 1, \quad \text{but} \quad c_i^{(l)} = 0 \quad \text{for every} \quad c^{(l)} \in S$$

(where $i$ is as in Section 16.1.2). So, the $i$th pool proves the latter claim: it contains blood from the $j$th person, but not from any ill person. In summary, we have a one-to-one correspondence between the medical state and the test results: a person is ill if and only if all pools that contain his/her blood turn out to be contaminated.

## 16.4 Exercises: Two-Level Pooling

### 16.4.1 Picking Parameters in Practice

1. In practice, pick the following parameters:

$$q \doteq 100$$
$$k = 3$$
$$n = q.$$

2. In this pick, could $q$ be even? Hint: no. After all, $q$ must be prime.
3. With this pick, is our $C$ a distinguishing matrix? Hint: since $q$ is odd,

$$k - 1 = 2 \nmid q = n,$$

as required in Section 16.1.5.
4. This pick is good enough for a population of a million or so, containing only 50 ill people or so. Why? Hint:

$$q^k \doteq 100^3 = 10^6$$
$$\frac{n}{k-1} = \frac{n}{2} = \frac{q}{2} \doteq \frac{100}{2} = 50.$$

5. In our pooling algorithm, how many pools are used? Hint:

$$nq = q^2 \doteq 10^4.$$

6. Thanks to our pooling algorithm, what is the save? Hint: instead of $10^6$ individual tests, we need only $10^4$ tests or so: one for each pool. This is 100 times as cheap.
7. Assume that we have a very big lab, which can carry out as many as $10^4$ tests concurrently, in one go. What is the save thanks to our pooling algorithm? Hint: to carry out $10^6$ individual tests, we need as many as 100 stages. Thanks to our pooling algorithm, we need one go only. This is 100 times as fast.

### 16.4.2 Two-Level Pooling

1. Consider now a naive two-level pooling method. Hint: see below.
2. Consider the same population: a million or so.
3. Split the population into $10^4$ disjoint groups.
4. The people in each group give blood to a different pool. This way, there are as many as $10^4$ pools.
5. This is the first level.
6. Test each and every pool.
7. Those pools that come clean contain blood from healthy people only.
8. Those that come contaminated, on the other hand, contain blood from an ill person, but we don't know who.
9. Thus, there are still up to $5,000$ people that could be ill.
10. Forward them to the second level.
11. These people can't be split into groups any more, because this could happen to be just the same as the original split.
12. Therefore, they must be tested individually.
13. This may require as many as $5,000$ more individual tests.
14. In total, the number of tests is more than in our original pooling strategy:

$$15,000 > 10,000.$$

15. Furthermore, in parallel, this requires two stages, which is twice as slow as our original pooling strategy.

Part VII


Towards
Quantum Computing

# Towards Quantum Computing

What is the key ingredient in set theory? This is the bit: 0 or 1. Now, take $k$ bits in a row. What do you get? This has a geometrical meaning: a hypercube of $n = 2^k$ nodes. On it, define a (complex) grid function. This makes a new quantum computer. Clearly, it has a great computational power. After all, $n$ grows exponentially with $k$. To use our quantum computer, we have another mathematical structure: a binary tree. This helps implement recursive algorithms like FFT [59].

Unfortunately, there is no quantum computer as yet: this is just science fiction. Still, the logic behind it could lead to a better understanding of both physics and computer science. Indeed, once implemented on a quantum computer, FFT takes a new form: quantum FFT. In it, the virtual binary tree is approached not from the top but from the bottom. This parses the entire binary tree node by node, leading to a simple loop, easy to run not only on a quantum computer but also on the good old digital computer.

Finally, quantum FFT could also be used to break the RSA code in cryptography. Fortunately, there is nothing to worry about: quantum computers are not available yet. Still, it is worthwhile to know how the code could be broken in theory.

# Quantum FFT

What is a quantum computer? In set theory, the elementary logic is based on a bit: 0 or 1. Now, $k$ bits in a row make a new geometrical structure: a hypercube, made of $n = 2^k$ nodes. On it, one can now define a new (complex) grid function: $v \equiv v_i$, where $i$ is made of the original $k$ bits, and indexes a node in the hypercube. This is our quantum computer, with its huge computational power: exponential in terms of the number of bits. To implement an algorithm like FFT on it, it makes sense to use a binary tree.

How to approach a binary tree? So far, we approached it from the top. This led to the recursive FFT algorithm. Here, on the other hand, we approach it from the bottom, and climb all the way up. This parses the entire binary tree, node by node. Instead of the original high-level recursion, we get a simple low-level loop, easy to run on the computer.

For this purpose, we imagine a quantum computer. This is just a dream: there is no quantum computer as yet. Still, it is worthwhile to imagine that we already have one. Indeed, on a quantum computer, FFT takes a new form: quantum FFT. Thanks to the new logic behind it, this (theoretical) algorithm leads to a new practical implementation: nonrecursively, using simple loops only. This is easy to run not only on a quantum computer but also on a digital (and even parallel) computer.

## 17.1 Qubit and Its Gates

### 17.1.1 Electron and Its Spin

A bit is a variable with just two possible values: either 0 or 1. A qubit, on the other hand, is a random variable: it could be 0 at probability $|v_0|^2$, or 1 at probability $|v_1|^2 = 1 - |v_0|^2$. For example, the electron could spin up at probability $|v_0|^2$, or down at probability $|v_1|^2 = 1 - |v_0|^2$. We'll never know for sure: after all, we're never going to look.

This is not the only way: there are many other ways to implement a qubit. Why do we use spin? Because this will teach us a lot about spin, and how it behaves in a magnetic field. This will make a gate: an automatic mechanism to change $v_0$ and $v_1$ implicitly. This way, nature will serve as our new computer: it will store $v_0$ and $v_1$ for us, and will also manipulate them for us, implicitly and indirectly.

### 17.1.2 Complex Degrees of Freedom

In the digital computer, the bit is the smallest unit of memory (Chapter 8, Section 8.11.2). Indeed, it stores just one bit: either 0 or 1. In our quantum computer, on the other hand, the electron stores much more: the entire state

$$v \equiv \begin{pmatrix} v_0 \\ v_1 \end{pmatrix} \in \mathbb{C}^2,$$

containing two complex numbers: $v_0$ and $v_1$. Still, these complex variables are not quite independent of each other. After all, they must satisfy a normalization condition:

$$\|v\|^2 = v_0^2 + v_1^2 = 1.$$

Later on, the state will get much more complicated: $v$ will get multidimensional, and will contain many more components. Only in the end will $v$ get normalized. Until then, we can use the components of $v$ to store and manipulate complex unknowns freely, even without normalization, and even without knowing what they really are. To do this, let's start from our two-dimensional $v$, and apply to it a $2 \times 2$ orthogonal matrix: a gate.

### 17.1.3 Orthogonal Pauli Matrix

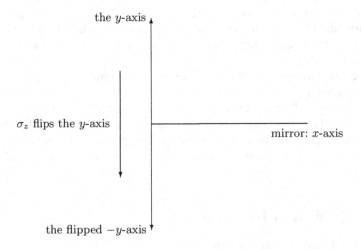

**Fig. 17.1.** The Pauli matrix $\sigma_z$ flips the $y$-axis. The $x$-axis acts like a mirror: it remains fixed, with eigenvalue 1.

Our first gate is the orthogonal Pauli matrix

$$\sigma_z \equiv \begin{pmatrix} 1 & \\ & -1 \end{pmatrix}$$

(Figure 17.1). (In the matrix, blank spaces stand for zero.) Let's apply it to $v$:

$$v \to \sigma_z v = \begin{pmatrix} v_0 \\ -v_1 \end{pmatrix}.$$

How to apply this in practice? Let our electron fly rightwards, through a magnetic field, oriented upwards (Figures 17.5–17.6). Thanks to the Schrodinger equation, $v_0$ will then precess at frequency $-\omega$, and $v_1$ at frequency $\omega$. Now, let's apply this from time $t = 0$ until time $t = t_0$, where

$$2\omega t_0 = \pi.$$

This way, $v_1$ will open a phaseshift of $\pi$ ahead of $v_0$. In other words, $v_1$ will indeed pick a minus sign, as required. Unfortunately, there is also a price to pay: an overall phaseshift of $-\pi/2$. This is global: it multiplies the entire vector $v$, as a whole. Physically, in quantum mechanics, this has no effect at all. Numerically, on the other hand, this global factor should be stored elsewhere, until the end. Throughout the algorithm, this global factor may change: it may get multiplied by more and more overall phaseshifts. Only in the end will the global factor come back to multiply $v$ as a whole.

### 17.1.4 The Orthogonal Hadamard Matrix

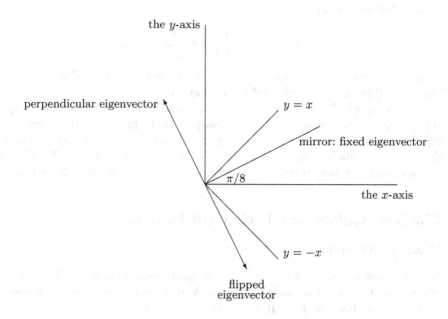

**Fig. 17.2.** The Hadamard matrix acts like a mirror too: the $x$-axis maps to the oblique line $y = x$, and the $y$-axis maps to the oblique line $y = -x$. This way, the mirror remains fixed (at angle $\pi/8$), while its perpendicular picks a minus sign.

So far, we've seen that the Pauli matrix $\sigma_z$ acts like a mirror, placed on the $x$-axis. This makes the eigenvectors: the $x$-axis remains fixed, at eigenvalue 1, while the

$y$-axis flips, at eigenvalue $-1$. The same principle works in yet another orthogonal matrix — the Hadamard matrix:

$$W_2 \equiv \frac{1}{\sqrt{2}} \begin{pmatrix} 1 & 1 \\ 1 & -1 \end{pmatrix}.$$

Is this familiar? This is actually the discrete Fourier transform, in the simple case of

$$k = 1, \quad n = 2^k = 2, \quad \text{and} \quad w = -1.$$

Still, there is a new thing here: normalization by $\sqrt{2}$. Thanks to this, $W_2$ is orthogonal.

What is the geometrical meaning of this transform? Well, it acts like a mirror too. Here, however, the mirror is oblique: it makes angle $\pi/8$ with the $x$-axis (Figure 17.2). As a matter of fact, you could look at things the other way around: the $x$-axis makes angle $-\pi/8$ with the mirror. Once mirrored, the $x$-axis maps to the oblique line $y = x$ that makes angle $\pi/8$ with the mirror. This is just what we wanted:

$$\begin{pmatrix} 1 \\ 0 \end{pmatrix} \to W_2 \begin{pmatrix} 1 \\ 0 \end{pmatrix} = \frac{1}{\sqrt{2}} \begin{pmatrix} 1 \\ 1 \end{pmatrix}.$$

The $y$-axis, on the other hand, makes angle $3\pi/8$ with the mirror. Once mirrored, it maps to the oblique line $y = -x$ that makes angle $-3\pi/8$ with the mirror. This is just what we wanted:

$$\begin{pmatrix} 0 \\ 1 \end{pmatrix} \to W_2 \begin{pmatrix} 0 \\ 1 \end{pmatrix} = \frac{1}{\sqrt{2}} \begin{pmatrix} 1 \\ -1 \end{pmatrix}.$$

Thus, the mirror marks the fixed eigenvector (of eigenvalue 1), and its perpendicular marks the flipping eigenvector (of eigenvalue $-1$).

How to apply $W_2$ in practice? Use a new magnetic field, oriented not upwards any more, but obliquely: at angle $\pi/4$ away. In 3-D, this is exactly in between two axes: the vertical $z$-axis (represented by state $(1,0)^t \in \mathbb{C}^2$) and a horizontal axis that is equally likely to spin up or down (represented by state $(1,1)^t \in \mathbb{C}^2$). Thus, our new magnetic field indeed aligns with the mirror in Figure 17.2, as required.

## 17.2 Two Qubits and Their Grid Function

### 17.2.1 Two Qubits

Consider now a system of two qubits: two electrons, which may spin up or down, at some probability. This makes a grid of four points, indexed by a binary index: $i = 00, 01$ (top row), and $i = 10, 11$ (bottom row in Figure 17.3).

### 17.2.2 Binary Index: a New Random Variable

Here, $i$ itself is a new random variable. In fact, $i$ could take either of these four values, each at some probability. More specifically, $i = 0$ or 1 or 2 or 3 if the electrons spin up-up or up-down or down-up or down-down, which may happen at probability $|v_{00}|^2$ or $|v_{01}|^2$ or $|v_{10}|^2$ or $|v_{11}|^2$, respectively. This way, the state $v$ is not only a vector (of norm 1) but also a grid function.

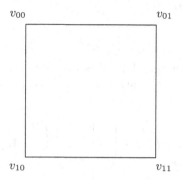

$$v_{00} \qquad\qquad\qquad\qquad v_{01}$$

$$v_{10} \qquad\qquad\qquad\qquad v_{11}$$

**Fig. 17.3.** Two (entangled) qubits make a grid of four points, indexed row by row by the binary index $i = 00, 01, 10, 11$, with the (normalized) grid function $v \equiv v_i$. How likely is the first electron to spin up (or 0), while the second spins down (or 1)? The probability for this is $|v_{01}|^2$.

### 17.2.3 Entanglement

This is indeed entanglement: the qubits depend on each other, and can never disregard one another. In other words, the values of $v$ on the top row are not necessarily proportional to those on the bottom row. Likewise, the values of $v$ on the left column are not necessarily proportional to those on the right column in the grid.

### 17.2.4 Two Qubits and Their Gate

Why is this efficient? Because one can now apply a gate ($2 \times 2$ unitary matrix) to the first qubit. As a result, the same gate will be applied to both columns at the same time. Likewise, one can also apply another gate to the second qubit only. As a result, this new gate will be applied to both rows at the same time. This is quite efficient: in just one go, one applies the same gate twice. In parallel computing, this is called SIMD: single instruction, multiple data. This is why quantum computing is so useful: it may help write and implement parallel algorithms in practice.

### 17.2.5 The Odd-Even Ordering

So far, our grid was ordered row by row. Still, it also makes sense to order it column by column: the left column (even index $i = 00, 10$), followed by the right column (odd index $i = 01, 11$). This is the odd-even ordering, used in FFT.

In this new ordering, the second qubit is an "ugly duckling:" it is no longer least significant, but most significant: at top priority in terms of indexing. We are now ready to apply FFT to our $v$. For this purpose, set

$$k = 2, \quad n = 2^k = 4, \quad \text{and} \quad w = \sqrt{-1}.$$

## 17.3 FFT on Two Qubits

### 17.3.1 Unitary FFT

Thanks to the odd-even ordering, FFT takes the form of a triple product of three $4 \times 4$ matrices. To have a unitary matrix, we also need to normalize by $\sqrt{2}$:

$$
W_4 \equiv \frac{1}{\sqrt{2}} \begin{pmatrix} I & I \\ I & -I \end{pmatrix} \begin{pmatrix} 1 & & & \\ & 1 & & \\ & & 1 & \\ & & & w \end{pmatrix} \begin{pmatrix} W_2 & \\ & W_2 \end{pmatrix},
$$

where $I$ is the $2 \times 2$ identity matrix. How to implement this?

### 17.3.2 Hadamard Transform on the First Qubit

What do we have here? This is a triple product of three $4 \times 4$ matrices. It is carried out from right to left. First, apply the rightmost matrix: the recursive call to $W_2$: the Hadamard transform on each column in the grid. (Thanks to the odd-even ordering, the grid is already ordered column by column.) To implement this, apply the Hadamard transform to the first qubit only. After all, this qubit stores the values of $v$ in each individual column. Thus, $W_2$ on the first qubit is the same as $W_2$ on both columns at the same time, as required.

### 17.3.3 Control Qubit

At the middle of our triple product, we have a diagonal unitary matrix. Let's call it $D_1$:

$$
D_1 \equiv \begin{pmatrix} 1 & & & \\ & 1 & & \\ & & 1 & \\ & & & w \end{pmatrix},
$$

where

$$
w = \sqrt{-1}.
$$

On its main diagonal, $D_1$ has 1's, except for the lower-right element, which is imaginary: $w = \sqrt{-1}$. How to apply $D_1$ to $v$? In a few stages:

- The second qubit tells us in what column we are in the grid. Apply to it the gate in Section 17.1.3, only this time use a lower frequency $\omega$, for which

$$
2\omega t_0 = \frac{\pi}{4}.
$$

  This opens a phaseshift: the second column gets ahead of the first one. In other words, the third and fourth components of $v$ (in the odd-even ordering) are now at phase $\pi/4$ ahead.

- Next, the second qubit can also serve as a control qubit. To see this, suppose that the quantum computer has an inner mechanism to measure the spin of the second qubit, without telling us whether it is up or down. This way, we'll never get to know the answer, hidden inside the computer. Moreover, the answer will soon be gone, leaving no mark or record whatsoever. This is good: the state won't get damaged.
- Now, based on the above measurement, the computer decides how to proceed. If the second qubit spins up (has value 0), then we are in the first column. In this case, do nothing: after all, the first and second main-diagonal elements are both 1.
- If, on the other hand, the second qubit spins down (has value 1), then we are in the second column. In this case, design a new magnetic field, and let the first electron pass through it, as follows.
- To the first electron, this new magnetic field should apply the gate

$$\begin{pmatrix} 1 & \\ & w \end{pmatrix}.$$

How to do this? Like before, only this time pick a bigger time step $t_1$, for which

$$2wt_1 = \frac{\pi}{2}.$$

This will change the third and fourth components in $v$ (the second column, in the odd-even ordering): the third component will lose phase $\pi/4$, and will fit the first and second components back again, as required. The fourth component, on the other hand, will gain phase $\pi/4$, and will open a phaseshift of $\pi/2$ ahead, as required.

In the above, the second qubit served as a control qubit for the first one. Indeed, the inner measurement carried out on the second qubit told the computer what to do to the first qubit. Still, this is not a must: we could also work the other way around, and interchange their roles. After all, $D_1$ doesn't distinguish between the qubits in any way.

### 17.3.4 Hadamard Transform on the Second Qubit

Finally, apply the leftmost matrix: the Hadamard transform $W_2$, applied to each individual row in the grid. To do this, apply $W_2$ to the second qubit only. After all, this qubit stores the values of $v$ in each individual row. Thus, applying $W_2$ to it actually applies the Hadamard transform to both rows at the same time, as required. This is the power of quantum computing.

## 17.4 Three Qubits: Discrete Cube

### 17.4.1 Three Entangled Qubits

Next, let's move on to a yet higher dimension:

$$k = 3, \quad n = 2^k = 8, \quad \text{and} \quad w = (-1)^{1/4}.$$

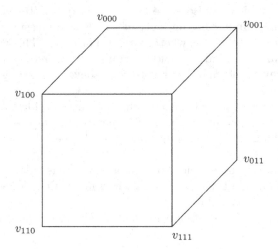

**Fig. 17.4.** Three qubits make a cube of eight points, indexed by the binary index $i = 000, 001, 010, 011, 100, 101, 110$, and $111$, with the (normalized) grid function $v \equiv v_i$. How likely is the first electron to spin up (or 0), while the second and third spin down (or 1)? The probability for this is $|v_{011}|^2$.

This way, we have not only two but three entangled qubits. Between them, they make a new random variable — the new index

$$i = 000, 001, 010, 011, 100, 101, 110, 111$$

(Figure 17.4). These have a geometrical meaning: corners of a discrete cube. On them, we can now define a new grid function (eight-dimensional complex vector of norm 1):

$$v \equiv v_i.$$

This way, $|v_i|^2$ is the probability to have index $i$. For example, $|v_{110}|^2$ is the probability to have the first and second electrons spin down, while the third spins up. (For numerical purposes, we also need to store an overall complex coefficient.)

This is our new quantum computer, with its great power: in just three qubits, nature stores eight degrees of freedom for us. To our new $v$, let's apply our unitary FFT.

### 17.4.2 Unitary FFT

To apply our unitary FFT, better reorder $v$, using a new odd-even ordering: start from the left part of the cube (indexed by an even $i$), and follow to the right part (indexed by an odd $i$). Thanks to this, FFT takes the form of a new triple product of three $8 \times 8$ matrices:

$$W_8 \equiv \frac{1}{\sqrt{2}} \begin{pmatrix} I & I \\ I & -I \end{pmatrix} \begin{pmatrix} I & & & \\ & 1 & & \\ & & w & \\ & & & w^2 \\ & & & & w^3 \end{pmatrix} \begin{pmatrix} W_4 & \\ & W_4 \end{pmatrix},$$

where $I$ is now the $4 \times 4$ identity matrix. How to apply this to $v$?

### 17.4.3 Recursion

How to apply this triple product to $v$? From right to left. Start from the rightmost matrix — call FFT recursively twice: on the even $i$'s (the left half of the cube), and also on the odd $i$'s (the right half of the cube). How to do this fast? Note that the third qubit is left out of the game. After all, the same is carried out, no matter whether it is 0 or 1. This is why, in the recursion, $W_4$ should be applied only once: to the first and second qubits only, leaving the third qubit idle. This is the power of quantum computing: the virtual binary tree is reduced to a new unary tree.

### 17.4.4 Diagonal Unitary Matrix

In the above triple product, in the middle, we have a diagonal unitary matrix. On its main diagonal, it has powers of $w$:

$$w = (-1)^{1/4}, \quad w^2 = \sqrt{-1}, \quad \text{and} \quad w^3 = (-1)^{3/4}.$$

Is this familiar? Well, this new $w^2$ was called $w$ in Section 17.3.3, and was used to define $D_1$. The same $D_1$ is used here too: let's apply it to the first and third qubits. This way, the second qubit remains idle, and only leads to duplication:

$$\begin{pmatrix} I & & & \\ & 1 & & \\ & & 1 & \\ & & & w^2 \\ & & & & w^2 \end{pmatrix}$$

(in the odd-even ordering). In Section 17.3.3, we could also define yet another matrix $D_2$, which is the same as $D_1$, except for one change: pick the frequency to be twice as low. This $D_2$ is useful here too: apply it to the second and third qubits. This time, the first qubit remains idle, and only leads to duplication:

$$\begin{pmatrix} I & & & \\ & 1 & & \\ & & w & \\ & & & 1 \\ & & & & w \end{pmatrix}.$$

Now, apply these diagonal matrices to $v$, one by one. It doesn't matter which one comes first: after all, diagonal matrices commute. Thus, their product is

$$\begin{pmatrix} I & & & \\ & 1 & & \\ & & w & \\ & & & w^2 & \\ & & & & w^3 \end{pmatrix},$$

as required.

### 17.4.5 Hadamard Transform on the Third Qubit

Finally, apply the Hadamard transform $W_2$ to the third qubit only. In our discrete cube, this leads to SIMD: the same operation is carried out row by row, in the same way. After all, the first and second qubits remain idle, and only lead to duplication: the same operation on four different rows, as required. This is the great power of quantum computing. This completes our unitary FFT on the cube. Let's extend this to a yet higher dimension: a hypercube.

## 17.5 $k$ Qubits: Hypercube

### 17.5.1 $k$ Entangled Qubits

Let's move on to a yet higher dimension, using a yet larger $k$:

$$n = 2^k, \quad \text{and} \quad w \equiv (-1)^{2/n}.$$

As $k$ increases by one, $n$ doubles, and $w$ takes a new phase, twice as small. Fortunately, $w$ could remain implicit: never stored or calculated explicitly. In fact, $w$ is only introduced to help make the presentation clearer.

### 17.5.2 Hypercube

Assume that our quantum computer contains $k$ (entangled) qubits. This way, $v$ will be much bigger: a now grid function, defined on a hypercube of $n = 2^k$ nodes. This is the great power of quantum computing: in just $k$ qubits, nature will store as many as $n = 2^k$ degrees of freedom for us. (For numerical purposes, we also need to store an external complex coefficient, to accumulate global phaseshifts of $v$.)

So far, we described our unitary FFT for $k \leq 3$ and $n \leq 8$. We are now ready to extend this to the general case as well. This will be done by mathematical induction on $k = 1, 2, 3, \ldots$. On the way, we'll also get rid of the recursion altogether, and use a simple loop instead.

### 17.5.3 Unitary FFT

For $k \leq 3$, we already wrote our unitary FFT as a triple product of three matrices. Let's extend this to a more general $k$, for which

$$n \equiv 2^k, \quad \text{and} \quad w \equiv (-1)^{2/n}.$$

In this case, our unitary FFT makes a new triple product of the form

$$
W_n = \frac{1}{\sqrt{2}} \begin{pmatrix} I & I \\ I & -I \end{pmatrix} \begin{pmatrix} I & & & & & & \\ & 1 & & & & & \\ & & w & & & & \\ & & & w^2 & & & \\ & & & & w^3 & & \\ & & & & & \ddots & \\ & & & & & & w^{n/2-2} & \\ & & & & & & & w^{n/2-1} \end{pmatrix} \begin{pmatrix} W_{n/2} & \\ & W_{n/2} \end{pmatrix},
$$

where $I$ is the $(n/2) \times (n/2)$ identity matrix.

### 17.5.4 The Odd-Even Ordering

In the rightmost matrix, $I$ appears four times, in four blocks. What do these $I$'s stand for?

Well, in the odd-even ordering, the $k$th qubit is an "ugly duckling:" it is no longer least significant, but most significant, prior to all other qubits. Thus, these four $I$'s tell us the state of the $k$th qubit: spin-up (as in the upper-left block) or down (as in the lower-right block).

## 17.6 Recursive vs. Nonrecursive FFT

### 17.6.1 Recursion

How to apply this triple product to $v$? Start from the right: apply the rightmost matrix. For this purpose, apply FFT recursively to the $k-1$ leading qubits. This leaves the $k$th qubit idle, leading to duplication only. This way, FFT is called recursively twice: not only to the even-numbered components in $v$ (the first block in the odd-even ordering), but also to the odd-numbered ones (the second subvector in $v$). This is done in just one recursive call: this is the power of quantum computing.

### 17.6.2 Recursion and Mathematical Induction

Recursion mirrors mathematical induction. Later on, we'll use mathematical induction to design a nonrecursive version and prove its correctness. In the induction step, we'll advance from $k-1$ to $k$. In the process, what happens to the dimension? It doubles from $n/2$ to $n$. In the process, what happens to the root of unity? It takes a new phase, twice as small:

$$
w \equiv (-1)^{2/n}.
$$

Indeed, in terms of this new $w$, the old root of unity was

$$
w^2 = (-1)^{4/n} = (-1)^{2/(n/2)}.
$$

The phase here is twice as big. This was the old root of unity, used in the previous step. This will be useful below.

### 17.6.3 Diagonal Unitary Matrix

In the above triple product, in the middle, we have an $n \times n$ diagonal unitary matrix. How to apply it to our up-to-date $v$? Apply $D_1$ to the first and $k$th qubits, $D_2$ to the second and $k$th qubits, ..., and $D_{k-2}$ to the $(k-2)$nd and $k$th qubits. (Ordering doesn't matter: diagonal matrices commute.) In summary, this actually applies the diagonal matrix

$$\begin{pmatrix} I & & & & & & \\ & 1 & & & & & \\ & & 1 & & & & \\ & & & w^2 & & & \\ & & & & w^2 & & \\ & & & & & \ddots & \\ & & & & & & w^{n/2-2} \\ & & & & & & & w^{n/2-2} \end{pmatrix}.$$

Indeed, the $(k-1)$st qubit remains idle, and only leads to duplication. (This is the least significant qubit, because the $k$th qubit already became most significant, thanks to our odd-even ordering.) Furthermore, in the previous step, we could use an induction hypothesis, with $w^2$ and $n/4$ playing the role of $w$ and $n/2$, respectively.

Next, define $D_{k-1}$ to be the same as $D_{k-2}$, except for one change: use a frequency twice as low. Once $D_{k-1}$ is ready, apply it to the $(k-1)$st and $k$th qubits. This will actually apply the new $n \times n$ diagonal unitary matrix

$$\begin{pmatrix} I & & & & & & \\ & 1 & & & & & \\ & & w & & & & \\ & & & 1 & & & \\ & & & & w & & \\ & & & & & \ddots & \\ & & & & & & 1 \\ & & & & & & & w \end{pmatrix}.$$

Indeed, here the $k-2$ leading qubits remain idle, and only lead to duplication. Being diagonal, these two matrices commute. Moreover, their product is

$$\begin{pmatrix} I & & & & & & \\ & 1 & & & & & \\ & & w & & & & \\ & & & w^2 & & & \\ & & & & w^3 & & \\ & & & & & \ddots & \\ & & & & & & w^{n/2-2} \\ & & & & & & & w^{n/2-1} \end{pmatrix}.$$

This is the middle term in our triple product, as required.

### 17.6.4 Hadamard Transform on the $k$th Qubit

Finally, apply the Hadamard transform to the $k$th qubit. This way, the $k-1$ leading qubits remain idle, and only lead to duplication. The only active qubit is the $k$th one, which distinguishes odd- from even-numbered components in $v$. As a result, the odd-numbered are added to (or subtracted from) the even-numbered, as required in the final stage in FFT.

### 17.6.5 Nonrecursive Version: Nested Loops

How to avoid recursion altogether, and replace it with simple loops? Use a new algorithm:

- Scan the qubits, one by one, as follows: for $j = 1, 2, \ldots, k$, look at the $j$th qubit, and do the following:
  - Use an inner loop to scan pairs of qubits, as follows: apply $D_1$ to the first and $j$th qubits, $D_2$ to the second and $j$th qubits, ..., and $D_{j-1}$ to the $(j-1)$st and $j$th qubits.
  - At the end of the inner loop, apply the Hadamard transform $W_2$ to the $j$th qubit.

Why is this correct? To prove this, use mathematical induction on $k = 1, 2, 3, \ldots$. For $k = 1$, this is easy: the inner loop is empty, and $W_2$ is the same as (unitary) FFT. Now, for $k \geq 2$, assume that the algorithm works for $k - 1$. (This is the induction hypothesis.) Now, look at the subloop $j = 1, 2, \ldots, k - 1$. Thanks to the induction hypothesis, this actually calls FFT recursively on the $k-1$ leading qubits. Thanks to quantum computing, this is done twice at the same time: not only for even- but also for odd-numbered components in $v$. After all, the $k$th qubit is not yet active, and only leads to duplication, as required.

We are not done yet. Now, the outer loop reaches the final qubit: $j = k$. This calls the inner loop for the last time. This applies the diagonal unitary matrix, as discussed above. Finally, $W_2$ is applied to the $k$th qubit, as required.

What have we done above? We carried out a nested loop over the qubits. This is indeed our nonrecursive version. Thanks to it, we can now estimate the total cost: $O(k^2)$. This is quite fast: much faster than any digital computer. Unfortunately, quantum computers are still only a dream.

## 17.7 Exercises: Spin

### 17.7.1 Electron and Its Spin

1. Let's model the spin of an electron.
2. For this purpose, we must first work in three dimensions. Assume that we have a detector, pointing at a fixed (normalized) direction $u$:

$$u \equiv \begin{pmatrix} u_1 \\ u_2 \\ u_3 \end{pmatrix} \in \mathbb{R}^3, \quad \|u\| = 1.$$

3. How likely is a given electron to spin (counterclockwise) around $u$? Hint: this depends on the angle $\theta$ between $u$ and $w$ (the direction around which the electron "really" spins in 3-D). In fact, the probability to spin (counterclockwise) around $u$ is $\cos^2(\theta/2)$.

4. For instance, assume that $u$ and $w$ are perpendicular to one another:

$$\theta = \frac{\pi}{2}.$$

5. What does this mean physically? Hint: the electron has a neutral state: it is equally likely to spin either clockwise or counterclockwise around $u$.

6. Confirm this algebraically as well. Hint:

$$\cos^2\left(\frac{\theta}{2}\right) = \cos^2\left(\frac{\pi}{4}\right) = \frac{1}{2}.$$

### 17.7.2 Magnetic Field and Spin

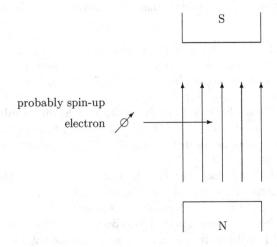

**Fig. 17.5.** The uniform magnetic field points upwards: $\mathbf{B} \equiv \|\mathbf{B}\|(0,0,1)^t$.

1. Consider a (uniform) magnetic field, pointing upwards, in the positive $z$-direction (Figure 17.5):

$$\mathbf{B} \equiv \|\mathbf{B}\| \begin{pmatrix} 0 \\ 0 \\ 1 \end{pmatrix}.$$

2. In this magnetic field, let's guess a dynamic state for the spin of our electron:

$$v \equiv v(t) \equiv \begin{pmatrix} v_0(t) \\ v_1(t) \end{pmatrix} \equiv \begin{pmatrix} \cos\left(\frac{\theta}{2}\right)\exp(-i\phi t) \\ \sin\left(\frac{\theta}{2}\right)\exp(i\phi t) \end{pmatrix},$$

where $i = \sqrt{-1}$ is the imaginary number.

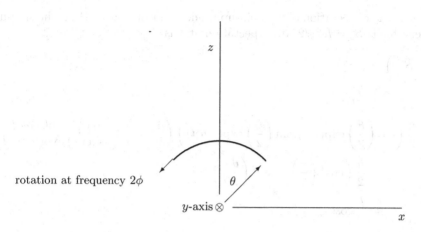

**Fig. 17.6.** Larmor precession: a view from the side. The expectations of $\tilde{S}_x$, $\tilde{S}_y$, and $\tilde{S}_z$ make a three-dimensional vector, rotating around the $z$-axis at frequency $2\phi$.

3. This is not the same $v$ as before. Here, $v$ is only mathematical, not physical. Indeed, here $v$ is two-dimensional, not three-dimensional. Furthermore, here $v$ is complex, not real.

4. As a matter of fact, $v(t)$ is not only a guess but also a solution to the Schrodinger equation.

5. Here, $\phi$ is the Larmor frequency. It is proportional to $\|\mathbf{B}\|$.

6. What is $\|v\|$? Hint: $\|v\| = 1$, as required.

7. What is the phaseshift between $v_0$ and $v_1$? Hint: $2\phi t$.

8. At what time is the phaseshift exactly $\pi$? Hint: at time $\pi/(2\phi)$.

9. During this time, what happened to $v$? Hint: $v_1$ was multiplied by $i$, and $v_0$ by $-i$.

10. Up to an overall phase, what happened to $v$? Hint: $v_1$ picked a minus sign.

11. Algebraically, what happened to $v$? Hint: $v$ was multiplied by $\sigma_z$ from the left:

$$v\left(\frac{\pi}{2\phi}\right) = \sigma_z v(0)$$

(up to an overall phase, which should be stored elsewhere).

12. In our magnetic field, how likely is the electron to spin up (counterclockwise around the $z$-axis)? Hint: the relevant observable is a diagonal $2 \times 2$ matrix:

$$\tilde{S}_z = \frac{\bar{h}}{2}\sigma_z = \frac{\bar{h}}{2}\begin{pmatrix} 1 & 0 \\ 0 & -1 \end{pmatrix}$$

(where $\bar{h}$ is Planck's constant). Its relevant eigenvector is $(1, 0)^t$. To have the probability, look at this eigenvector, take its inner product with $v$, calculate the absolute value, and square it up:

$$|(1,0)v|^2 = |v_0|^2 = \left|\cos\left(\frac{\theta}{2}\right)\right|^2 = \cos^2\left(\frac{\theta}{2}\right).$$

13. What is the expectation of the spin-up random variable at $v$? Hint: the spin-up observable is $\tilde{S}_z = \hbar\sigma_z/2$. Its expectation at $v$ is

$$
\begin{aligned}
\left(v, \tilde{S}_z v\right) &= \bar{v}^t \tilde{S}_z v \\[2mm]
&= \frac{\hbar}{2} \bar{v}^t \sigma_z v \\[2mm]
&= \frac{\hbar}{2} \left( \cos\left(\frac{\theta}{2}\right) \exp(i\phi t), \sin\left(\frac{\theta}{2}\right) \exp(-i\phi t) \right) \begin{pmatrix} 1 & 0 \\ 0 & -1 \end{pmatrix} \begin{pmatrix} \cos\left(\frac{\theta}{2}\right) \exp(-i\phi t) \\ \sin\left(\frac{\theta}{2}\right) \exp(i\phi t) \end{pmatrix} \\[2mm]
&= \frac{\hbar}{2} \left( \cos^2\left(\frac{\theta}{2}\right) - \sin^2\left(\frac{\theta}{2}\right) \right) \\[2mm]
&= \frac{\hbar}{2} \cos(\theta).
\end{aligned}
$$

14. Next, look at another random variable: spin-right (counterclockwise around the $x$-axis). What is its expectation at $v$? Hint: the relevant observable is now

$$
\tilde{S}_x = \frac{\hbar}{2}\sigma_x = \frac{\hbar}{2}\begin{pmatrix} 0 & 1 \\ 1 & 0 \end{pmatrix}.
$$

Thus,

$$
\begin{aligned}
\left(v, \tilde{S}_x v\right) &= \bar{v}^t \tilde{S}_x v \\[2mm]
&= \frac{\hbar}{2} \bar{v}^t \sigma_x v \\[2mm]
&= \frac{\hbar}{2} \left( \cos\left(\frac{\theta}{2}\right) \exp(i\phi t), \sin\left(\frac{\theta}{2}\right) \exp(-i\phi t) \right) \begin{pmatrix} 0 & 1 \\ 1 & 0 \end{pmatrix} \begin{pmatrix} \cos\left(\frac{\theta}{2}\right) \exp(-i\phi t) \\ \sin\left(\frac{\theta}{2}\right) \exp(i\phi t) \end{pmatrix} \\[2mm]
&= \frac{\hbar}{2} \left( \cos\left(\frac{\theta}{2}\right) \sin\left(\frac{\theta}{2}\right) \exp(2i\phi t) + \sin\left(\frac{\theta}{2}\right) \cos\left(\frac{\theta}{2}\right) \exp(-2i\phi t) \right) \\[2mm]
&= \frac{\hbar}{2} 2 \sin\left(\frac{\theta}{2}\right) \cos\left(\frac{\theta}{2}\right) \frac{\exp(2i\phi t) + \exp(-2i\phi t)}{2} \\[2mm]
&= \frac{\hbar}{2} \sin(\theta) \cos(2\phi t).
\end{aligned}
$$

15. Next, look at the spin-in random variable (counterclockwise around the $y$-axis). What is its expectation at $v$? Hint: the relevant observable is now

$$
\tilde{S}_y = \frac{\hbar}{2}\sigma_y = \frac{\hbar}{2}\begin{pmatrix} 0 & -i \\ i & 0 \end{pmatrix}.
$$

Therefore,

$$
\begin{aligned}
\left(v, \tilde{S}_y v\right) &= \bar{v}^t \tilde{S}_y v \\[2mm]
&= \frac{\hbar}{2} \bar{v}^t \sigma_y v \\[2mm]
&= \frac{\hbar}{2} \left( \cos\left(\frac{\theta}{2}\right) \exp(i\phi t), \sin\left(\frac{\theta}{2}\right) \exp(-i\phi t) \right) \begin{pmatrix} 0 & -i \\ i & 0 \end{pmatrix} \begin{pmatrix} \cos\left(\frac{\theta}{2}\right) \exp(-i\phi t) \\ \sin\left(\frac{\theta}{2}\right) \exp(i\phi t) \end{pmatrix}
\end{aligned}
$$

$$= \frac{\bar{h}}{2} \left( \cos\left(\frac{\theta}{2}\right) \sin\left(\frac{\theta}{2}\right) (-i) \exp(2i\phi t) + \sin\left(\frac{\theta}{2}\right) \cos\left(\frac{\theta}{2}\right) i \exp(-2i\phi t) \right)$$

$$= \frac{\bar{h}}{2} 2 \sin\left(\frac{\theta}{2}\right) \cos\left(\frac{\theta}{2}\right) \frac{\exp(2i\phi t) - \exp(-2i\phi t)}{2i}$$

$$= \frac{\bar{h}}{2} \sin(\theta) \sin(2\phi t).$$

16. What is the mathematical meaning of $2\phi t$? Hint: the phaseshift between $v_0$ and $v_1$.

17. By now, we have three expectations. Put them in one three-dimensional vector:

$$\frac{\bar{h}}{2} \left( \sin(\theta) \cos(2\phi t), \sin(\theta) \sin(2\phi t), \cos(\theta) \right).$$

Hint: see Figure 17.6.

18. How does this vector look like? Hint: at frequency $2\phi$, it rotates around the vertical $z$-axis (at angle $\theta$ from it).

19. This is Larmor's precession.

20. In classical electromagnetics, a charged ball also spins in the same way. Why? Hint: this is a special case of a more general theorem: the expectations in quantum mechanics obey the same laws as in classical physics.

21. Simulate quantum FFT on your digital computer.

22. Does it work?

23. To confirm that it works, apply inverse quantum FFT. Do you get the original $v$ back again?

# 18

## Shor's Factoring Algorithm

Our quantum computer is very simple: a list of $k$ qubits. At some probability, each qubit can spin up (value 0) or down (value 1). This makes a hypercube of $2^k$ nodes. On it, we have a (complex) grid function: the probabilities to have a certain configuration of 0's and 1's. There are as many as $2^k$ different configurations, each at some probability. This is the great power of quantum computing: with as little as $k$ qubits, we get as many as $2^k$ (complex) degrees of freedom.

Unfortunately, this is still science fiction. Still, nothing can stop us from "playing" with this in our imagination. This may lead to a lot of insight in both computer science and physics.

By now, we've already seen one quantum algorithm: quantum FFT. This is based on yet another fundamental structure from set theory: a binary tree. Let's use quantum FFT in cryptography.

The RSA key exchange uses two long prime numbers: $p < q$, both kept secret. Their product $pq$, on the other hand, is not secret at all, but placed in the public domain. How to break the code? In other words, how to factorize $pq$ in terms of its inner (unknown) ingredients, $p$ and $q$? This is a "salad" problem: separate the salad back to its original ingredients.

This is quite difficult: much more difficult than preparing the salad in the first place... This is the factoring problem: the job of the hacker. Once the hacker uncovers $p$ and $q$ explicitly, he can easily read our secret messages. It is important to understand how he could do this. This is the key to closing the gap, and fixing and improving the code in the future.

On the good old digital computer, it is practically impossible to factor $p$ and $q$ out. To break the code, the hacker must use a new quantum computer. On it, he should run quantum FFT, as described above. This remains implicit: there is no need to read the results. Even so, this is good enough to uncover $p$ and $q$, and break the RSA code. Fortunately, no hacker has a quantum computer as yet (or at least we hope not). As a matter of fact, quantum computer is still science fiction. Still, we better get ready for a future day, when quantum computers may become available. For this purpose, we better understand well how the code could break. Only then could we close the gap, and design new safer codes.

## 18.1 FFT and RSA

### 18.1.1 Implicit Quantum FFT

By now, our quantum FFT remains implicit: we can never read the results, or we might spoil the entire state forever. How to use quantum FFT in practice? Only implicitly! How, and for what purpose? To break the RSA code.

### 18.1.2 The Factoring Problem

In the RSA code, in the public domain, we only have the product $pq$. How to uncover the individual (very long) prime factors $p$ and $q$? This is the factoring problem: it actually breaks the RSA code. On the digital computer, this is practically impossible. On the (theoretical) quantum computer, on the other hand, there is a practical algorithm: Shor's factoring algorithm.

## 18.2 Hypercube and Its Index

### 18.2.1 Quantum Computer: Hypercube

To help solve the factoring problem, assume that we already have a quantum computer of $k$ qubits. How does this look like? Well, each qubit could be either 0 or 1. Together, the qubits make a new geometrical structure: a hypercube of $2^k$ nodes, indexed by

$$i = 0, 1, 2, 3, \ldots, 2^k - 1.$$

This hypercube is completely nonphysical: it is in our heads only. To index the nodes in it, $i$ must take its binary representation (or configuration): a list of $k$ binary digits. For this purpose, each qubit makes a little random variable, with two possible values: either 0 or 1. Together, the qubits make a bigger random variable: $i$ itself. Still, this is not as simple as a system of coins: it is much more complicated.

### 18.2.2 Binary Index: a New Random Variable

Here, $i$ itself is a random variable too in its own right. What is the value of $i$? Well, we don't know for sure: this is uncertain, and can be predicted at some probability only.

What is the probability? Well, this depends on each individual qubit. If it spins up, then it has value 0. If, on the other hand, it spins down, then it has value 1. This is uncertain: each qubit makes a little random variable. Together, these $k$ qubits might spin in such a configuration that makes $i$. The probability for this is $|v_i|^2$, where $v$ is the state: a normalized (complex) $2^k$-dimensional vector (or grid function), defined on all $0 \leq i < 2^k$.

This is not as simple as a system of coins. Indeed, unlike coins, here the qubits are entangled: not quite independent. More precisely, their little states depend on each other. This is why $v$ *cannot* be written as a product of $k$ little states: it contains not only $k$ but $2^k$ degrees of freedom. This is what gives our quantum computer its exponential power. Let's see a few examples.

## 18.3 Transformation: from State to State

### 18.3.1 The Initial State

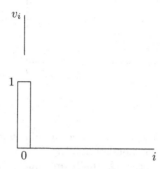

**Fig. 18.1.** Initially, $i$ is not random but deterministic: $i = 0$ at probability 1 (a must), and $i \neq 0$ at probability 0 (no chance at all). In other words, we know for sure that all qubits spin up, not down.

Let's prepare our quantum computer in a very simple (deterministic) state: all qubits spin up (and have value 0). This way, $i$ is actually deterministic. After all, we already know its value for sure:

$$i = 0.$$

This is why the state takes a particularly simple form: only the first component is nonzero, and all the rest vanish:

$$v = (1, 0, 0, 0, 0, \ldots, 0)^t$$

(Figure 18.1).

### 18.3.2 Determinism

This $2^k$-dimensional vector tells us the whole story. Indeed, it contains all the probabilities. Only one component is nonzero: the first one. Thus, $i$ can take only one value: the first one. This means that $i$ can be 0 only: no other value is possible. Indeed, the probability for this is as high as $|v_0|^2 = 1$. Could $i$ take any other value? No! After all, the probability to have $i \neq 0$ is as low as $|v_i|^2 = 0$.

This is indeed determinism. Still, this is going to change soon: thanks to quantum FFT, $v$ will soon become as nondeterministic as ever: a constant vector of all 1's (up to a normalization factor, which we disregard).

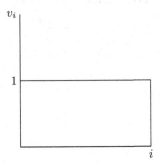

**Fig. 18.2.** After FFT, $v$ gets constant and uniform: all components are the same: $v_i = 1$ (up to normalization). This means nondeterminism: $i$ is equally likely to take any value between 0 and $2^k - 1$. In other words, all qubits are equally likely to spin up or down.

### 18.3.3 The One State: Uniform Distribution

So far, our state $v$ is deterministic: it has only one nonzero component. But this is not what we want. Let's go ahead and apply quantum FFT. In other words, to the above $v$, let's apply the $2^k \times 2^k$ matrix $W_{2^k}$:

$$v \to (W_{2^k})\, v = (1, 1, 1, 1, \ldots, 1)^t$$

(Figure 18.2).

### 18.3.4 Nondeterminism

What is this new $v$? It is as nondeterministic as ever. In fact, each qubit is equally likely to spin up or down. This way, $i$ is really random: equally likely to take any value from 0 to $2^k - 1$.

## 18.4 Period Finding

### 18.4.1 Modular Power

But this is still not quite what we want. We really want the state to have a "periodic zebra" pattern. To have this, recall that $p$ and $q$ are the long prime number that we need to uncover. Fortunately, their product $pq$ is available. Let's pick a new natural number $1 < a < pq - 1$ at random. Now, is $a$ coprime to $pq$? If not, then $a$ must be a multiple of $p$ or $q$ (but not of $pq$). In this case, we're done: just calculate

$$\text{GCD}\,(a, pq)\,,$$

and you get both $p$ and $q$ explicitly, as required.

So, assume that $a$ is coprime to $pq$. In other words, $a$ is *not* a multiple of $p$ or $q$:

$$\mathrm{GCD}\,(a, pq) = 1.$$

With this new $a$ at hand, let's calculate a new random variable: the modular power

$$a^i \bmod pq$$

(Chapter 13, Section 13.4.3).

### 18.4.2 Quantum Computing

This is done inside our quantum computer, without telling us what the precise value of $i$ really is. This way, we never spoil our state, but only transform it into a new "zebra" pattern.

### 18.4.3 "Periodic Zebra" State

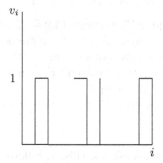

**Fig. 18.3.** After calculating the modular power, the state takes a "zebra" pattern: $v_i = 1$ at $i$'s at spacing $o$ from one to another, and $v_i = 0$ in between.

After this calculation, what is the new state? Well, $v$ has transformed from its old uniform pattern to a new "zebra" pattern:

$$\begin{pmatrix} 1 \\ 1 \\ 1 \\ 1 \\ 1 \\ \vdots \end{pmatrix} \rightarrow \begin{pmatrix} 1 \\ 0 \\ 1 \\ 0 \\ 1 \\ \vdots \end{pmatrix}$$

(Figure 18.3). On the left-hand side, the old $v$ had a (boring) uniform pattern: 1's everywhere. On the right-hand side, $v$ got a much more interesting pattern: 1's remain only at those $i$'s for which $(a^i \bmod pq)$ agrees with the result of our quantum calculation:

$$v_i = \begin{cases} 1 & \text{if } (a^i \bmod pq) = \text{result of our quantum calculation} \\ 0 & \text{otherwise.} \end{cases}$$

After all, other $i$'s that disagree with our quantum calculation could never take place: they have probability 0 — no chance at all.

### 18.4.4 The Order of $a$

Why is this a zebra pattern? To see this, recall the order of $a$:

$$o = o_{pq}(a).$$

This is the minimal natural number for which

$$a^o \equiv 1 \bmod pq.$$

Fortunately,

$$o \mid \phi = (p-1)(q-1).$$

This can be proved as in Chapter 13, Sections 13.9.5–13.10.1. Unfortunately, we have no access to $\phi$, which is kept secret. (After all, we are now the hackers...) Thus, we have a great task: to uncover $o$, without knowing $\phi$. Why is $o$ so important? Later on, we'll use $o$ to solve the factoring problem. Fortunately, we already have a clue about $o$: it makes the spacing in our up-to-date state.

### 18.4.5 Spacing and Period

Indeed, our up-to-date $v$ should look like this: 1, followed by $o-1$ zeroes, followed by 1, followed by $o-1$ zeroes, and so on. As a matter of fact, this pattern is not quite unique: it could also shift, provided that the same spacing is kept: $o-1$ zeroes between the 1's. (The actual shift depends on the precise result $a^i$ in our quantum calculation above.) Fortunately, such a shift is completely immaterial: it would only lead to a local phaseshift in our quantum FFT below, with no effect whatsoever.

## 18.5 Period Finding: Use Quantum FFT

### 18.5.1 Applying Quantum FFT

Unfortunately, the period $o$ is still unknown. How to find it? Apply quantum FFT! In the new state $v$, we'll then have a new (approximate) zebra pattern:

$$v \to \begin{pmatrix} 1 \\ 0 \\ 1 \\ 0 \\ 1 \\ \vdots \end{pmatrix}.$$

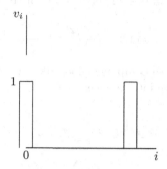

**Fig. 18.4.** After quantum FFT, the state is near-zebra, with spacing as large as $2^k/o$.

### 18.5.2 Near Zebra State

This is a new near-zebra pattern: 1, followed by $2^k/o - 1$ zeroes, followed by 1, followed by $2^k/o - 1$ zeroes, and so on (Figure 18.4). Why is this only an approximation? Because $2^k/o$ is no longer integer:

$$o \nmid 2^k.$$

As a matter of fact, $2^k/o$ could be written as an integer minus a fraction:

$$\frac{2^k}{o} = \left\lceil \frac{2^k}{o} \right\rceil - \alpha,$$

for some $0 < \alpha < 1$. Fortunately, we can still pick $k$ so large that

$$2^k \gg (pq)^2 > \phi^2 > o^2.$$

This way, the new spacing $2^k/o$ is much bigger than the old spacing $o$. This way, the new random variable $o$ is not as sensitive as the old random variable $i$: even if we have an error in $i$, this is still unlikely to affect $o$.

## 18.6 Interference: Constructive or Destructive?

### 18.6.1 Constructive Interference

What produces the near-zebra pattern? To see this, consider the old state $v$ in Figure 18.3, and apply quantum FFT to it. How should the new state look like? It should exhibit interference: constructive, followed by destructive, followed by constructive again, followed by destructive again, and so on. Indeed, consider an $i$ that is roughly a multiple of $2^k/o$:

$$i \doteq l\frac{2^k}{o},$$

for some $0 \le l \le o$. In other words, the error is small (in magnitude):

$$e = e(i) = i - l\frac{2^k}{o} \ll \frac{2^k}{o}.$$

After FFT, how should the $i$th component look like? Well, during FFT, $v$ is highly likely to undergo constructive interference:

$$2^{k/2}\left((W_{2^k})\,v\right)_i = \sum_{j=0}^{2^k-1} w^{ij} v_j$$

$$= \sum_{0 \le j < 2^k/o} w^{ioj}$$

$$\doteq \sum_{0 \le j < 2^k/o} w^{l2^k j}$$

$$= 1 + 1 + 1 + \cdots + 1$$

$$= \frac{2^k}{o} + \alpha.$$

This is indeed constructive: in the above sum, all terms are positive, and help increase the total sum.

### 18.6.2 Interference: Constructive vs. Destructive

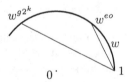

**Fig. 18.5.** In the complex plane, $w$ lies on the unit circle, at phase $2\pi/2^k$. Now, if $e \ll g2^k/o$, then $eo \ll g2^k$. In this case, $w^{g2^k}$ has a much bigger phase than $w^{eo}$, and a much longer distance from 1. In other words, $1 - w^{g2^k} \gg 1 - w^{eo}$. In summary, the numerator is much bigger than the denominator (in absolute value). This is constructive interference, with high probability $|v_i|^2$.

And what about a more general $i$, which could be quite different from any multiple of $2^k/o$? Well, for such an $i$, during FFT, $v$ undergoes a different kind of interference:

not constructive but destructive. How to analyze both kinds at the same time? For this purpose, write $e$ as an integer plus a roundoff error:

$$e = M + g,$$

for some

$$|g| \le \frac{1}{2}.$$

This way, we can now estimate the above sum asymptotically:

$$2^{k/2} \left( (W_{2^k}) v \right)_i$$

$$= \sum_{j=0}^{2^k-1} w^{ij} v_j$$

$$= \sum_{0 \le j < 2^k/o} w^{ioj}$$

$$= \frac{1 - w^{io(2^k/o+\alpha)}}{1 - w^{io}}$$

$$= \frac{1 - w^{eo(2^k/o+\alpha)}}{1 - w^{eo}}$$

$$= \frac{1 - w^{eo\alpha+eo2^k/o}}{1 - w^{eo}}$$

$$= \frac{1 - w^{eo\alpha+g2^k}}{1 - w^{eo}}$$

$$\sim \begin{cases} \frac{\sqrt{(1-\cos(2\pi g))^2+\sin^2(2\pi g)}}{2\pi eo/2^k} = \frac{\sqrt{2(1-\cos(2\pi g))}}{2\pi e} \cdot \frac{2^k}{o} = \frac{\sin(\pi g)}{\pi e} \cdot \frac{2^k}{o} & \text{if } e \ll g\frac{2^k}{o} \\ O(1) & \text{otherwise,} \end{cases}$$

as can be seen geometrically on the unit circle in the complex plane (Figure 18.5).

### 18.6.3 Low Error — High Probability

Thus, those $i$'s with error as low as

$$e(i) \ll g\frac{2^k}{o}$$

win: after FFT, they have probability as high as

$$|v_i|^2 \sim \left( \frac{\sin(\pi g)}{\pi e} \cdot \frac{2^k}{o} \right)^2$$

(up to normalization). When is this maximal? When the error is as low as

$$e = g.$$

This $i$ is indeed nearly a multiple of $2^k/o$:

$$i \doteq l\frac{2^k}{o}.$$

This $i$ is indeed most likely: after FFT, it has probability as high as

$$|v_i|^2 \sim \left(\frac{\sin(\pi g)}{\pi g} \cdot \frac{2^k}{o}\right)^2$$

(up to normalization). Is this result familiar? We've already seen it before: in the special case of $e = g = 0$, we even saw an exact equation (Section 18.6.1).

### 18.6.4 High Error — Low Probability

And what about those $i$'s that exhibit a high error? For these, things are the other way around: the interference is not constructive but destructive. This leads to a much lower probability (up to normalization again):

$$|v_i|^2 = O(1) \quad \text{if} \quad g2^k = O(eo) \text{ and } e \neq 0.$$

In particular, this is relevant to an $i$ with a big error:

$$|v_i|^2 = O(1) \quad \text{if} \quad 2^k = O(eo).$$

(Since $|g| \leq 1/2$, this condition tells us that the previous formula also holds.) Let's use this in practice.

## 18.7 Finding the Order of $a$

### 18.7.1 Measuring $i$: Back to Determinism

How to uncover $o$? By now, we have a near-zebra state. It is time to change it for the last time. For this purpose, just read (or measure) $i$. This makes $v$ as simple as in the beginning: just one nonzero component:

$$v_j = \begin{cases} 1 & \text{if} \quad j = i \\ 0 & \text{if} \quad j \neq i. \end{cases}$$

This is the end of our quantum computation. Now, we have determinism again: we know $i$ for sure, at probability $|v_i|^2 = 1$. Indeed, after measuring $i$, $v_i$ is the only nonzero component. The other $v_j$'s (for any $j \neq i$), on the other hand, vanish. After all, these $j$'s disagree with $i$, and can't be true any more.

Moreover, thanks to our asymptotic estimate, $i$ is highly likely to be nearly a multiple of $2^k/o$:

$$i \doteq l\frac{2^k}{o}$$

(for some unknown integer $0 \leq l \leq o$). To uncover $o$, we can now use the good old digital computer again.

## 18.7.2 Continued Fraction

Unfortunately, $l$ is still unknown. As a result, $o$ is unknown as well. Although we don't care about $l$, we do care about $o$. How to uncover it? By approximating

$$\frac{i}{2^k} \doteq \frac{l}{o}$$

in terms of a continued fraction. This is done on a digital computer: no need to use a quantum computer any more.

How to do this? Define a new sequence of natural numbers $o_1, o_2, o_3, o_4, \ldots$, which are optimal in the following sense: $o_1$ is the maximal integer number that makes a lower bound:

$$o_1 \le \frac{i}{2^k}.$$

(In our case, $o_1 = 0$.) Next, $o_2$ is the maximal natural number that produces an upper bound of the form

$$o_1 + \frac{1}{o_2} \ge \frac{i}{2^k}.$$

Next, $o_3$ is the maximal natural number that produces a new lower bound of the form

$$o_1 + \frac{1}{o_2 + \frac{1}{o_3}} \le \frac{i}{2^k}.$$

Next, $o_4$ is the maximal natural number that produces a new upper bound of the form

$$o_1 + \frac{1}{o_2 + \frac{1}{o_3 + 1/o_4}} \ge \frac{i}{2^k},$$

and so on. On the left-hand side, we obtain a list of alternating (lower or upper) bounds, each could also be written simply as $l/o$, where $o$ increases monotonically down the list, yielding better and better accuracy. Each such $o$ could be used to solve the factoring problem (see below). Better yet, run the above quantum algorithm a few times, and obtain a few $i$'s. To solve the factoring problem, use only an $o$ obtained in all runs (provided that $o < pq$).

# 18.8 Breaking the RSA Code

## 18.8.1 Nontrivial Square Root

What is so good about $o$ (the order of $a$)? Well, thanks to a lemma in [59], at probability $1/2$ or more, $o$ is even, and $a^{o/2}$ makes a nontrivial square root of 1 modulus $pq$. What does this mean? Well, to be nontrivial, this square root mustn't be $\pm 1$ modulus $pq$:

$$a^{o/2} \not\equiv \pm 1 \bmod pq.$$

In other words, $pq$ mustn't divide the following two numbers:

$$pq \nmid a^{o/2} \pm 1.$$

But $pq$ must divide their product:

$$pq \mid a^o - 1 = \left(a^{o/2} - 1\right)\left(a^{o/2} + 1\right).$$

After all, $o$ was originally defined as the order of $a$.

### 18.8.2 Factorization

Is this situation familiar? Well, this is as in Chapter 13, Section 13.11.4, only with a much lower modular power:

$$p \mid a^{o/2} - 1 \quad \text{and} \quad q \mid a^{o/2} + 1,$$

or vice versa. This is what we wanted. To have $p$ and $q$ explicitly, just calculate

$$\text{GCD}\left(a^{o/2} - 1, pq\right).$$

Thanks to the above lemma, at probability 1/2 or more, this should indeed work. Otherwise, pick a new $a$, and try all over again.

## 18.9 Exercises: FFT of Any Order

### 18.9.1 How to Truncate a State?

1. Look again at quantum FFT.
2. So far, the dimension was a power of 2: $2^k$. In this case, we already know how to apply FFT (the unitary matrix $W_{2^k}$).
3. But what if the dimension is a new natural number $L$ that may not be a power of 2 any more? In this case, $v$ is an $L$-dimensional vector. How to apply $W_L$ to $v$? Hint: see below.
4. Let $v$ be a $2^k$-dimensional state. How to truncate it? In other words, how to modify $v$ to leave only $L$ nonzero components, followed by $2^k - L$ zeroes? Hint: the following algorithm tells us whether the present experiment could be accepted (since $i < L$), or must be rejected (since $i \geq L$) and restarted all over again:
   - Look at $L$, in its binary representation.
   - Add leading zeroes to it (if necessary), until it has $k$ bits too.
   - In our quantum computer, use the random variable $i$: the binary index. In this process, be careful to keep $i$ hidden inside the quantum computer: never look what the precise value of $i$ really is. Moreover, $i$ will be gone soon, with no trace whatsoever. This is important to avoid spoiling the entire state.
   - Consider the (nondeterministic) binary representation of $i$, qubit by qubit.
   - Start from the most significant qubit. Never look whether it spins up (value 0) or down (value 1), but only compare it to the corresponding bit in $L$.
   - If it is smaller, then accept.
   - If, on the other hand, it is larger, then reject.
   - If, on the other hand, it is the same, then:
     - If this is the last qubit, then reject.
     - Otherwise, move on to the next (less significant) qubit, and compare it too, as above, and so on.
5. Does this truncate $v$, as required? Hint: by the end of the above algorithm, we still don't know what $i$ really is. We only know whether $i < L$ or not. Only if $i < L$ do we get to continue our experiment. Thus, in this case, the state was indeed truncated:

$$v \to \begin{pmatrix} \mathbf{v} \\ \mathbf{0} \end{pmatrix},$$

where $\mathbf{v}$ is the top $L$-dimensional subvector of $v$, and $\mathbf{0}$ is a new $(2^k - L)$-dimensional zero subvector. Why? Because there is now no chance to have $i \geq L$ any more.

6. How to apply $W_L$ to the top subvector $\mathbf{v}$? Hint: truncate $v$ as above, and then apply $w_{2^k}$ to it, with a few changes: the new $w$ should have a new phase, $2^k/L$ times as big. For this purpose, start with a new initial frequency, $2^k/L$ times as high as in the original algorithm. In the end, truncate again, and multiply by $2^{k/2}/\sqrt{L}$. This way, the final state will be the same as $W_L\mathbf{v}$, as required.

# 19

# Towards
# Feynman Diagrams

Set theory is also the basis for graph theory. Indeed, what is a graph? It is a set
of nodes. On top of this, the graph is also a set of edges: every two nodes may be
connected by an edge. In an oriented graph, each edge has a direction: it leads from
one node to another, but not the other way around. In this kind of graph, the edge
can be illustrated as an arrow. Finally, in a weighted graph, each edge also has a
weight: a (complex) number, telling us how likely the edge is to get active. This
could be viewed as a flow: the weight tells us how much mass flows along the arrow.

Thanks to graph theory, we can now introduce Feynman diagrams [56]. These are
elementary pieces of graph, containing one node (or vertex), and three edges issuing
from it. Thanks to their edges, Feynman diagrams can be tied to one another, and
assemble into bigger and bigger graphs.

What is the physical meaning of this? Well, a Feynman diagram illustrates
an elementary quantum-mechanical process, in which two particles scatter from
one another, and emit a new particle. This is quite useful in particle physics and
quantum field theory. Thanks to their elegant rules, Feynman diagrams form a
complete algebraic structure, easy to visualize and comprehend. This way, modern
physics gets as easy and accessible as ever, just like a fun game.

## 19.1 System: Particle and Anti-Particle

### 19.1.1 Particle and Its Anti-Particle

In quantum mechanics, consider some given particle (say, an electron). Assume
that it could be in $m$ possible states (each at some probability), where $m$ is a fixed
natural number. For an electron, for example, there are only two states: spin up or
down ($m = 2$, Figure 19.1).

Usually, the particle has a positive energy. Still, in theory, it might also have a
negative energy. In this case, its anti-particle has a positive energy. For example,
look at an electron with a negative energy. What is its anti-particle? This is a
positron, with positive energy and charge.

For a more general particle, what is its anti-particle? This is its opposite image:

$$\text{anti-particle} = \text{missing particle}.$$

This way, it has the opposite energy, charge, and spin: all three pick a minus sign.

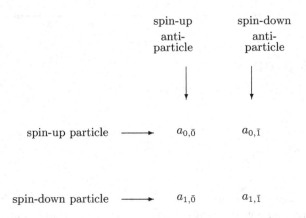

**Fig. 19.1.** The simple case $m = 2$. The system contains a particle and an anti-particle, each with spin, up or down, as indicated by their indices: $0 \leq i, j < m = 2$. The $i\bar{j}$ possibility has complex amplitude $a_{i,\bar{j}}$, and probability $|a_{i,\bar{j}}|^2$ (known as the fine structure constant).

### 19.1.2 Complex Amplitude and Probability

Together, our original particle and the anti-particle make a new system, with the following $m \times m$ matrix:

$$A \equiv \left(a_{i,\bar{j}}\right)_{0 \leq i, j < m}$$

(Figure 19.1). (The bar reminds us that this is an anti-particle.) What do we have in $A$? Each element is called a complex amplitude. In quantum mechanics, this allows a nondeterministic state: a superposition (or sum) of a few different possibilities, each at its own complex amplitude and probability. How to calculate the probability? Easy: take the complex amplitude, calculate its absolute value, and square it up. Let's see a simple example.

### 19.1.3 Example: Electron and Positron

For example, if our original particle is an electron, then the anti-particle is a positron. This is quite a simple case: $m = 2$, so the state of the electron-positron pair could be indexed by $0 \leq i, j < m = 2$. In particular, the electron could spin either up ($i = 0$) or down ($i = 1$). Likewise, the positron could spin either up ($j = 0$) or down ($j = 1$). This way, the entire electron-positron system has four possible states: either up-up or up-down or down-up or down-down.

Each of these four is a deterministic state, with no ambiguity or uncertainty. Still, in quantum mechanics, we often have a nondeterministic state, in which all these four are possible, each at its own complex amplitude and probability. This makes a new $2 \times 2$ matrix:

$$A \equiv \begin{pmatrix} a_{0,\bar{0}} & a_{0,\bar{1}} \\ a_{1,\bar{0}} & a_{1,\bar{1}} \end{pmatrix}$$

(Figure 19.1). In $A$, each element is a complex amplitude. For example, up-down has complex amplitude $a_{0,\bar{1}}$. This tells us the probability to see up-down: $|a_{0,\bar{1}}|^2$. Clearly, these four probabilities must have a sum of 1. To guarantee this, we also assume that $A$ has already been normalized:

$$\sum_{i,j=0}^{m-1} |a_{i,\bar{j}}|^2 = 1.$$

### 19.1.4 Virtual Gauge Boson

Imagine now a theoretical (nonphysical) situation: the electron "sits" on top of the positron. How come they don't cancel each other? Well, this is a virtual particle: it fluctuates ever so frequently to an electron-positron pair and back, and so on. This is a virtual process, in an infinitesimal time. The joint (united) particle is called a gauge boson: a photon.

In terms of spin, what is the state of the united photon? Clearly, this depends on the original spins of the electron-positron. Still, could it be deterministic: say, up-up? No! It could only be nondeterministic: a superposition of up-up *minus* down-down. This means uncertainty: either up-up or down-down, at equal probability. In this virtual particle, the up-electron can indeed fluctuate to a down-electron and back:

$$\begin{matrix} \text{electron} & - \\ \text{positron} & - \end{matrix} \quad \begin{pmatrix} 0 \\ \bar{0} \end{pmatrix} \quad \rightleftarrows \quad \begin{pmatrix} 1 \\ \bar{1} \end{pmatrix} \quad \begin{matrix} - & \text{electron} \\ - & \text{positron.} \end{matrix}$$

At the same time, the down-positron fluctuates to an up-positron and back. This, however, runs the other way around: back-in-time. (After all, this takes only an infinitesimal time.)

To do this, the top arrow emits a $0\bar{1}$-photon (to be absorbed soon). The bottom arrow emits the same, but back-in-time. Or, in terms of standard time, it absorbs a $0\bar{1}$-photon (to be emitted soon). This is why the net action of both arrows is zero: nothing really happens, as required.

Thanks to the frequent fluctuations, we have uncertainty: at a given time, we can't tell whether it is up-up or down-down. After all, both states are equally likely.

## 19.2 Matrix of Complex Amplitudes

### 19.2.1 Traceless Matrix

What is so good about this kind of nondeterminism? Well, thanks to it, $A$ takes the form

$$A = \frac{\sqrt{-1}}{\sqrt{2}} \begin{pmatrix} 1 & 0 \\ 0 & -1 \end{pmatrix}.$$

Look at the main-diagonal elements: what is their sum? Zero! This means that $A$ is traceless: it has zero trace. This is indeed what we want.

## 19.2.2 Anti-Hermitian Matrix

Thanks to the same logic, up-down is no longer allowed on its own: it is too deterministic. Only a nondeterministic state is allowed: up-down $\pm$ down-up. More precisely, what is allowed is

$$A = \frac{1}{\sqrt{2}} \begin{pmatrix} 0 & 1 \\ -1 & 0 \end{pmatrix} \quad \text{or} \quad \frac{\sqrt{-1}}{\sqrt{2}} \begin{pmatrix} 0 & 1 \\ 1 & 0 \end{pmatrix}.$$

Are these matrices familiar? They are the well-known Pauli matrices: $\sigma_z$, $\sigma_y$, and $\sigma_x$, with a little change: the new coefficient $\sqrt{-1}/\sqrt{2}$ makes them anti-Hermitian rather than Hermitian.

## 19.2.3 Why Anti-Hermitian?

Why should $A$ be anti-Hermitian? Consider an $i\bar{j}$-particle, with complex amplitude $a_{i,\bar{j}}$. Now, interchange $i$ and $j$. What is the $j\bar{i}$-particle? What is its complex amplitude?

To find out, start from the $i\bar{j}$-particle again. Its complex amplitude tells us how likely it is to be emitted (not only virtually but also physically). This can also be viewed the other way around: the original particle being emitted is the same as the anti-particle being absorbed. How to have the anti-particle? For this, we also need a little change — an extra bar on top: not $a_{i,\bar{j}}$ any more, but $\bar{a}_{i,\bar{j}}$. (This bar is on top of the complex amplitude, not the indices.) This way, the imaginary part picks a minus sign:

$$\Im \bar{a}_{i,\bar{j}} = -\Im a_{i,\bar{j}}.$$

Why is this good? Because the imaginary part tells us the frequency, which is proportional to the energy (Planck's law). Now, in the anti-particle, the energy should indeed pick a minus sign.

So far, we talked about the absorbed anti-particle. But this is not quite what we want: we want it emitted, not absorbed. For this, we need to reverse time, and pick a minus sign:

$$a_{j,\bar{i}} \equiv -\bar{a}_{i,\bar{j}}$$

(in agreement with $A$ being anti-Hermitian). This is what we wanted: the new $j\bar{i}$-particle is the anti-$i\bar{j}$-particle, emitted back-in-time. Its new complex amplitude can now be placed in $A$, to mirror $a_{i,\bar{j}}$. It can now be used to reverse time.

## 19.2.4 Transition

Thanks to the above, our original $i\bar{j}$-particle can now be used in a few (equivalent) ways. In the simple way, start from an $i$-particle, and convert it into a new $j$-particle:

$$i \to j.$$

In this transition, a $i\bar{j}$-particle is emitted. In other words, its anti-particle is absorbed. Or, back-in-time, this anti-$i\bar{j}$-particle is emitted (leftward rather than rightward):

$$i \leftarrow j.$$

But this arrow could also be used by the $j\bar{i}$-particle, emitted leftward in the same way. This shows once again that it was indeed a good idea to make $A$ anti-Hermitian.

Still, in what follows, we consider even a more general $A$, not necessarily anti-Hermitian or traceless. This makes the $j\bar{i}$-particle more independent: it is not necessarily the anti-particle of the $i\bar{j}$-particle any more. This way, we can imagine even a deterministic boson, and study it.

## 19.3 Emitting a Gauge Boson

### 19.3.1 Scattering

Feynman diagrams help visualize an elementary physical process. For example, Figure 19.2 illustrates scattering: an $i$-particle hits an object, converts into a new $j$-particle, and bounces off. To change in this way, it must also emit a new $i\bar{j}$-gauge boson. This preserves the total amount of particles:

$$i = j + i\bar{j}.$$

This process has complex amplitude $a_{i,\bar{j}}$, and probability $|a_{i,\bar{j}}|^2$ (known as the fine structure constant).

This can also be written in terms of anti-particles (Figure 19.2, second picture): an incoming $\bar{j}$-particle converts into an outgoing $\bar{i}$-particle, emitting a new $i\bar{j}$-boson (as before). This still preserves the total amount of particles:

$$\bar{j} = \bar{i} + i\bar{j}.$$

Next, drop the bars. This is no longer equivalent, but different: an incoming $j$-particle, converting into a new $i$-particle, emitting a new $j\bar{i}$-boson, with a new complex amplitude: $a_{j,\bar{i}}$. The total amount of particles is still preserved:

$$j = i + j\bar{i}.$$

Finally, rewrite this in terms of an anti-$j\bar{i}$-boson (or indeed an $i\bar{j}$-particle), not emitted but absorbed, helping the incoming $j$-particle convert into a new $i$-particle (bottom of Figure 19.2). The total amount of particles is still preserved:

$$j + i\bar{j} = i.$$

What is its complex amplitude? On one hand, this is the anti-boson of the $j\bar{i}$-boson, so it must have complex amplitude $\bar{a}_{j,\bar{i}}$. On the other hand, once read back-in-time (downward rather than upward), this mirrors the top picture, giving $-a_{i,\bar{j}}$. Fortunately, both are the same:

$$-a_{i,\bar{j}} = \bar{a}_{j,\bar{i}}.$$

(provided that $A$ is indeed anti-Hermitian).

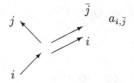

$\Updownarrow$   equivalent

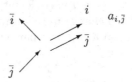

$\downarrow$   anti-equivalent

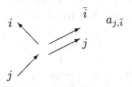

$\Updownarrow$   absorbe

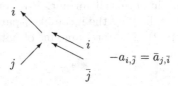

**Fig. 19.2.** An incoming $i$-particle turns into an outgoing $j$-particle, emitting an $i\bar{j}$-particle. Next, this is equivalent to an incoming $\bar{j}$-particle that turns into an outgoing $\bar{i}$-particle. Next, drop the bars. This is equivalent not to the former pictures but to the bottom picture. After all, emitting a $j\bar{i}$-particle is the same as absorbing its anti-particle: an $i\bar{j}$-particle. Once read downwards, this mirrors the top picture, picking a minus sign to reverse time.

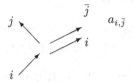

$$\Updownarrow \quad \text{absorbe back-in-time}$$

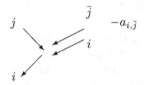

$$\Updownarrow \quad \text{emit back-in-time}$$

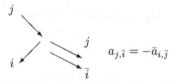

**Fig. 19.3.** Convert an incoming $i$-particle into an outgoing $j$-particle, emitting an $i\bar{j}$-particle. Next, read this in the opposite direction, back-in-time: convert an incoming $j$-particle back into an outgoing $i$-particle, by absorbing an $i\bar{j}$-particle back-in-time. For this, the complex amplitude must pick a minus sign. Next, instead of absorbing an $i\bar{j}$-particle, emit its anti-particle: a new $j\bar{i}$-particle (back-in-time). For this, the complex amplitude must also pick a bar on top.

## 19.4 How Does $A$ Act?

### 19.4.1 Transition and Matrix

In the top picture in Figures 19.2–19.3, the transition $i \to j$ is illustrated geometrically. This could happen at complex amplitude $a_{i,\bar{j}}$ and probability $|a_{i,\bar{j}}|^2$, emitting a new $i\bar{j}$-particle. How to represent this algebraically? For simplicity, look at one electron. Let $0 \le i, j < 2$ be fixed. For example, assume that $i = 0$ (spin-up) and $j = 1$ (spin-down). Assume also that $A$ has zero main diagonal:

$$A = \begin{pmatrix} & a_{0,\bar{1}} \\ a_{1,\bar{0}} & \end{pmatrix}.$$

### 19.4.2 *A* Acts from the Right

Now, represent the states of $i$ and $j$ as row vectors: $i = (1,0)$ (spin-up) and $j = (0,1)$ (spin-down). This way, the transition $i \to j$ can be represented by multiplying by $A$ from the right:

$$i = (1,0) \to (1,0)A = a_{0,\bar{1}}(0,1).$$

This way, we indeed obtain $j$ at complex amplitude $a_{0,\bar{1}}$ and probability $|a_{0,\bar{1}}|^2$, as required.

### 19.4.3 *A* Acts from the Left

Recall that $i$ and $j$ are fixed numbers. Next, interpret $i$ as missing $j$ (or $\bar{j}$), and $j$ as missing $i$ (or $\bar{i}$). In these terms, $i \to j$ is equivalent to $\bar{j} \to \bar{i}$. We've already seen this geometrically. How to write this algebraically as well? Easy: represent the state of $\bar{j}$ by the column vector $(0,1)^t$ (positron spinning down), and the state of $\bar{i}$ by $(1,0)^t$ (positron spinning up). Multiply by $A$ from the left:

$$\bar{j} = \begin{pmatrix} 0 \\ 1 \end{pmatrix} \to A \begin{pmatrix} 0 \\ 1 \end{pmatrix} = a_{0,\bar{1}} \begin{pmatrix} 1 \\ 0 \end{pmatrix}.$$

This has the same complex amplitude and probability as before.

### 19.4.4 Inverse Transition

How about the inverse transition $j \to i$? This is represented by

$$j = (0,1) \to (0,1)A = a_{1,\bar{0}}(1,0).$$

### 19.4.5 Back-In-Time

How is this related to our original transition? Well, it is actually $i \leftarrow j$: back-in-time (Figure 19.3). Instead of emitting an $i\bar{j}$-particle, we now need to emit an anti-$i\bar{j}$-particle back in time. This has a complex amplitude $-\bar{a}_{i,\bar{j}}$: a minus sign to pay for time reversal, and a bar on top to pay for energy, which is now negative, not positive. But this is the same as in a $j\bar{i}$-particle:

$$a_{j,\bar{i}} = -\bar{a}_{i,\bar{j}}$$

(provided that $A$ is anti-Hermitian). Once a $j\bar{i}$-particle is emitted, we indeed get the inverse transition: $j \to i$, as required. In our example, in particular, the new complex amplitude is

$$a_{1,\bar{0}} = -\bar{a}_{0,\bar{1}}$$

(if $A$ is indeed anti-Hermitian).

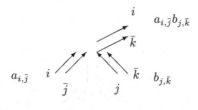

**Fig. 19.4.** Incoming $i\bar{j}$- and $j\bar{k}$-particles emit a new $i\bar{k}$-particle, with a new complex amplitude: the product of the original complex amplitudes.

## 19.5 Two Bosons Interact

### 19.5.1 Two Bosons Emit a New Boson

Consider now a new system of two incoming bosons: an $i\bar{j}$-boson of matrix

$$A \equiv \left(a_{i,\bar{j}}\right)_{0 \leq i,j < m},$$

and a $j\bar{k}$-boson of matrix

$$B \equiv \left(b_{j,\bar{k}}\right)_{0 \leq j,k < m}$$

(Figure 19.4). Once they meet, they interact, and emit a new $i\bar{k}$-boson. This still preserves the total amount of particles:

$$i + \bar{j} + j + \bar{k} = i + \bar{k}.$$

Assume now that $i$ and $k$ are specific, but $j$ is not: it runs from $j = 0, 1, 2, \ldots, m-1$. In this case, what is the new complex amplitude? It is the sum of products:

$$\sum_{j=0}^{m-1} a_{i,\bar{j}} b_{j,\bar{k}} = (AB)_{i,\bar{k}}.$$

This will be useful later.

### 19.5.2 A Boson Splits Into Two New Bosons

But this is not the only way to have an $i\bar{k}$-boson. In both incoming bosons, interchange the roles of $i$ and $k$ (Figure 19.5). What do you get? Again, there are two incoming bosons: a $k\bar{j}$-boson of matrix $A$, and a $j\bar{i}$-boson of matrix $B$. Once they meet, they interact, and emit a new $k\bar{i}$-boson. This still preserves the total amount of particles:

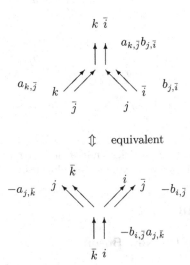

**Fig. 19.5.** In the previous diagram, interchange $i$ and $k$. This way, we have incoming $k\bar{j}$- and $j\bar{i}$-particles, which emit a new $k\bar{i}$-particle. In the bottom picture, this is rewritten in terms of the corresponding anti-particles. Thanks to the minus sign, the absorbed $i\bar{k}$-particle has complex amplitude $-b_{i,\bar{j}}a_{j,\bar{k}}$.

$$k + \bar{j} + j + \bar{i} = k + \bar{i}.$$

Next, rewrite this the other way around: replace each boson by its anti-boson. What do you get? The outgoing $k\bar{i}$-boson makes an incoming $i\bar{k}$-boson. The incoming $k\bar{j}$- and $j\bar{i}$-bosons make outgoing $j\bar{k}$- and $i\bar{j}$-bosons. In the process, the complex amplitudes pick a minus sign. Once read downwards, this mirrors Figure 19.4, with a new minus sign, to account for the reversed time.

### 19.5.3 Commutator: New Complex Amplitudes

Again, assume now that $i$ and $k$ are specific, but $j$ is not: it runs from $j = 0, 1, 2, \ldots, m-1$. This way, the new $i\bar{k}$-boson takes a new contribution to its complex amplitude:

$$-\sum_{j=0}^{m-1} a_{j,\bar{k}}b_{i,\bar{j}} = -\sum_{j=0}^{m-1} b_{i,\bar{j}}a_{j,\bar{k}} = -(BA)_{i,\bar{k}}.$$

In summary, thanks to Figures 19.4–19.5, the total complex amplitude of the new $i\bar{k}$-boson is

$$[A, B]_{i, \bar{k}} = (AB)_{i, \bar{k}} - (BA)_{i, \bar{k}},$$

where $[A, B]$ is the commutator:

$$[A, B] = AB - BA.$$

In summary, the new boson has a new matrix: the commutator of the original matrices. Here, "new" can take both meanings: either in time, or in reversed time.

## 19.6 Exercises: Weak and Strong Interactions

### 19.6.1 Weak Interaction: Quarks and $W$-Bosons

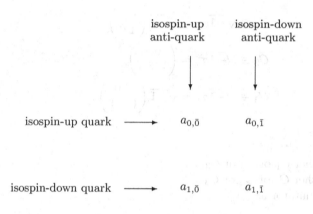

**Fig. 19.6.** Weak interaction: quark and anti-quark, whose isospin could be up or down, as indicated by their index: $0 \leq i, j < m = 2$. The $i\bar{j}$-gauge boson ($W$- or $Z$-boson) has complex amplitude $a_{i\bar{j}}$, and probability $|a_{i\bar{j}}|^2$ to be emitted (which is its fine structure constant).

1. In the atom, in the nucleus, there are protons and neutrons.
2. Each proton (or neutron) contains three quarks.
3. Each quark has isospin (which mirrors spin): either up (0) or down (1).
4. A quark and an anti-quark make a gauge boson: a $W$-boson. It may have isospin: up-up ($0\bar{0}$), up-down ($0\bar{1}$), down-up ($1\bar{0}$), or down-down ($1\bar{1}$) (Figure 19.6).
5. In the nucleus, what keeps the protons and neutrons together? Hint: this is the weak interaction: an up-quark from one proton turns into a down-quark, emitting a new $0\bar{1}$-$W$-boson (mirroring Figure 19.2). This is only a virtual boson, which is soon absorbed back, and so on. This is indeed nondeterminism: the proton turns into a neutron, and so on, back and forth.
6. Could two $W$-bosons attract each other as well? Hint: yes, as in Figures 19.4–19.5.

## 19.6.2 Weak Structure Constants

1. Recall the Pauli matrices:

$$\sigma_x \equiv \begin{pmatrix} 0 & 1 \\ 1 & 0 \end{pmatrix}$$

$$\sigma_y \equiv \sqrt{-1} \begin{pmatrix} 0 & -1 \\ 1 & 0 \end{pmatrix}$$

$$\sigma_z \equiv \begin{pmatrix} 1 & 0 \\ 0 & -1 \end{pmatrix}.$$

2. Are they traceless?
3. Are they Hermitian?
4. How to make them anti-Hermitian? Hint: multiply by $\sqrt{-1}$, and define three new matrices:

$$C_1 \equiv \sqrt{-1}\sigma_x = \sqrt{-1} \begin{pmatrix} 0 & 1 \\ 1 & 0 \end{pmatrix}$$

$$C_2 \equiv \sqrt{-1}\sigma_y = \begin{pmatrix} 0 & 1 \\ -1 & 0 \end{pmatrix}$$

$$C_3 \equiv \sqrt{-1}\sigma_z = \sqrt{-1} \begin{pmatrix} 1 & 0 \\ 0 & -1 \end{pmatrix}.$$

5. Are they traceless?
6. Are they anti-Hermitian?
7. Pick $A$ to be either $C_1$ or $C_2$ or $C_3$.
8. Pick $B$ to be either $C_1$ or $C_2$ or $C_3$.
9. Span their commutator as

$$[A, B] = c_1 C_1 + c_2 C_2 + c_3 C_3,$$

for some real coefficients $c_1$, $c_2$, and $c_3$.

10. $c_1$, $c_2$, and $c_3$ are called structure constants. They depend on $A$, $B$, and the original definition of $C_1$, $C_2$, and $C_3$:

$$c_1 \equiv c_1 (A, B, C_1, C_2, C_3)$$
$$c_2 \equiv c_2 (A, B, C_1, C_2, C_3)$$
$$c_3 \equiv c_3 (A, B, C_1, C_2, C_3).$$

11. Physically, what is $|c_1|^2 + |c_2|^2$? Hint: this is the probability to attract (weakly) $W$-boson $A$ to $W$-boson $B$ by emitting a new (virtual) $W$-boson (Figure 19.4), to be absorbed soon. This converts an isospin-up quark in $A$ to an isospin-down quark in $B$, back and forth (Figure 19.3).

12. Look at $c_3$. What does it tell us? Hint: this is the complex amplitude to have the following nondeterministic state: emitting either $0\bar{0}$-$W$-boson and $1\bar{1}$-anti-$W$-boson or $0\bar{0}$-anti-$W$-boson and $1\bar{1}$-$W$-boson (at equal probability). In other words, this is a virtual fluctuation:

$$W\text{-boson} \quad - \quad \begin{pmatrix} 0 \\ \bar{0} \end{pmatrix} \rightleftarrows \begin{pmatrix} 1 \\ \bar{1} \end{pmatrix} \quad - \quad W\text{-boson}.$$

Here, the top arrow emits a $0\bar{1}$-anti-$W$-boson (to be absorbed soon). The bottom arrow works the same, but back-in-time. After all, this is a virtual process, taking place at an infinitesimal time.

13. What is the probability to have this nondeterministic state? Hint: $|c_3|^2$.
14. What is the net effect of this virtual process? Hint: zero: nothing happens, as required.
15. Does this preserve energy? Hint: yes! The total energy remains the same.

### 19.6.3 Quantum Chromodynamics: Quarks and Gluons

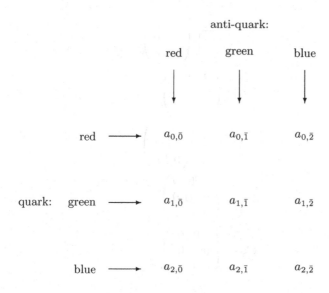

**Fig. 19.7.** Strong interaction: quark and anti-quark, whose color is indexed by $0 \le i, j < m = 3$. The $i\bar{j}$-gluon has complex amplitude $a_{i\bar{j}}$, and probability $|a_{i\bar{j}}|^2$ to be emitted (which is its fine structure constant).

1. So far, we used $m = 2$. In this context, a quark has only two options: either isospin up or down.
2. Next, let's move on to a more complicated system, in which the quark has three options to be colored: either red (0) or green (1) or blue (2), at some probability each (Figure 19.7).
3. In a proton (or a neutron), there are three quarks. What attracts them to each other? Hint: this is the strong interaction. For example, a red quark changes color to green, emitting a new $0\bar{1}$-gluon (mirroring Figure 19.2). This is a virtual

particle, which is soon absorbed back again, returning to the original state, and so on, ever so frequently, attracting the quarks to each other.

4. Could two gluons attract each other as well? Hint: yes, as in Figures 19.4–19.5.
5. Extend the Pauli matrices to this case as well. Hint: design eight new independent $3 \times 3$ matrices:

$$C_1 \equiv \begin{pmatrix} 0 & 1 & 0 \\ -1 & 0 & 0 \\ 0 & 0 & 0 \end{pmatrix}$$

$$C_2 \equiv \sqrt{-1} \begin{pmatrix} 0 & 1 & 0 \\ 1 & 0 & 0 \\ 0 & 0 & 0 \end{pmatrix}$$

$$C_3 \equiv \begin{pmatrix} 0 & 0 & -1 \\ 0 & 0 & 0 \\ 1 & 0 & 0 \end{pmatrix}$$

$$C_4 \equiv \sqrt{-1} \begin{pmatrix} 0 & 0 & 1 \\ 0 & 0 & 0 \\ 1 & 0 & 0 \end{pmatrix}$$

$$C_5 \equiv \begin{pmatrix} 0 & 0 & 0 \\ 0 & 0 & 1 \\ 0 & -1 & 0 \end{pmatrix}$$

$$C_6 \equiv \sqrt{-1} \begin{pmatrix} 0 & 0 & 0 \\ 0 & 0 & 1 \\ 0 & 1 & 0 \end{pmatrix}$$

$$C_7 \equiv \begin{pmatrix} 1 & 0 & 0 \\ 0 & -1 & 0 \\ 0 & 0 & 0 \end{pmatrix}$$

$$C_8 \equiv \begin{pmatrix} 0 & 0 & 0 \\ 0 & 1 & 0 \\ 0 & 0 & -1 \end{pmatrix}.$$

6. Are they traceless?
7. Are they anti-Hermitian?
8. Pick $A$ to be one of these eight matrices.
9. Pick $B$ to be one of these eight matrices.
10. Span their commutator as

$$[A, B] = c_1 C_1 + c_2 C_2 + c_3 C_3 + \cdots + c_8 C_8,$$

for some real coefficients $c_1, c_2, c_3, \ldots, c_8$.
11. The coefficients $c_1, c_2, c_3, \ldots, c_8$ are called structure constants.
12. They depend on $A$, $B$, and the original definition of the matrices $C_1, C_2, C_3, \ldots, C_8$.

### 19.6.4 Strong Structure Constants

1. So far, $i$ and $j$ indexed the quarks. Next, let $1 \le i, j \le 8$ index the matrices $C_1, C_2, C_3, \ldots, C_8$.

2. Let $A$ be fixed and specific, as above: one of the above eight matrices.
3. $B$, on the other hand, remains unspecified: either one of the above eight matrices.
4. The transformation $[A, \cdot]$ maps $B$ to $[A, B]$:

$$[A, \cdot] : B \to [A, B].$$

5. How does this transformation look like? Hint: see below.
6. Use the above basis $C_1, C_2, C_3, \ldots, C_8$ to write the transformation explicitly, as an $8 \times 8$ matrix. Hint: to filter out its $j$th column, apply it to some $B \equiv C_j$. This filters out its $(i, \bar{j})$th element:

$$[A, \cdot]_{i, \bar{j}} = [A, C_j]_i$$
$$= c_i$$
$$\equiv c_i (A, B, C_1, C_2, C_3, \ldots, C_8)$$

$(1 \le i, j \le 8)$.

7. Physically, what is $|c_1|^2 + |c_2|^2$? Hint: this is the probability to attract (strongly) gluon $A$ to gluon $B$ by a new (virtual) gluon (Figure 19.4), which converts a red quark in $A$ to a green quark in $B$, back and forth (Figure 19.3).
8. Likewise, what is $|c_3|^2 + |c_4|^2$? Hint: this is the probability to attract (strongly) gluon $A$ to gluon $B$ by a new (virtual) gluon (Figure 19.4), which converts a red quark from $A$ to a blue quark in $B$, back and forth (Figure 19.3).
9. Likewise, what is $|c_5|^2 + |c_6|^2$? Hint: this is the probability to attract (strongly) gluon $A$ to gluon $B$ by emitting a new (virtual) gluon (Figure 19.4), which converts a green quark from $A$ to a blue quark in $B$, back and forth in time (Figure 19.3).

### 19.6.5 Killing Form

1. Thus, $[A, \cdot]$ contains the (strong) structure constants corresponding to $A$ (with $B \equiv C_j$ at the $j$th column).
2. Next, let both $A$ and $B$ be fixed, specific, and more general: linear combinations of the above eight matrices.
3. This defines two $8 \times 8$ matrices: $[A, \cdot]$ and $[B, \cdot]$.
4. This way, $[B, \cdot]$ contains the structure constants corresponding to $B$.
5. Multiply $[A, \cdot]$ by $[B, \cdot]$, and look at the trace of this product:

$$\text{trace} ([A, \cdot] [B, \cdot]) = \sum_{i=1}^{8} ([A, \cdot] [B, \cdot])_{i, \bar{i}}$$

$$= \sum_{i,j=1}^{8} [A, \cdot]_{i, \bar{j}} [B, \cdot]_{j, \bar{i}}.$$

6. This is the Killing form.
7. What is the physical meaning of this? Hint: Figure 19.8 illustrates a strong interaction: a gluon in state $A$ is attracted to an anti-gluon in state $\bar{B}$. This is done virtually: converting gluon $j$ to $i$, and then going back-in-time, converting gluon $i$ back to $j$ in the past, and so on. This could be done for all indices $1 \le i, j \le 8$, leading to the complex amplitude

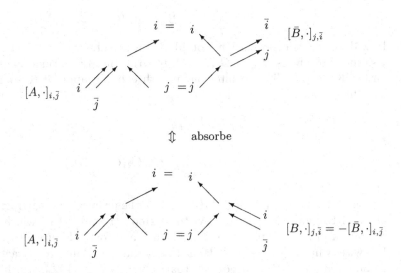

Fig. 19.8. Strong interaction: gluon $A$ is attracted to anti-gluon $\bar{B}$. Gluon $A$ is absorbed, converting gluon $j$ to $i$. Then, going back-in-time, gluon $i$ converts back to $j$, emitting anti-gluon $\bar{B}$ back in the past. Then, the process restarts symmetrically, from $\bar{B}$ back to $A$, and so on.

$$\text{trace}\left([A, \cdot]\,[B, \cdot]\right) = \sum_{i,j=1}^{8} [A, \cdot]_{i,\bar{j}} [B, \cdot]_{j,\bar{i}} ..$$

8. Look at this the other way around: the anti-gluon at state $\bar{B}$ is also attracted to the gluon in state $A$.
9. Is this as in classical physics? Hint: in classical mechanics, a force triggers a counter-force in the opposite direction, which picks a minus sign. This way, in a closed system, the total force is zero. (See exercises at the end of Chapter 8.)

# Part VIII

Appendix:
Applications in C++

# Appendix: Applications in C++

Set theory and mathematical logic form the theoretical basis for computer science. Indeed, in Chapters 7–8, we've already seen a fundamental mathematical structure: a binary tree, useful in recursive algorithms. In a binary tree, each node has two branches: either leftward or rightward. This is indeed the most elementary ingredient in set theory: a bit, which can take two possible values: either 0 or 1. This is indeed the basis for the digital computer. In a quantum computer, on the other hand, the little bit develops into a new random variable: a qubit, which could be either 0 (at some probability) or 1 (at some other probability, as in Chapters 14–18).

The binary tree has a more general version: an N-ary tree. In it, each node has not only two but N branches. In general relativity, for example, we often set N= 4. In Chapter 11, we used this to implement dynamic tensors, including their arithmetic operations.

Here, we do this in practice, in C++. The code is based on the framework in [45]. Still, it stands on its own. After all, it follows the original logic in Chapter 11. This is indeed the power of an object-oriented language like C++: it lets you design new mathematical structures, as in the original math.

In the final chapter, we also use C++ in numerical analysis. After all, in Chapter 3, we've already used Zron's lemma to prove the Han-Banach theorems: the basis for the finite-element method. We are thus ready to implement a mesh of finite elements, and assemble the stiffness-mass matrix on it. In particular, we can now linearize and assemble the Maxwell system.

To understand the C++ code in detail, one better look at [45] first. Still, one can also read the present code on its own, and get some idea about how it works. After all, the idea is quite simple: scan the finite elements one by one, and assemble the stiffness and mass matrices on them. The stiffness matrix discretizes the derivatives: the curl of the electric field $E$ (the unknown in the system). The mass matrix, on the other hand, is a bit trickier: it also contains a nonlinear part, which must be linearized time and again (at each and every Newton iteration) at the current approximation to $E$. Once the linear system is solved numerically, the approximate solution gets improved, ready for the next iteration.

# Dynamic Trees
# and Tensors
# in C++

Thanks to set theory, we already have a useful mathematical structure: a binary tree. In it, each node has two branches: leftward or rightward. At its bottom, however, the binary tree has a different kind of nodes: leaves, which have no branches at all.

The bottom level could be mirrored by a new geometrical structure: a hypercube. Each leaf in the original tree is mirrored by a corner in the hypercube. We've already seen this in a quantum computer, where the corners are indexed by 0 and 1.

To see this mirroring, use mathematical induction. Indeed, as the binary tree grows by one level, the number of leaves doubles. Likewise, as the hypercube grows by one dimension, the number of corners doubles. What is so good about the hypercube? Well, it is not only geometrical but also algebraic: a tensor.

This is true not only for a binary but also for an N-ary tree, where each node has not only two but N branches. This makes a bigger tensor: in each dimension, we have now more numbers: not only 0 and 1 but also 2, 3, 4, ..., N−1. In general relativity, for example, we often use N= 4, so each dimension is indexed by $0, 1, 2, 3$. This leads to spacetime and its metric (Chapter 11).

In this chapter, we use an N-ary tree to implement tensors, and their arithmetic operations. This makes the tensor highly dynamic: it may be as big as you like, with as many indices (or dimensions) as you like.

On the digital computer, this is done best in an object-oriented language like C++. This way, the implementation follows the original math: it is designed as in the original formulation. This leads to an efficient implementation of arithmetic operations between tensors: addition, multiplication, and contraction. Since our N-ary tree has no limit on the number of levels, our tensor has no limit on the number of indices (or dimensions).

## 20.1 Dynamic Tensor: N-ary Tree

### 20.1.1 Dynamic Tensor

In Chapter 11, we often use $4 \times 4$ matrices, indexed by two indices: $m, n = 0, 1, 2, 3$. Furthermore, we also use $4 \times 4 \times 4$ tensors, indexed by three indices: $\mu, \nu, \alpha = 0, 1, 2, 3$. Moreover, we even use bigger tensors that contain $4^4$ entries, indexed by four indices: $\rho, \mu, \sigma, \nu = 0, 1, 2, 3$. How to implement all these tensors in one go? Best implement

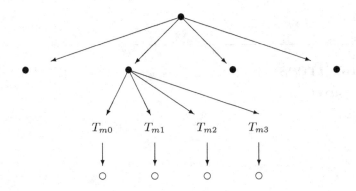

**Fig. 20.1.** A $4 \times 4$ tensor, implemented as a two-level tree. From each node, four branches issue to the next lower level. The leaves at the bottom, on the other hand, point to nothing: they only contain the individual entries $T_{mn}$ ($0 \leq m, n \leq 3$), which could be numbers, or even more complicated objects like polynomials.

dynamic tensors, with as many indices as you like. This way, the algebraic operators can be written once and for all, using recursion.

### 20.1.2 Full N-ary Tree

For this purpose, we need a full tree: each level should have exactly 'N' subtrees, where 'N' is still unspecified. (The quotation marks indicate that N will be a C++ term.) In general relativity, we'll often set 'N' = 4. After all, this is the dimension of spacetime. How many levels should the tree have? As many as the number of indices in the tensor. Indeed, each leaf will store one entry. Since the total number of leaves is a power of 4, they will be just enough to store all the entries in the tensor. In Figure 20.1, for instance, we use two levels to store a matrix: a tensor with two indices, containing 16 entries.

What is the individual entry? In an object-oriented language like C++, it could be denoted by 'T': an abstract object, which could be just anything. Later on, 'T' will be specified concretely: it could stand for a simple number, or even for a more complicated object, like polynomial. After all, we don't care what the entries are, so long as they have arithmetic operations, and can be added to each other, and multiplied by one another.

## 20.2 The Implementation

### 20.2.1 The Tensor Class

The tensor class contains two fields, to store data. The first one is called "entry": it will store the entry. Still, only the leaves will store a meaningful entry. The other nodes, on the other hand, will store no entry at all: their "entry" field will remain empty.

The second field, "next", will point to the subtensors. As discussed above, there will be 'N' subtensors. Each subtensor will be allocated memory later on:

```
template<class T, int N> class tensor{
  protected:
    T entry;
    tensor* next[N];
  public:
    tensor(const T&s=0, tensor* p=0):entry(s){
      for(int i=0; i<N; i++)
        next[i] = p;
    } // default constructor
```

## 20.3 Constructors and Destructor

### 20.3.1 Default Constructor

The above constructor takes two arguments: 's', to serve as the entry, and 'p', to point to a subtensor. If the constructor is called with no arguments at all, then it would use the default values: zero entry in the head, and no subtensors at all. This would make a very small "tree:" a dangling leaf, ready to be placed in a bigger tree later on.

If, on the other hand, the constructor is called with some concrete arguments 's' and 'p', then all subtensors would coincide, and duplicate each other. This would make little sense, and should be avoided.

### 20.3.2 Copy Constructor

The copy constructor, on the other hand, takes another kind of argument: an existing tensor 't':

```
tensor(const tensor&t):entry(t.entry){
  for(int i=0; i<N; i++)
    if(t.next[i])
```

If 't' is nontrivial, then it has subtensors, which should be copied recursively, one by one:

```
      next[i] = new tensor(*t.next[i]);
    else
      next[i] = 0;
} // copy constructor
```

### 20.3.3 Destructor

How to remove the subtrees, and release the memory they occupy for future use? In the following destructor, their addresses are deleted recursively:

```
~tensor(){
  for(int i=0; i<N; i++){
    delete next[i];
    next[i] = 0;
  }
} // destructor
```

## 20.4 Member Functions

### 20.4.1 Number of Levels

Recursion is also useful to count the total number of levels in the tree (or the total number of indices in the tensor):

```
int indices() const{
  return next[0] ? next[0]->indices() + 1 : 0;
} // total number of indices
```

### 20.4.2 Read an Entry

Recall that the entries are stored in the leaves only. The rest of the nodes, on the other hand, remain empty. How to read an entry?

```
const T& operator()() const{
  return entry;;
} // read an entry
```

Because of the word "const", the entry can only be read, but not changed.

### 20.4.3 Read a Subtensor

The subtensors, on the other hand, can be read in two ways: either by

```
const tensor* readNext(int i) const{
  return next[i];
} // read ith subtensor
```

(read only), or by

```
tensor* Next(int i){
  return next[i];
} // read/write ith subtensor
```

(read/write, because the word "const" is now missing). Finally, we also declare some more member functions, to be defined explicitly later:

```
const tensor& operator=(const tensor&);

const tensor& operator+=(const tensor&);

const tensor& operator*=(const T&);

const tensor operator*(const T&) const;

const tensor& operator*=(const tensor&);

void replace(int, int);

void contractFirst(int);

void contract(int, int);
};
```

## 20.5 Member Operators

### 20.5.1 Assignment Operator

This completes the body of the class. Next, let's define those functions that were only declared above. The assignment operator is defined recursively:

```
template<class T, int N>
const tensor<T,N>&
tensor<T,N>::operator=(const tensor<T,N>&t){
  if(this != &t){
    entry = t.entry;
    for(int i=0; i<N; i++)
      if(next[i])
        if(t.next[i])
```

Indeed, the subtensors are assigned recursively, one by one:

```
        *next[i] = *t.next[i];
      else{
        delete next[i];
        next[i] = 0;
      }
    else
      if(t.next[i])
```

If, however, the current tensor has no subtensors as yet, then they must be constructed from scratch by the copy constructor:

```
        next[i] = new tensor(*t.next[i]);
  }
  return *this;
} // assignment operator
```

### 20.5.2 Addition

Thanks to recursion, arithmetic operators can also be defined easily. Let's start from addition (Chapter 11, Section 11.15.1):

```
template<class T, int N>
const tensor<T,N>&
tensor<T,N>::operator+=(const tensor<T,N>&t){
```

Here, the argument 't' is added to the current tensor. For this, they must have the same number of indices. (Otherwise, issue an error message, which is left as an exercise.) For instance, if there are no indices at all (no subtrees), then the dangling leaf should be added:

```
if(!next[0])
   entry += t.entry;
```

If, on the other hand, there are a few subtensors, then add them recursively, one by one:

```
else
   for(int i=0; i<N; i++)
      *next[i] += *t.next[i];
return *this;
} //  add a tensor
```

## 20.6 Multiplication

### 20.6.1 Tensor Times Scalar

Recursion can also be used to multiply by a scalar 's':

```
template<class T, int N>
const tensor<T,N>&
tensor<T,N>::operator*=(const T&s){
   if(!next[0])
      entry *= s;
```

Indeed, multiply the subtensors recursively, one by one:

```
else
   for(int i=0; i<N; i++)
      *next[i] *= s;
return *this;
} //  multiply by a scalar
```

This way, the user can now write "Q *= q" to multiply the tensor 'Q' by the scalar 'q', and store the result back in 'Q'. Still, some users might want to write simply "Q * q", keeping 'Q' unchanged. To make them happy, let's write another operator:

```
template<class T, int N>
const tensor<T,N>
tensor<T,N>::operator*(const T&s) const{
  return tensor(*this) *= s;
} // tensor times scalar
```

### 20.6.2 Tensor Times Tensor

Thanks to this new function, we can now implement yet another operation — tensor-times-tensor (Chapter 11, Section 11.15.3):

```
template<class T, int N>
const tensor<T,N>&
tensor<T,N>::operator*=(const tensor<T,N>&t){
```

As before, there are two possibilities: if the current tensor is just a dangling leaf (no subtrees), then this entry should multiply 't':

```
if(!next[0])
  *this = t * entry;
```

Otherwise, scan the current subtensors, one by one:

```
else
  for(int i=0; i<N; i++)
```

and multiply each of them recursively by 't':

```
      *next[i] *= t;
  return *this;
} // multiply by another tensor
```

This way, in the innermost recursive call, all leaves will be multiplied by 't', as required. The result will be the tensor product: a new big tensor, containing all possible products of an entry from the original tensor times an entry from 't'.

## 20.7 Contraction

### 20.7.1 Replace a Tree by a Subtree

Look at the Riemann tensor $R^{\rho}_{\mu\sigma\nu}$. It uses four indices: $\rho, \mu, \sigma, \nu$. Let's contract $\rho$ with $\sigma$, to yield the Ricci tensor $R_{\mu\nu}$. For this purpose, we don't need all 16 instances $0 \le \rho, \sigma \le 3$, but only the four diagonal ones: $\sigma = \rho = 0, 1, 2, 3$. In other words, for our purpose, $\sigma$ is no longer an independent index: it depends on $\rho$ quite explicitly. Therefore, the original index $\sigma$ can drop, leaving only those $\sigma$'s that are the same as $\rho$.

In the Riemann tensor, $\sigma$ is the third index. This makes our task a bit difficult. Let's start from an easier task (Chapter 11, Section 11.16.1).

Consider a $4 \times 4$ tensor $T_{\mu\nu}$. Suppose that we need only four entries in it: say, $T_{0\nu}$ ($0 \le \nu \le 3$). This is the leftmost subtree in Figure 20.2.

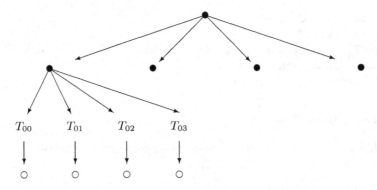

**Fig. 20.2.** In the "replace" function, the original $4 \times 4$ tensor $T_{\mu\nu}$, which contains 16 entries, is replaced by its leftmost subtree, which contains four entries only.

So, we actually don't need $\mu$ any more. We can safely drop it, to obtain a new smaller tensor, with one index only. To do this, we need to replace the original tree by its own (leftmost) subtree. This is done in the following member function:

```
template<class T, int N>
void
tensor<T,N>::replace(int pick, int index){
```

Here, there are two arguments: "index" tells us what index we want to drop. In the above example, $\mu$ is the first index in $T$, so "index" $= 1$. Furthermore, the argument "pick" tells us which son to pick. Since we only want to pick those entries with $\mu = 0$, we use "pick" $= 0$ (firstborn son).

So, in our example, we only need to keep the leftmost subtree. (Firstborn son replaces father.) The other three, on the other hand, may safely drop:

```
if(index==1){
  entry = next[pick]->entry;
  for(int i=0; i<N; i++)
    if(i != pick){
      delete next[i];
      next[i] = next[pick]->next[i];
    }
```

By now, the leftmost subtree is nearly copied. Indeed, its head has already been copied to the head of the current tree. (Individual son replaced individual father.) Furthermore, three subtrees of it have already been copied to the corresponding subtrees in the current tree. (Three grandsons replaced three uncles, respectively. This has been done by address, to avoid copying big 'T' objects.) There is just one more subtree to copy:

```
  next[pick]->entry.~T();
  next[pick] = next[pick]->next[pick];
}
```

(Grandson replaced his own father.) But what about our original task? Well, since $\sigma$ is the third index in the Riemann tensor, we'll have to call this function with "index" = 3, not 1. For this, we'll need recursion:

```
    else
      for(int i=0; i<N; i++)
        next[i]->replace(pick, index-1);
  }  // replace the current tensor by the "pick"th subtensor
```

### 20.7.2 Contract with the First Index

This will help complete our original task: in the Riemann tensor $R^{\rho}_{\mu\sigma\nu}$, contract the first index $\rho$ with the third index $\sigma$ (Chapter 11, Section 11.16.5):

```
    template<class T, int N>
    void
    tensor<T,N>::contractFirst(int index){
```

Indeed, this function contracts with the first index: $\rho$. To contract $\rho$ with $\sigma$, it will be called with "index" = 3. This way, $\sigma$ will drop, and will take the dependent value $\sigma = \rho =$ 'i':

```
    for(int i=0; i<N; i++)
      next[i]->replace(i, index-1);
```

To contract with $\rho$, just sum the subtensors, and store the sum in the leftmost subtree:

```
    for(int i=1; i<N; i++)
      *next[0] += *next[i];
```

Finally, put this in the head:

```
    replace(0, 1);
  }  // contract with the first index
```

### 20.7.3 Contract with Any Index

Our original task is now complete: contracting with the first index: $\rho$. Still, what about a more difficult task, say, contracting the second and fourth indices with each other? For this purpose, we need to call the following function, with "index1" = 2 and "index2" = 4:

```
    template<class T, int N>
    void
    tensor<T,N>::contract(int index1, int index2){
      if(index1 == 1)
```

If "index1" = 1, then this is a contraction with the first index, which we already know how to do:

```
        contractFirst(index2);
    else
```

If, on the other hand, "index1" > 1, then use recursion:

```
        for(int i=0; i<N; i++)
            next[i]->contract(index1-1, index2-1);
    } // contract
```

### 20.7.4 Print the Tensor

Finally, let's define an ordinary (nonmember) function to print the entries onto the screen:

```
    template<class T, int N>
    void print(const tensor<T,N>& t){
```

To increase efficiency, the function takes the tensor 't' not by name but by a (constant) reference. Now, distinguish between two cases: if 't' is just a leaf, then print the entry:

```
        if(!t.readNext(0))
            print(t());
```

Otherwise, use recursion:

```
        else
            for(int i=0; i<N; i++)
                print(*t.readNext(i));
    } // print the leaves
```

# 21

# Nonlinear
# Maxwell Solver
# in C++

Finally, we also use C++ in numerical analysis: to linearize and assemble the Maxwell system on a mesh of finite elements. The logic is quite simple: scan the finite elements one by one, and assemble the stiffness and mass matrices on them, using nested loops. The stiffness matrix discretizes the derivatives: the curl of the electric field $E$ (the unknown in the system). The mass matrix, on the other hand, is a bit trickier: it also contains a nonlinear part, which must be linearized time and again, at the current approximation to $E$. Once the linear system is solved numerically, the approximate solution gets improved, ready for the next iteration.

The code is based on Chapter 29 in [45]. Still, it improves on it in two aspects: first, it discretizes the curl directly. This way, one can easily introduce variable coefficients, as in dialectric materials. Furthermore, the nonlinear term is linearized not only in 1-D but also in 3-D. This is done in Newton's iteration.

## 21.1 Maxwell's Equations

### 21.1.1 The Maxwell System

In this chapter, we use the same logic to discretize the Maxwell system on a mesh of (linear) finite elements. How does the Maxwell system look like? This is an integral equation, defined in a three-dimensional domain. The left-hand side is integrated in the entire domain, whereas the right-hand side is integrated on part of its boundary, say, its left edge:

$$\int\int\int \left( (\nabla \times E)^t \left( \nabla \times \tilde{E} \right) + K E^t \tilde{E} + K_1 \|E\|^2 E^t \tilde{E} \right) dxdydz$$

$$= \sqrt{K} \int\int (\mathbf{n} \times B)^t \tilde{E} dxdy,$$

where $K$ and $K_1$ are given negative parameters, $E$ is the unknown electric field, $\tilde{E}$ is a test vector function, $B$ is the unknown magnetic field, and $\mathbf{n}$ is the outer normal vector at the boundary.

### 21.1.2 The Finite-Element Mesh

The domain is then approximated by a mesh of tetrahedra. This way, $E$ can be approximated as a linear combination of (linear) nodal basis functions $\psi_J$, with unknown (complex) coefficients $c_J$, where $J$ indexes the mesh nodes:

$$E \doteq \begin{pmatrix} \sum_J c_J^{(1)} \psi_J \\ \sum_J c_J^{(2)} \psi_J \\ \sum_J c_J^{(3)} \psi_J \end{pmatrix}.$$

The test functions $\tilde{E}$ take a similar form. In other words, they could be just any nodal basis function:

$$\tilde{E} = \begin{pmatrix} \psi_I \\ 0 \\ 0 \end{pmatrix} \quad \text{or} \quad \begin{pmatrix} 0 \\ \psi_I \\ 0 \end{pmatrix} \quad \text{or} \quad \begin{pmatrix} 0 \\ 0 \\ \psi_I \end{pmatrix},$$

where $I$ indexes the mesh nodes. How to linearize and assemble this into a discrete system, ready to solve numerically on the digital computer?

### 21.1.3 The Stiffness and Mass Matrices

How to discretize the Maxwell system? For this purpose, we need to assemble the coefficient matrix, which contains two parts: the stiffness and mass matrices.

The stiffness matrix discretizes the curl: $\nabla \times E$ (where $E$ is the unknown electric field). The mass matrix, on the other hand, contains two terms: the linear term, with the coefficient $K$ (stored in the external parameter "HELM"), and the nonlinear term, with the coefficient $K_1$ (stored in "HELMNonlin"). In our numerical examples, we often pick $K = K_1 = -20$.

## 21.2 How to Linearize?

### 21.2.1 Linearization

Because of this nonlinear term, the entire system takes the form:

$$A(\tilde{x}) = \tilde{f},$$

where $A()$ is the original nonlinear mapping that maps the (unknown) solution $\tilde{x}$ to the given right-hand side $\tilde{f}$. This will produce the new (linearized) system

$$Ae = f,$$

where $x$ is the current approximation to $\tilde{x}$, $f$ is its residual

$$f \equiv A(x) - \tilde{f},$$

$e$ is the (unknown) approximate error

$$e \doteq x - \tilde{x},$$

and $A$ is the Jacobian

$$A \equiv A'(x)$$

at $x$. The (sparse) coefficient matrix $A$ will be constructed in the constructor: a new C++ function, which will take three arguments: the mesh 'm', and two vectors (or grid functions), defined on the mesh nodes: 'x' (to store the current approximation to $\tilde{x}$, which approximates the electric field $E$), and 'f', to store its residual:

$$f = A(x) - \tilde{f}.$$

(The quotation marks tell us that $m$, $x$, and $f$ are placed in a C++ code.)

### 21.2.2 Newton's Iteration

This makes one Newton iteration: solving for $e$, and using it to update and improve $x$. In each Newton iteration, $A$ is different: it must be updated beforehand. After all, the linearization is carried out at the new (improved) $x$. In other words, $A'(x)$ must be recalculated (at the new $x$) before being used at the present iteration. Once the linear system is solved for the error $e$, $x$ can get improved, ready for the next iteration:

$$x \leftarrow x - e.$$

Thus, the constructor will be called time and again. After all, in each Newton iteration, $A$ must be recalculated. More precisely, only the nonlinear term must be linearized again at the new $x$, and assembled again on the finite elements. On top of this, there is a little more work: the new residual $f$ must also be recalculated at the up-to-date $x$, and reassembled on the mesh. This way, everything gets ready to solve the linear system, and complete one iteration.

## 21.3 The Implementation

### 21.3.1 The Matrix Elements

In the coefficient matrix $A$, the elements could actually be not only scalars but also more complicated. This is why they are declared as abstract 'T' objects, which could be just anything. Later on, 'T' should be specified as a "matrix2": a little $2 \times 2$ matrix, ready to store a complex number. This will help linearize the nonlinear term in real arithmetic: since the real and imaginary parts are stored separately, they are ready to differentiate, to produce $A \equiv A'(x)$ in each Newton iteration.

### 21.3.2 The Initial Guess

In summary, how to solve for $\tilde{x}$? Start from an initial guess 'x', with the initial residual

$$'f' = A('x') - \tilde{f}.$$

For example, 'x' could be picked as the zero vector. Most often, $A(\mathbf{0}) = \mathbf{0}$, so 'f' is initially $-\tilde{f}$.

### 21.3.3 Vectors and Their Components

Now, in Newton's iteration, we improve on 'x' time and again:

$$x \leftarrow x - e,$$

until it approximates $\tilde{x}$ well. Here, both 'x' and 'f' are grid functions, defined on the mesh nodes. On each node, 'x' actually contains three scalars, to form the electric field $E$ at this point. Each scalar is by itself a complex number, stored as a pair: the real part, followed by the imaginary part. Later on, this will help differentiate in real arithmetic.

### 21.3.4 The Mesh Nodes

We use linear finite elements: each tetrahedron contains four degrees of freedom, at its four corners. In the mesh 'm', the total number of nodes will be stored in the parameter "nodes" (to be specified later). Since the electric field contains three spatial components, $A$ has a $3 \times 3$ block form. Each block is of order 2 times "nodes" (to store the real and imaginary parts of each complex number).

## 21.4 The Constructor of the Matrix

### 21.4.1 The Constructor

To produce the matrix $A$, the constructor takes three arguments: the mesh 'm', and the vectors 'x' and 'f':

```
template<class T>
sparseMatrix<T>::sparseMatrix(mesh<tetrahedron>&m,
    const runVector<point>&x,
    runVector<point>&f){
  int nodes = m.indexing();
```

(See [45] for more details.) The boundary conditions will be of Neumann-mixed type. The vector "xBoundary" will tell us where mixed conditions are imposed. For this purpose, "xBoundary" will be nonzero only at the left edge of the domain.

```
runVector<int> xBoundary(nodes,0);
```

### 21.4.2 Initialization to Zero

In the beginning, the matrix contains no rows at all:

```
this->item = new row<T>*[this->number = 3 * nodes];
for(int i=0; i<this->number; i++)
  this->item[i] = 0;
```

Later on, we'll also calculate the residual 'f' = $A('x') - \tilde{f}$. For this purpose, 'f' is initialized to be the zero (complex) vector:

```
f = point(0.,0.);
```

# 21.5 The Unit Tetrahedron

## 21.5.1 Nodal Basis Functions

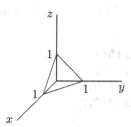

**Fig. 21.1.** The unit tetrahedron $T$.

The nodal basis functions $p_0$, $p_1$, $p_2$, and $p_3$ are now defined as polynomials of three independent variables in the unit tetrahedron (Figure 21.1). Each basis function is linear, and has the value 1 at just one corner, and zero at the other three. To define them, we need a few polynomials of one variable (with a low-case letter) or two variables (with a capital letter):

```
polynomial<double> zero(1,0.);
polynomial<polynomial<double> > Zero(1,zero);
polynomial<polynomial<polynomial<double> > >
    ZZero(1,Zero);
polynomial<double> one(1,1.);
polynomial<polynomial<double> > One(1,one);
polynomial<double> minus1(1,-1.);
polynomial<polynomial<double> > Minus1(1,minus1);
polynomial<double> oneMinusx(1.,-1.);
polynomial<polynomial<double> >
    oneMinusxMinusy(oneMinusx,minus1);
polynomial<polynomial<double> > yy(zero,one);
polynomial<double> x1(0.,1.);
polynomial<polynomial<double> > xx(1,x1);
list<polynomial<polynomial<polynomial<double> > > >
    P(4,ZZero);
P(0) =
    polynomial<polynomial<polynomial<double> > >
    (oneMinusxMinusy,Minus1);
P(1) =
    polynomial<polynomial<polynomial<double> > >
    (1,xx);
P(2) =
    polynomial<polynomial<polynomial<double> > >
    (1,yy);
P(3) =
```

```
polynomial<polynomial<polynomial<double> > >
(Zero, One);
```

## 21.5.2 Nodal Basis Functions and Their Gradients

The constant gradients $\nabla p_0$, $\nabla p_1$, $\nabla p_2$, and $\nabla p_3$ are also stored, for further use:

```
point3 gradient[4];
gradient[0] = point3(-1,-1,-1);
gradient[1] = point3(1,0,0);
gradient[2] = point3(0,1,0);
gradient[3] = point3(0,0,1);
```

## 21.6 Scanning the Finite Elements

### 21.6.1 Scanning the Tetrahedra

We are now ready to scan the tetrahedra in our mesh 'm'. Each tetrahedron will be denoted by $t$:

```
for(const mesh<tetrahedron>* runner = &m;
    runner; runner =
    (const mesh<tetrahedron>*)runner->readNext()){
```

What to do in this loop? To see this, we need to see how an individual tetrahedron $t$ looks like.

### 21.6.2 Mapping to a General Tetrahedron $t$

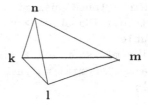

**Fig. 21.2.** A general tetrahedron $t$, vertexed at **k**, **l**, **m**, and **n**.

Look at $t$: an individual tetrahedron in the mesh (Figure 21.2). How to map the unit tetrahedron $T$ onto it? This is done by the affine mapping

$$E_t : T \to t.$$

(Don't confuse $E_t$ with the electric field $E$.) Its Jacobian, $S_t$, is now stored in 'S':

```
matrix3 S((*runner)()[1]() - (*runner)()[0](),
          (*runner)()[2]() - (*runner)()[0](),
          (*runner)()[3]() - (*runner)()[0]());
matrix3 Sinverse = inverse(S);
matrix3 SinverseT = transpose(Sinverse);
```

Since $|\det(S_t)|$ will be used often later, it makes sense to calculate it here once and for all:

```
double detS = fabs(det(S));
```

Furthermore, we also define a few polynomials of three variables, and initialize them to zero:

```
list<polynomial<polynomial<polynomial<double> > > >
    X(3,ZZero);
list<polynomial<polynomial<polynomial<double> > > >
    Y(3,ZZero);
polynomial<polynomial<polynomial<double> > >
    E2(1,Zero);
```

"E2" is now ready to store $\|E\|^2$ in $t$. This will be useful in the nonlinear term.

## 21.7 Scanning Blocks and Corners

### 21.7.1 Scanning the Blocks

Since the matrix has $3 \times 3$ blocks, we can scan the blocks one by one, using the new index "iblock" $= 0, 1, 2$:

```
for(int iblock=0;iblock<3;iblock++){
```

### 21.7.2 Scanning the Corners

Inside this loop, we also scan the corners in $t$, corner by corner. In this inner loop, 'i' is the local index of the corner, whereas 'I' is its global index in the entire mesh:

```
for(int i=0; i<4; i++){
    int I = (*runner)()[i].getIndex();
```

### 21.7.3 The Electric Field $E$

To get ready for real arithmetic in $t$, the electric field $E$ should better be written explicitly, in terms of its real and imaginary parts:

$$E_1 = \left( \text{``X[0]''} + \sqrt{-1}\,\text{``Y[0]''} \right) \circ E_t^{-1}$$
$$E_2 = \left( \text{``X[1]''} + \sqrt{-1}\,\text{``Y[1]''} \right) \circ E_t^{-1}$$
$$E_3 = \left( \text{``X[2]''} + \sqrt{-1}\,\text{``Y[2]''} \right) \circ E_t^{-1}.$$

(Clearly, this is not yet the exact electric field, but only the best approximation we currently have.) For this purpose, we must sum up four polynomials of three variables: four degrees of freedom from 'x' (at four corners in $t$) times the relevant nodal basis function in 'P':

```
X(iblock) += x[I+iblock*nodes][0] * P[i];
Y(iblock) += x[I+iblock*nodes][1] * P[i];
```

In this form, $E$ is ready for real arithmetic later on.

### 21.7.4 Mixed Boundary Conditions

Moreover, let's go ahead and define "xBoundary" at the four corners of $t$: 1 at those corners that lie on the left edge of the domain (where mixed boundary condition should be imposed), and 0 elsewhere:

```
if((*runner)()[i]()[0] <= 1.e-6)
    xBoundary(I) = 1;
}
```

In this form, "xBoundary" will be used later to assemble the boundary term. This completes the inner loop over the corners of $t$.

### 21.7.5 The Nonlinear Term

Next, in $t$, let's go ahead and calculate the quadratic polynomial

$$\text{``E2''} = \|E\|^2 = |E_1|^2 + |E_2|^2 + |E_3|^2 :$$

```
E2 += X[iblock] * X[iblock];
E2 += Y[iblock] * Y[iblock];
}
```

This will be useful later. This completes the loop over "iblock" $= 0, 1, 2$. This index is now free for further use.

## 21.8 Curl and Gradient

### 21.8.1 Scanning the Blocks Again

Now, how to construct the individual rows in $A$? Recall that $A$ has $3 \times 3$ blocks. Let's scan them once again, using the same index "iblock" $= 0, 1, 2$:

```
for(int iblock=0;iblock<3;iblock++){
```

In this loop, we'll often need the new parameter $K_1 |\det(S_t)|$:

```
double helmNonlin = HELMNonlin * detS;
```

### 21.8.2 Scanning the Corners Again

In this loop, let's scan the corners in $t$ once again, corner by corner, as before:

```
for(int i=0; i<4; i++){
    int I = (*runner)()[i].getIndex();
```

Recall that 'I' is the global index of the present corner in the entire mesh. This way, at the 'I'th node in 'm', we can now consider a nodal basis function:

$$\psi_I = p_i \circ E_t^{-1}.$$

This is a composite function in $t$. In fact, $E_t^{-1}$ maps from $t$ back to the unit tetra-hedron, where $p_t$ is indeed a nodal basis function. This way, $\psi_I$ is indeed a nodal basis function in $t$: it is 1 at the present corner, and 0 at the other three.

### 21.8.3 The Gradient

Thanks to the chain rule, $\psi_I$ has a constant gradient in $t$:

$$\nabla \psi_I = \nabla \left( p_i \circ E_t^{-1} \right) = S_t^{-t} \nabla p_i \circ E_t^{-1}.$$

Let's go ahead and store it for further use:

```
point3 gradi = SinverseT * gradient[i];
```

### 21.8.4 The Curl

The electric field $E$ contains three scalar components. Let's focus on one component, say the first one, indexed by "iblock" $= 0$. To approximate it, define the new vector function

$$\psi_I^{(1)} = \begin{pmatrix} \psi_I \\ 0 \\ 0 \end{pmatrix}.$$

Its curl is

$$\nabla \times \psi_I^{(1)} = \nabla \times \begin{pmatrix} \psi_I \\ 0 \\ 0 \end{pmatrix} = \begin{pmatrix} 0 \\ (\psi_I)_z \\ -(\psi_I)_y \end{pmatrix}.$$

Fortunately, this curl is constant in $t$. Let's go ahead and store it for future use:

```
point3 curli = iblock == 0 ?
  point3(0.,gradi[2],-gradi[1])
:
  iblock == 1 ?
    point3(-gradi[2],0.,gradi[0])
:
  point3(gradi[1],-gradi[0],0.);
```

This is how the curl looks like in vacuum: a constant. In dialectric materials, on the other hand, the curl could vary. In this case, "curli" should also be multiplied by the inverse permeability matrix $\mu^{-1}$ from the left. For simplicity, however, we stick to vacuum, where $\mu$ is the identity matrix, as before.

## 21.9 Assembling the Residual

### 21.9.1 The Nonlinear Term

Now, let's assemble the contribution from the nonlinear term to 'f'. For this purpose, we need to integrate $K_1\|E\|^2 E\psi_I$ in $t$. Fortunately, this integral could be calculated in the unit tetrahedron $T$:

```
f(I+iblock*nodes) += helmNonlin *
    point(integral(E2 * X[iblock] * P[i]),
        integral(E2 * Y[iblock] * P[i]));
```

(because "helmNonlin" $= K_1|\det(S_t)|$, where $S_t$ is the Jacobian of $E_t : T \to t$).

## 21.10  The Matrix Element

### 21.10.1  Inner Loop on the Corners

We are now ready to assemble the contributions from $t$ to the ('I','J')th element in the present block in $A$. For this purpose, we need an inner loop over the corners in $t$:

```
for(int j=0; j<4; j++){
    int J = (*runner)()[j].getIndex();
    polynomial<polynomial<polynomial<double> > >
        Pij = P[i] * P[j];
```

This polynomial will be useful later.

### 21.10.2  The Linear Mass Term

Let's start with the linear mass term. Its coefficient is

```
double helm = HELM * detS / 120.;
if(i==j)
    helm *= 2.;
```

(where "HELM" $= K$). The coefficient $1/120$ follows from the integral over the tetrahedron:

$$\int\int\int_t (p_i p_j)\circ E_t^{-1}dxdydz = |\det(S_t)| \int\int\int_T p_i p_j dxdydz = \begin{cases} \frac{|\det(S_t)|}{60} & \text{if} \quad i = j \\ \frac{|\det(S_t)|}{120} & \text{if} \quad i \neq j. \end{cases}$$

### 21.10.3  Inner Loop on the Blocks

In general, there may be a contribution to all $3 \times 3$ blocks. For this purpose, we must loop not only over "iblock" $= 0, 1, 2$ but also over "jblock" $= 0, 1, 2$:

```
for(int jblock=0; jblock<3; jblock++){
```

In vacuum, on the other hand, things are simpler: the linear mass term contributes to the (1,1)th, (2,2)th, and (3,3)th blocks only:

```
double helmDiag = iblock == jblock ?
   helm
:
   0.;
```

The linear mass term is now ready to assemble. Let's leave it until later.

### 21.10.4  The Stiffness Term

Next, let's calculate the stiffness term as well. As we did for $\psi_I$, let's go ahead and calculate the gradient and curl of $\psi_J$ as well:

```
point3 gradj = SinverseT * gradient[j];
point3 curlj = jblock == 0 ?
   point3(0.,gradj[2],-gradj[1])
:
   jblock == 1 ?
     point3(-gradj[2],0.,gradj[0])
:
     point3(gradj[1],-gradj[0],0.);
```

By now, we already have two curls in $t$:

$$\text{``curli''} = \nabla \times \psi_I$$
$$\text{``curlj''} = \nabla \times \psi_J.$$

## 21.11  Assemble the Linear Terms

### 21.11.1  Assemble the Curl and the Linear Mass Term

Fortunately, these curls are constant vectors in $t$. We can now calculate their inner product, add the linear mass term (stored in "helmDiag" above), and integrate in $t$ (which is actually calculated in $T$):

```
T addIJ = helmDiag +
     detS / 6. * (curlj * curli);
```

(because $T$ has volume 1/6). This way, the linear contributions are stored in "addIJ". Later on, when 'T' is specified as "matrix2", "addIJ" should become a multiple of the identity matrix of order 2. This is now added to the relevant matrix element, using the "assemble" function (to be written later):

```
assemble(I+iblock*nodes, J+jblock*nodes, addIJ);
```

This is the linear part of $A$. There is no need to recalculate it in each and every Newton iteration, because it remains the same.

### 21.11.2  Assemble the Residual

Moreover, "addIJ" contributes to 'f' as well:

```
f(I+iblock*nodes) +=
    addIJ * x[J+jblock*nodes];
```

This assembles the linear stiffness-mass terms, as required.

## 21.12  Linearization

### 21.12.1  The Linearized Term

Now, let's linearize and assemble the nonlinear mass term as well. For this purpose, we need to differentiate with respect to the coefficient $c_J$ of the nodal basis function $\psi_J$. For instance, consider the (1,1)th block, indexed by "iblock" = 0 and "jblock" = 0. In this block, we need the partial derivative

$$\frac{\partial \left( \|E\|^2 E_1 \right)}{\partial c_J^{(1)}}.$$

### 21.12.2  Differentiation in Real Arithmetic

Unfortunately, in complex arithmetic, this is not well-defined: $\|E\|^2$ is not differentiable! Fortunately, in real arithmetic, it is. Indeed, let's write the real and imaginary parts explicitly:

$$c_J^{(1)} \equiv \left( \Re c_J^{(1)}, \Im c_J^{(1)} \right)^t \quad \text{and} \quad E_1 \equiv (\Re E_1, \Im E_1)^t.$$

In this form, we have the $2 \times 2$ Jacobian

$$\frac{\partial \left( \|E\|^2 \Re E_1, \|E\|^2 \Im E_1 \right)}{\partial \left( \Re c_J^{(1)}, \Im c_J^{(1)} \right)}$$

$$= \begin{pmatrix} \Re E_1 \\ \Im E_1 \end{pmatrix} \left( \frac{\partial \left( \|E\|^2 \right)}{\partial \Re c_J^{(1)}}, \frac{\partial \left( \|E\|^2 \right)}{\partial \Im c_J^{(1)}} \right) + \|E\|^2 \frac{\partial \left( \Re E_1, \Im E_1 \right)}{\partial \left( \Re c_J^{(1)}, \Im c_J^{(1)} \right)}$$

$$= E_1 \left( \frac{\partial \left( \|E\|^2 \right)}{\partial \Re E_1}, \frac{\partial \left( \|E\|^2 \right)}{\partial \Im E_1} \right) \frac{\partial \left( \Re E_1, \Im E_1 \right)}{\partial \left( \Re c_J^{(1)}, \Im c_J^{(1)} \right)} + \|E\|^2 \frac{\partial \left( \Re E_1, \Im E_1 \right)}{\partial \left( \Re c_J^{(1)}, \Im c_J^{(1)} \right)}$$

$$= E_1 \left( 2\Re E_1, 2\Im E_1 \right) \begin{pmatrix} \psi_J & 0 \\ 0 & \psi_J \end{pmatrix} + \|E\|^2 \begin{pmatrix} \psi_J & 0 \\ 0 & \psi_J \end{pmatrix}$$

$$= \left( 2 E_1 E_1^t + \|E\|^2 \begin{pmatrix} 1 & 0 \\ 0 & 1 \end{pmatrix} \right) p_j \circ E_t^{-1}.$$

To assemble into the $(1,1)$th block, multiply this by $\psi_I = p_i \circ E_t^{-1}$, and integrate in $t$ (which can actually be done in $T$). Let's start with the latter term, which contains $\|E\|^2$. As a matter of fact, such a term should assemble not only in the $(1,1)$th but also in the $(2,2)$th and $(3,3)$th blocks:

```
if(iblock==jblock){
    point J1 = point(integral(E2 * Pij),0.);
    point J2 = point(0.,J1[0]);
    matrix2 Jacobian = matrix2(J1,J2);
    assemble(I+iblock*nodes,J+jblock*nodes,
        helmNonlin * Jacobian);
}
```

The former term, on the other hand, is trickier: it is relevant in the off-diagonal blocks as well. For instance, in the $(1,2)$th block,

$$
\frac{\partial\left(\|E\|^2\Re E_1, \|E\|^2\Im E_1\right)}{\partial\left(\Re c_J^{(2)}, \Im c_J^{(2)}\right)}
$$

$$
= \begin{pmatrix}\Re E_1 \\ \Im E_1\end{pmatrix}\left(\frac{\partial\left(\|E\|^2\right)}{\partial\Re c_J^{(2)}}, \frac{\partial\left(\|E\|^2\right)}{\partial\Im c_J^{(2)}}\right) + \|E\|^2\frac{\partial\left(\Re E_1, \Im E_1\right)}{\partial\left(\Re c_J^{(2)}, \Im c_J^{(2)}\right)}
$$

$$
= E_1\left(\frac{\partial\left(\|E\|^2\right)}{\partial\Re E_2}, \frac{\partial\left(\|E\|^2\right)}{\partial\Im E_2}\right)\frac{\partial\left(\Re E_2, \Im E_2\right)}{\partial\left(\Re c_J^{(2)}, \Im c_J^{(2)}\right)} + (0)
$$

$$
= E_1\left(2\Re E_2, 2\Im E_2\right)\begin{pmatrix}\psi_J & 0 \\ 0 & \psi_J\end{pmatrix}
$$

$$
= 2\left(E_1 E_2^t\right)p_j \circ E_t^{-1}.
$$

To assemble into the ('I','J')th element in the $(1,2)$th block, multiply by $\psi_I = p_i \circ E_t^{-1}$, and integrate in $t$ (which is actually done in $T$). In general, such a term should assemble not only in the $(1,2)$th but also in the ("iblock","jblock")th block:

```
point J1(integral(X[iblock] * X[jblock] * Pij),
    integral(X[iblock] * Y[jblock] * Pij));
point J2(J1[1], integral(Y[iblock] * Y[jblock] * Pij));
matrix2 Jacobian(J1,J2);
Jacobian *= 2.;
assemble(I+iblock*nodes,J+jblock*nodes,
    helmNonlin * Jacobian);
}
```

(To avoid duplicate calculations, one might want to modify this code a little, and store products like "X[iblock]" times "X[jblock]" (or "Y[jblock]") earlier, right after 'X' and 'Y' were calculated in $t$.) This closes the inner loop over "jblock" $= 0, 1, 2$. Instead, we'll soon open a yet inner loop over the corners.

## 21.13  Assemble the Boundary Term

### 21.13.1  Mixed Boundary Conditions

On the left edge of the domain, we also have a mixed boundary condition:

$$\mathbf{n} \times B = E \quad \text{plus a given free vector function}$$

(where $B$ is the unknown magnetic field, and $\mathbf{n}$ is the outer normal vector, pointing from the left edge leftwards). On the right-hand side, we have two terms. The former will be thrown to the left-hand side of Maxwell's system, and contribute to $A$. On its way, it will pick a minus sign. The latter, on the other hand, will remain on the right-hand side, and contribute to $\tilde{f}$.

How to assemble these terms? For this purpose, we must first detect boundary triangles, vertexed at boundary points: 'I', 'J', and 'K' that lie on the left edge of the domain.

### 21.13.2  Yet Inner Loop on the Corners

For this purpose, we need a yet inner loop over the corners:

```
for(int k=0; k<4; k++){
    int K = (*runner)()[k].getIndex();
    if((i!=j)&&(j!=k)&&(k!=i)
        &&(xBoundary[I] *
        xBoundary[J] * xBoundary[K]==1)){
```

This is a boundary triangle, with three distinct nodal basis functions, vertexed at 'i' $\neq$ 'j' $\neq$ 'k'. Its area is half the norm of the vector product of the edges:

```
point3 jMinusi = (*runner)()[j]()
    - (*runner)()[i]();
point3 kMinusi = (*runner)()[k]()
    - (*runner)()[i]();
double radiation = sqrt(fabs(HELM))
    * l2norm(jMinusi & kMinusi) / 24.;
```

(where "HELM" $= K$, and '&' stands for '$\times$': vector product in 3-D). The factor $1/24$ follows from the integral over the unit triangle:

$$\int_0^1 \int_0^{1-x} p_i p_j \, dy \, dx = \begin{cases} 1/12 & \text{if} \quad i = j \\ 1/24 & \text{if} \quad i \neq j. \end{cases}$$

### 21.13.3  Contribute to Matrix Elements

The mixed boundary term is imaginary: since $K < 0$, it has the coefficient $\sqrt{-1}\sqrt{|K|}$. This is why it is placed in the off-diagonal elements of a little $2 \times 2$ matrix, representing a complex number in real arithmetic. Furthermore, as discussed in Section 21.13.1, it also picks a minus sign:

```
T radiationMatrix(
    point(0.,radiation),
    point(-radiation,0.));
```

This is now ready to contribute to the ('I','I')th and ('I','J')th elements in the (1,1)th, (2,2)th, and (3,3)th blocks:

```
assemble(I+iblock*nodes, I+iblock*nodes,
    radiationMatrix);
assemble(I+iblock*nodes, J+iblock*nodes,
    radiationMatrix);
```

This way, thanks to 'j'↔'k', the main-diagonal element gets a contribution twice as big. This is as required: it has coefficient 1/12, not just 1/24. (To see this, look at the ('I','I')th element in each block. It gets the same contribution twice from this triangle.)

### 21.13.4  Contribute to the Residual

Furthermore, the same boundary term contributes to 'f' as well:

```
f(I+iblock*nodes) +=
    radiationMatrix
    * x[I+iblock*nodes];
f(I+iblock*nodes) +=
    radiationMatrix
    * x[J+iblock*nodes];
```

### 21.13.5  Assemble the Free Function

On its right-hand side, the boundary condition in Section 21.13.1 may also contain a free vector function. (Say, the constant vector function $(1,1,1)^t$ on the left edge of the domain.) In $\tilde{f}$, this should be integrated over the above boundary triangle. In the residual 'f' $= A(\text{'x'}) - \tilde{f}$, on the other hand, this picks a minus sign:

```
f(I+iblock*nodes) -=
    point(0.,2. * radiation);
}
```

As discussed above, this is imaginary. This is why it is placed in the second component in real arithmetic. Why factor 2? Because we already divided by 24 above. As discussed above, at corner 'i' in the boundary triangle, we have yet another factor 2 (thanks to 'j'↔'k'). This leads to

$$2 \cdot 2 \cdot \frac{1}{24} = \frac{1}{6},$$

in agreement with the integral over the unit triangle:

$$\int_0^1 \int_0^{1-x} p_i \, dy \, dx = \frac{1}{6}.$$

It is now time to close:

```
        }
```

This closes the innermost loop over 'k' = 0, 1, 2, 3. This assembles the boundary term as well. The stiffness-mass matrix $A$ is now complete. It is now time to close the loops over 'j', 'i', "iblock", and "runner":

```
        }
      }
     }
    }
  } // linearize and assemble Maxwell system
```

### 21.13.6 The "Assemble" Function

In the above, we often called the "assemble" function to add a contribution to the ('I','J')th matrix element. Like the above constructor, this function should also be written in the sparse-matrix class in [45]:

```
void assemble(int I, int J, const T& contribution){
```

There are no blocks here: we want to add to the 'I'th row, at the 'J'th column in our matrix. Now, there are two possibilities. If the 'I'th row already exists, then the new contribution is placed in a new little row, and added:

```
    if(this->item[I]){
      row<T> r(contribution,J);
      *this->item[I] += r;
    }
```

If, on the other hand, the 'I'th row doesn't exist as yet, then it must be allocated dynamic memory, and constructed from scratch:

```
    else
      this->item[I] = new row<T>(contribution,J);
  } // assemble to the (I,J)th matrix element
```

# References

1. Adamowicz, Z. and Zbierski, P.: *Logic of Mathematics: A Modern Course of Classical Logic* (third edition). John Wiley and Sons, N.Y., 1997.
2. Aharoni, R.: *Mathematics, Poetry, and Beauty*. World Scientific Publishing Company, 2015.
3. Aharoni, R.: *Lecture Notes on Set Theory*. Technion–Israel Institute of Technology.
4. Alcubierre, M.: *Introduction to 3+1 Numerical Relativity*. Oxford Univ. Press (2012).
5. Baumgarte, T.W., and Shapiro, S.L.: *Numerical Relativity: Solving Einstein Equations on the Computer*. Cambridge Univ. Press (2010).
6. Bransden, B.H., and Joachain, C.J.: *Quantum Mechanics*. Prentice-Hall, 2000.
7. Cameron, P.J.: *Sets, Logic and Categories*. Springer, N.Y., 1999.
8. Carlson, B.: *Lecture Notes on Quantum Mechanics*, 2019.
9. Carroll, S.: *Something Deeply Hidden: Quantum Worlds and the Emergence of Spacetime*. Oneworld Publications (2019).
10. Cenzer, D., Larson, J., Porter, C., and Zapletal, J.: *Set Theory and Foundations of Mathematics: An Introduction to Mathematical Logic*. world Scientific, Singapore, 2020.
11. Chirvasa, M.: *Finite Difference Methods in Numerical Relativity*. VDM Verlag (2010).
12. Courant, R. and John, F.: *Introduction to Calculus and Analysis* (vol. 1–2). Springer, N.Y., 1998–1999.
13. Drucker, T.: *Perspectives on the History of Mathematical Logic*. American Mathematical Society, Birkhauser, 2008.
14. Edwards, R.E.: *Functional Analysis: Theory and Applications*. Dover Publications, NY, 2012.
15. Einstein, A.: *Relativity: The Special and the General Theory*. Martino Fine Books, 2010.
16. Eves, H.: *Foundations and Fundamental Concepts of Mathematics*. Dover Publications, N.Y., 1997.
17. Fraenkel, A.A., Bar-Hilel, Y., and Levy, A.: *Foundations of Set Theory*. Elsevier, 1973.
18. Gibson, C.C.: *Elementary Euclidean Geometry: An Introduction*. Cambridge University Press, 2003.
19. Glendinning, P.: *Stability, Instability and Chaos: An Introduction to the Theory of Nonlinear Differential Equations*. Cambridge University Press (1994).
20. Gowers, T.: *Mathematics: A Very Short Introduction*. Oxford University Press, Oxford, UK, 2002.
21. Griffiths, D.J., and Schroeter, D.F.: *Introduction to Quantum Mechanics* (3rd edition). Cambridge University Press, 2018.

22. Hausdorff, F.: *Set Theory (Third Edition)*. AMS Bookstore, 1978.
23. Hrabovsky, C.: *Classical Mechanics: The Theoretical Minimum*. 2014.
24. d'Inverno, R.: *Approaches to Numerical Relativity*. Cambridge Univ. Press (2008).
25. Jech, T.: *Set Theory: The Third Millennium Edition*. Springer, NU, 2006.
26. Jones, H.F.: *Groups, Representations and Physics (Second Edition)*. CRC Press, N.Y., 1998.
27. Kadanoff, L.P.: *Quantum Statistical Mechanics*. Taylor and Francis, 2019.
28. Karatzas, I. and Shreve, S.E.: *Brownian Motion and Stochastic Calculus* (second edition). Springer, New York, 1991.
29. Kibble, T.: *Classical Mechanics (5th Edition)*. Imperial College Press, London, 2004.
30. Klein, F.: *Elementary Mathematics from an Advanced Standpoint: Arithmetic, Algebra, Analysis*. Dover Publications, N.Y., 2004.
31. Kuratowski, K.: *Introduction to Set Theory and Topology (Second Edition)*. Elsevier, 2014.
32. Meloney, A.: *Lecture Notes on Classical Mechanics*, 2010.
33. Mcloncy, A.: *Lecture Notes on General Relativity*, 2019.
34. Newman, J.R.: *The World of Mathematics*. Courier Dover Publications, N.Y., 2000.
35. Miller, G. L.: Riemann's hypothesis and tests for primality. *J. Comput. Sys. Sci.* 13 (1976), pp. 300–317.
36. Pinter, C.: *A Book of Set Theory*. Dover Publications, N.Y., 2014.
37. Pollard, S.: *Philosophical Introduction to Set Theory*. Dover Publications, N.Y., 2015.
38. Potter, M.: *Set Theory and Its Philosophy: A Critical Introduction*. Oxford University Press, Oxford, 2004.
39. Rabin, M. O.: Probabilistic algorithm for testing primality. *J. Number Theory* 12 (1980), pp. 128–138.
40. Rivest, R., Shamir, A., and Adleman, L.: A method for obtaining digital signatures and public-key cryptosystems. *Communications ACM* 21 (1978), pp. 120–126.
41. Robinett, R.: *Quantum Mechanics: Classical Results, Modern Systems, and Visualized Examples (Second Edition)*. Oxford Univ. Press, 2006.
42. Romano, A., and Furnari, M.M.: *The Physical and Mathematical Foundations of the Theory of Relativity: A Critical Analysis*. Birkhouser, NY, 2019.
43. Schieve, W.C., and Horwitz, L.P.: *Quantum Statistical Mechanics*. Cambridge Univ. Press, 2009.
44. Schuller, M.F.: *Lectures on the Geometrical Anatomy of Theoretical Physics and Quantum Theory*. Lecture notes, independently published, 2019.
45. Shapira, Y.: *Solving PDEs in C++: Numerical Methods in a Unified Object–Oriented Approach (Second Edition)*. SIAM, Philadelphia, PA, 2012.
46. Shapira, Y.: *Linear Algebra and Group Theory for Physicists and Engineers*. Birkhouser, Springer Nature, N.Y. (2019).
47. Shapira, Y.: *Classical and Quantum Mechanics with Lie Algebras*. WS, Singapore (2021).
48. Shibata, M.: *Numerical Relativity*. World Scientific Publishing (2016).
49. Sibley, T.Q.: *The Foundations of Mathematics*. Wiley, 2008.
50. Smullyan, R.M., and Fitting, M.: *Set Theory and the Continuum Problem*. Dover Publications, N.Y., 2009.
51. Sobczyk, G.: *New Foundations in Mathematics: The Geometric Concept of Number*. Birkhauser, 2013.
52. Steinbauer, R.: *The Penrose and Hawking Singularity Theorems Revisited*. Math Faculty, Vienna Univ., 2006.
53. Strang, G.: *Introduction to Linear Algebra* (third edition). SIAM, Philadelphia, 2003
54. Strogatz, S.H.: *Nonlinear Dynamics and Chaos: With Applications to Physics, Biology, Chemistry, and Engineering (Second Edition)*. Westview Press (2014).

55. Susskind, L., and Friedman, A.: *Quantum Mechanics: The Theoretical Minimum.* Penguin, 2015.
56. Susskind, L.: *Lecture Notes on Quantum Field Theory and Particle Physics.* Stanford Univ., 2015.
57. Taylor, J.R.: *Classical Mechanics.* University Science Books, Herndon, VA, 2005.
58. Vaisman, I.: *Analytical Geometry.* World Scientific, 1997.
59. Vazirani, U.V.: *Lecture Notes on Quantum Mechanics and Quantum Computation.* UC, Berkeley.
60. Wilder, R.L.: *Introduction to the Foundations of Mathematics: Second Edition.* Dover Publications, N.Y., 2012.
61. Woodhouse, N.M.J.: *Special Relativity. (2nd edition).* Springer, London, 2007.
62. Wootters, M.: *Lecture Notes on Algebraic Coding Theory.* Stanford Univ., 2021.

# Index

Printed in the United States
by Baker & Taylor Publisher Services